Biological Neural Networks in Invertebrate Neuroethology and Robotics

This is a volume in
NEURAL NETWORKS: FOUNDATIONS TO APPLICATIONS

Edited by Steven F. Zornetzer, Joel Davis, Clifford Lau, and Thomas McKenna

BIOLOGICAL NEURAL NETWORKS IN INVERTEBRATE NEUROETHOLOGY AND ROBOTICS

Edited by

RANDALL D. BEER
Department of Computer Engineering and Science and Department of Biology
Case Western Reserve University
Cleveland, Ohio

ROY E. RITZMANN
Department of Biology and Department of Neuroscience
Case Western Reserve University
Cleveland, Ohio

THOMAS MCKENNA
Computational Neuroscience Program
Office of Naval Research
Arlington, Virginia

ACADEMIC PRESS, INC.
Harcourt Brace Jovanovich, Publishers
Boston   San Diego   New York
London   Sydney   Tokyo   Toronto

This book is printed on acid-free paper. ∞

COPYRIGHT © 1993 BY ACADEMIC PRESS, INC.

All rights reserved.
No part of this publication may be reproduced or
transmitted in any form or by any means, electronic
or mechanical, including photocopy, recording, or
any information storage and retrieval system, without
permission in writing from the publisher.

ACADEMIC PRESS, INC.
1250 Sixth Avenue, San Diego, CA 92101-4311

United Kingdom Edition published by
ACADEMIC PRESS LIMITED
24-28 Oval Road, London NW1 7DX

Library of Congress Cataloging-in-Publication Data
Biological neural networks in invertebrate neuroethology and robotics
 / edited by Randall D. Beer, Roy E. Ritzmann. Thomas McKenna.
       p. cm. — (Neural networks, foundations to applications)
   Papers originally presented at a workshop held at the National
Academy of Sciences Study Center in Woods Hole, Massachusetts, in
the fall of 1991, sponsored by the Office of Naval Research.
   Includes bibliographical references and index.
   ISBN 0-12-084728-0 (acid-free paper)
   1. Neural circuitry—Congresses. 2. Invertebrates—Physiology-
-Congresses. 3. Neural networks (Computer science)—Congresses.
 I. Beer, Randall D. II. Ritzmann, Roy. III. McKenna, Thomas M.
 IV. Series.
QP363.3.B56 1992   92-26266
591.1'88—dc20   CIP

PRINTED IN THE UNITED STATES OF AMERICA

92 93 94 95   BC   9 8 7 6 5 4 3 2 1

Contents

Contributors   vii
Preface   ix

I. NEUROETHOLOGY I: CONTROL OF LEG MOVEMENT   1

Chapter I. Integration of Individual Leg Dynamics with Whole Body Movement in Arthropod Locomotion   3
Robert J. Full

Chapter II. The Walking of Cockroaches—Deceptive Simplicity   21
Fred Delcomyn

Chapter III. Load Compensatory Reactions in Insects: Swaying and Stepping Strategies in Posture and Locomotion   43
Sasha N. Zill

Chapter IV. Integration by Spiking and Nonspiking Local Neurons in the Locust Central Nervous System: The Importance of Cellular and Synaptic Properties for Network Function   69
Gilles Laurent

II. NEUROETHOLOGY II: CONTROL OF ORIENTATION   87

Chapter V. Multisensory Processing for Movement: Antennal and Cercal Mediation of Escape Turning in the Cockroach   89
Christopher M. Comer and John P. Dowd

Chapter VI. The Neural Organization of Cockroach Escape and Its Role in Context-Dependent Orientation   113
Roy E. Ritzmann

Chapter VII. Neuroethology of Acoustic Startle and Escape in Insects   139
Ronald R. Hoy

| Chapter VIII. | Organization of Goal-Oriented Locomotion: Pheromone-Modulated Flight Behavior of Moths | 159 |

Edmund A. Arbas, Mark A. Willis, and Ryohei Kanzaki

| Chapter IX. | A New Role for the Insect Mushroom Bodies: Place Memory and Motor Control | 199 |

Makoto Mizunami, Josette M. Weibrecht and Nicholas J. Strausfeld

III. COMPUTER MODELING   227

| Chapter X. | Modeling a Reprogrammable Central Pattern Generating Network | 229 |

A. I. Selverston, P. Rowat, and M. E. T. Boyle

| Chapter XI. | Voyages Through Weight Space: Network Models of an Escape Reflex in the Leech | 251 |

Shawn R. Lockery and Terrence J. Sejnowski

| Chapter XII. | Simulations of Cockroach Locomotion and Escape | 267 |

Randall D. Beer and Hillel J. Chiel

| Chapter XIII. | Lobster Walking as a Model for an Omnidirectional Robotic Ambulation Architecture | 287 |

Joseph Ayers and Jill Crisman

IV. ROBOTICS   317

| Chapter XIV. | Legged Robots | 319 |

Marc H. Raibert and Jessica K. Hodgins

| Chapter XV. | A Robot that Walks; Emergent Behaviors from a Carefully Evolved Network | 355 |

Rodney A. Brooks

| Chapter XVI. | Control of a Hexapod Robot Using a Biologically Inspired Neural Network | 365 |

Roger D. Quinn and Kenneth S. Espenschied

| Chapter XVII. | Modeling Neural Function at the Schema Level: Implications and Results for Robotic Control | 383 |

Ronald C. Arkin

Index   411

Contributors

Numbers in parantheses refer to the pages on which the authors' contributions begin.

Arbas, Edmund (159), Arizona Research Laboratory, Division of Neurobiology, 611 Gould-Simpson Building, University of Arizona, Tucson, Arizona 85721.

Arkin, Ronald (383), Mobile Robot Laboratory, College of Computing, Georgia Institute of Technology, Atlanta, Georgia 30332-0280.

Ayers, Joseph (287), Department of Biology and Marine Science Center, Northeastern University, East Point, Nahant, Massachusetts 01908.

Beer, Randall (267), Departments of Computer Engineering and Science and Department of Biology, Case Western Reserve University, 10900 Euclid Avenue, Cleveland, Ohio 44106.

Boyle, M. E. T. (229), Department of Biology, University of California—San Diego, La Jolla, California 92093.

Brooks, Rodney, A. (355), Artificial Intelligence Laboratory, Massachusetts Institute of Technology, 545 Technology Square, Cambridge, Massachusetts 02139.

Chiel, Hillel (267), Department of Biology and Department of Neuroscience, Case Western Reserve University, 10900 Euclid Avenue, Cleveland, Ohio 44106.

Comer, Christopher M. (89), Department of Biology, Box 4348, University of Illinois, Chicago, Illinois 60680.

Crisman, Jill (287), Department of Electrical and Computer Engineering, Northeastern University; Boston, Massachusetts 02115-5095.

Delcomyn, Fred (21), Department of Entomology, University of Illinois—Urbana, 505 South Goodwin Street, Urbana, Illinois 61801.

Dowd, John (89), Department of Biology, University of Illinois—Chicago, Chicago, Illinois 60680.

Espenschied, Kenneth (365), Department of Mechanical and Aerospace Engineering, Case Western Reserve University, 10900 Euclid Avenue, Cleveland, Ohio 44106.

Full, Robert (3), Department of Integrative Biology, University of California—Berkeley, Berkeley, California 94720.

Hodgins, Jessica K. (319), IBM T. J. Watson Research Center, P.O. Box 704, Yorktown Heights, New York 10598.

Hoy, Ronald R. (139), Section of Neurobiology and Behavior, Cornell University, Ithaca, New York 14853.

Kanzaki, Ryohei (159), Institute of Biological Studies, University of Tsukuba, 1-1-1 Tennodei, Tsukuba-shi, Ibaraki 305, Japan.

Laurent, Gilles (69), Division of Biology, Computation and Neural Systems Program, California Institute of Technology, Pasadena, California 91125.

Lockery, Shawn R. (251), Computational Neurobiology Laboratory, Salk Institute for Biological Studies and Howard Hughes Medical Institute, P.O. Box 85800, San Diego, California 92186-5800.

Mizunami, Makoto (199), Department of Biology, Faculty of Sciences, Kyushu University, Fukuoka 812, Japan.

Quinn, Roger (365), Department of Mechanical and Aerospace Engineering, Case Western Reserve University, 10900 Euclid Avenue, Cleveland, Ohio 44106.

Raibert, Marc (319), Artificial Intelligence Laboratory, Massachusetts Institute of Technology, 545 Technology Square, Cambridge, Massachusetts 02139.

Ritzmann, Roy E. (113), Department of Biology and Department of Neuroscience, Case Western Reserve University, 10900 Euclid Avenue, Cleveland, Ohio 44106.

Rowat, Peter (229), Department of Biology, University of California—San Diego, La Jolla, California 92093.

Sejnowski, Terrence (251), Computational Neurobiology Laboratory, Salk Institute for Biological Studies and Howard Hughes Medical Institute, P.O. Box 85800, San Diego, California 92186-5800.

Selverston, A. I. (229), Department of Biology, University of California—San Diego, La Jolla, California 92093.

Strausfeld, Nicholas, J. (199), Arizona Research Laboratory, Division of Neurobiology, 611 Gould–Simpson Building, University of Arizona, Tucson, Arizona 85721.

Weibrecht, Josette M. (199), Arizona Research Laboratories, Division of Neurobiology, University of Arizona, Tucson, Arizona 85721.

Willis, Mark A. (159), Arizona Research Laboratories, Division of Neurobiology, University of Arizona, Tucson, Arizona 85721.

Zill, Sasha (43), Department of Anatomy, School of Medicine, Marshall University, 1542 Spring Valley Drive, Huntington, West Virginia 25755.

Preface

Historically, engineers have often turned to biological systems for inspiration. Velcro was inspired by the manner in which burrs attach themselves to clothing. Early attempts to develop flying machines were inspired by observations of birds and other flying animals. Although flying machines that attempted to mimic bird flight in every detail were ultimately unsuccessful, many design features of the modern airplane, such as wing shape, were derived from these observations. Today the development of autonomous robots could similarly benefit from a study of the behavioral control mechanisms that have evolved in animals.

In the fall of 1991, the Office of Naval Research sponsored a workshop entitled "Locomotion Control in Legged Invertebrates" that was held at the National Academy of Sciences Study Center in Woods Hole, Massachusetts. The purpose of this workshop was to bring together invertebrate neuroethologists interested in sensorimotor control of movement with engineers interested in the design of autonomous mobile robots. This turned out to be a highly stimulating workshop. With this book, we hope to extend the interdisciplinary interactions that it fostered beyond the immediate participants. While each chapter primarily reviews the authors' own work, authors have also been encouraged to place their work in the broader context of interdisciplinary themes explored by the workshop.

Of what benefit might invertebrate neuroethology be to roboticists? A mobile robot that is to accomplish some task autonomously in a realistic environment faces many of the same challenges that animals do. In order to move throughout their environments, robots must generate locomotory rhythms, maintain stable gaits at different speeds, adjust to changes in terrain and maintain an upright posture. Animals accomplish this with a combination of pattern generation and reflex control that is described in the chapters by Delcomyn, Zill and Laurent. As indicated in both Raibert's and Full's chapters, the solutions to these problems do not simply come from considerations of neural control, but must also factor in the mechanical properties of musculoskeletal structures. While the biological solutions to these problems may not be unique, or even necessarily the best ones, some roboticists have begun to utilize biological principles in the design of their robotic systems. This is clearly indicated in the chapters by Raibert, Brooks, Quinn and Arkin.

Useful mobile robots must be able to do more than just locomote. They must interact with their environment. Similarly, animals must seek out goals and avoid predators in order to survive. Moreover, animals must accomplish these tasks efficiently in order to compete with other animals for resources and to survive interactions with predators that have developed efficient mechanisms of capture. Finally, animals must perform these tasks in the context of ever changing environmental conditions. Certainly, a robot that could interact with its environment on the level of the animals described in the chapters by Comer, Ritzmann, Hoy and Arbas would be a significant advance.

We believe that computer modeling can play an important role in fostering these interactions between biologists and engineers. The development of a computer model can pose specific experimental questions that must be addressed before a quantitative understanding of the system can be achieved. The success or failure of a computer model serves as a very strong test for some qualitative theory of the system's operation. Once a model has been developed, its manipulation can suggest additional experiments which can either verify the model's predictions or serve as the basis for refining the model. In addition, computer models can aid in abstracting biological control principles for robotic application. The chapters by Selverston, Lockery, Beer and Ayers illustrate these various uses of computer models.

The benefits of increased interaction between biologists and engineers is not a one-way street. Although the organization of the book implies a movement from neuroethology to computer modeling and finally to robotics, we see a much more complex interaction. Attempts to implement in hardware theoretical models based upon biological data can actually drive biological experimentation. It is easy for the biologist to get caught in the trap of ignoring many properties of the system. Often this is necessary in order to understand one area very well. For example, in studying a neural circuit in the central nervous system, the neurobiologist may not concentrate on sensory inputs that have only subthreshold effects and may completely ignore the mechanical properties of the peripheral structures. This can result in a model that appears to account for the behavior, but simply does not work when placed in simulation or in hardware. Factors such as stability and posture may not seem important to a locomotion controller until one tries to implement them in a moving robot. When faced with the problem that the biologically-inspired machine falls over, one recourse for the engineer is to ask the biologist to determine how the animal solves stability problems. In this way, robotics and neuroethology tend to drive each other, with computer modeling providing an excellent environment in which to create and test hypotheses from complex biological data prior to implementation in hardware. Ultimately, this results in a more complete understanding of the biological system. The result is a positive synergy from which all parties benefit.

The biological systems that are described in this volume are concentrated in the invertebrate phyla. The reason for this reflects the primary goal of the work. Many

Preface

neuroethologists have exploited the technological advantages of invertebrates to achieve a very detailed understanding of control systems. The number of neurons involved is orders of magnitude smaller than in many vertebrate systems. Individual neurons can be filled with dyes and unambiguously identified based upon morphological characteristics, and mechanical properties of leg joints are often less complex than vertebrate joints. The work on the stomatogastric ganglion of the lobster described by Selverston is an excellent example of just how far one can go in understanding complex circuits in invertebrate systems.

These animals have certainly provided useful models for understanding more complex systems such as those found in vertebrates and even in mammalian systems such as our own. Indeed, the mechanisms involved in the production of action potentials were first understood in the squid giant axon. However, we must not lose sight of the fact that our primary goal here is not to develop models for understanding human systems; rather, it is to provide insight to solutions for engineering problems. For this purpose, the technical advantages of working with invertebrates provide the most appropriate and efficient sources of relatively quick information.

One criticism that might be raised is that invertebrate systems are too simple to provide much useful information to the engineers. Hopefully the biological chapters will lay this concern to rest. Although technically approachable, the invertebrate systems are far from simple. We find populations of 100 or more neurons controlling complex movements, context dependent behavior, interactions between pattern generators and peripheral sensory structures, and sophisticated sensory processing. Perhaps the best example of the complex nature of these systems is provided by Strausfeld's chapter. Here we find that insects are fully capable of solving problems involving associative learning. Even more comforting, this work provides evidence that these and other higher control processes found in insect brains can be understood using the same techniques that have been exploited by others in the lower ganglia.

In summary, we hope that this volume will communicate some of the enthusiasm that was shared by the participants in the ONR workshop at Woods Hole. We fully expect that interactions among biologists, modelers and roboticists, of the sort illustrated here, will result in exciting discoveries and technological breakthroughs. It is our hope that this volume provides a catalyst for such interactions.

We would like to thank the Office of Naval Research for supporting the workshop that made this book possible. We would also like to thank Kathy Tibbetts at Academic Press for supervising the process that actually brought this book into existence.

Randall D. Beer
Roy Ritzmann
Thomas McKenna

PART I

NEUROETHOLOGY I: CONTROL OF LEG MOVEMENT

Chapter I

Integration of Individual Leg Dynamics with Whole Body Movement in Arthropod Locomotion

ROBERT J. FULL
Department of Integrative Biology
University of California at Berkeley
Berkeley, California

I. Introduction

This chapter highlights recent advances in the mechanics of invertebrate legged locomotion to show why neural control studies must consider musculoskeletal dynamics. In contrast to studies of neural control, little is known about the mechanics of locomotion or, more specifically, leg dynamics. In a 1985 symposium on insect locomotion Delcomyn stated that

> [The mechanics of locomotion] is an area of research that has never attracted many adherents, yet which provides much information that is essential in order for progress to be made in our understanding of the physiological basis of locomotion.

In addition to the issue of approach, I will propose concepts that can be transferred from the mechanics of invertebrate legged locomotion to the design of more versatile legged robots.

II. Forward versus Inverse Dynamics

The last 20 years has seen substantial and significant research on the neural control of arthropod legs (see reviews by Bassler, 1987; Clarac, 1981; Cruse, 1985, 1990;

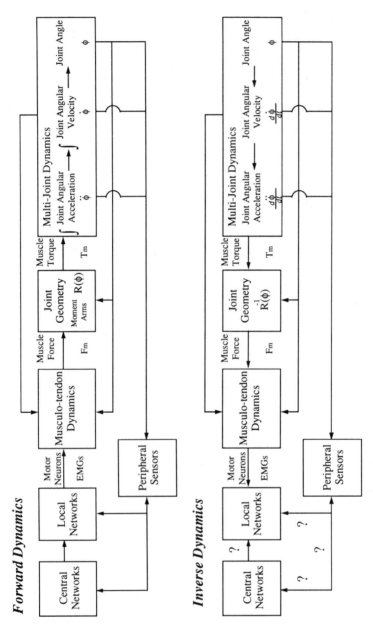

FIGURE 1. Forward versus inverse approach to dynamics. Forward dynamics characterizes the "controller" to predict leg movements, the "plant." Inverse dynamics uses leg movement and force production to make inferences about the "controller." Most of the research on arthropod locomotion has taken the forward approach. Research using the inverse approach is necessary in the future. Adapted from Zajac, F. E., and Gordon, M. E., *Exerc. Sport Sci. Rev.* **17**, p. 187. © 1989 ACSM.

Chapter I Integration of Individual Leg Dynamics

Delcomyn, 1981, 1984, 1985; Graham, 1985; Pearson, 1985; Zill, 1985). Considerable progress toward understanding the control of legged locomotion has been achieved by characterizing the "controller" (i.e., the associated central and sensory neural networks; Fig. 1). Such a determination of the pathways and information flow between sensory input and motor output layers is essential to the understanding of locomotor dynamics. The characterization of central pattern generators and relaxation oscillators; the identification of servo systems based on velocity, load, and position; and the description of ipsilateral and contralateral leg coupling are evidence of the advancements.

Despite this progress, attempting to characterize the controller before or in isolation from the musculoskeletal system (i.e., the "plant") is extremely problematic for several reasons (Zajac and Gordon, 1989). First, the flow of information associated with neural and musculoskeletal systems is not unidirectional but is a closed loop. Muscle and joint sensors feed back information about limb and body position, velocity, and force. In this sense, the musculoskeletal system can be considered the controller, and the neural system becomes the plant. Second, a single motor neuron's activity pattern can result in several completely different musculoskeletal responses, depending on the context within which the neural activity is generated. For example, the activation of an extensor muscle while joint angle is decreasing can result in absorption of energy by that muscle, whereas identical activation of the same extensor muscle while a joint angle is increasing can result in generation of energy. If limb segments on both sides of the joint have the same angular velocity, the activation of the same extensor muscle can result in a near-isometric contraction and energy transfer. In this sense, "feedback" can occur within the musculoskeletal system, and joint position and velocity can determine musculoskeletal dynamics. Likewise, joint angle can affect joint geometry by altering moment arms that will, in turn, affect joint dynamics through torque development. The third major difficulty with conventional "forward" dynamics has been that force determination for two or more legs on the ground is an indeterminate problem.

An alternative approach, termed *inverse dynamics*, uses the output to make inferences concerning the input (Fig. 1). Muscle torque at a joint can be determined from joint dynamics. Muscle force at a joint can be estimated from joint geometry. Muscle activity patterns can be predicted from muscle force estimates. Neural function can then be inferred from muscle activity patterns. With the inverse approach, quantified behavior (i.e., leg movements and force development) can be related unambiguously to neural control. The more that is learned about the plant (i.e., the musculoskeletal system), the better the neural controller can be defined and the better the whole system can be understood from an engineering and design standpoint. This approach is complementary to the more conventional *forward dynamics* and neuronal *circuit breaking* approaches.

Few investigations have used inverse dynamics and correlated motor neuron output, muscle activity, and kinematics (i.e., description of stepping patterns) with the actual kinetics (i.e., force development) involved in generating locomotion (Delcomyn, 1985). Studies that have been conducted underscore the diversity of leg function. Ground reaction forces of standing and slow-walking spiders differ depending on the leg measured (Blickhan and Barth, 1985). Cruse (1976) demonstrated that pairs of legs in a walking stick insect each generate a distinct ground reaction force pattern. In rock lobsters walking under water, leg four appears to control movement, whereas leg five functions as a strut (Clarac and Cruse, 1982; Cruse *et al.*, 1983). In crayfish, leg four produces most of the propulsive force, whereas leg three exerts the largest vertical force (Klarner and Barnes, 1986). Vertical force patterns are distinct in the second and third legs of crickets (Harris and Ghiradella, 1980).

Even though several of these studies have demonstrated the importance of load or force in feedback control, they have not produced an adequate model of leg function. It is not clear what types of leg dynamics neural activity patterns actually generate. For several reasons, this gap prevents the study of information flow from the nervous system to behavior.

First, in most neural studies, leg function has been oversimplified. Function is typically divided into a power stroke (or stance phase) and a return stroke (e.g., Cruse, 1990). This dichotomy assumes that all legs in a stance phase are equivalent. Yet, muscle of joints in different legs or even the same leg can be shortening to produce energy, lengthening while absorbing energy, or contracting isometrically to transfer energy. The assumption that extensors only extend and flexors only flex is unrealistic. In 1952 Hughes proposed a simple model for the function of insect legs. Legs could function as levers, inclined struts, or both, depending on the direction of the ground reaction force vector. If a significant horizontal accelerating force is observed which directs the ground reaction force in an anterior or forward direction, then the leg functions as a lever in that direction. A leg functions as a strut if the ground reaction force vector is directed back toward the joint. Hughes (1952) reported that the second and third legs function as inclined struts, while the first leg of cockroaches functions as a lever. Cruse (1976) showed that the legs of stick insects do not necessarily follow this model. First and second legs in stick insects can act as levers or inclined struts depending on the orientation of movement considered.

Second, the dynamic passive functions of the leg have been largely ignored. Legs can function as inverted pendulums and springs that may demand less, or simpler, neural control.

Third, many of the studies have focused on animals (i.e., stick insects, crayfish, and lobsters) moving at very low speeds. These animals lack the speed and maneuverability desirable in a model of legged locomotion. Moreover, leg control undoubtedly changes significantly when little or no time is available for feedback during medium to fast locomotion (Delcomyn, 1991; Zill, 1985).

III. Motion of the Body or Center of Mass

For most multilegged animals, it is difficult to predict the movement of the whole body even when the dynamics of isolated legs are known. Numerous studies have documented the enormous variation in leg position and phase (Delcomyn, 1985; Delcomyn and Cocatre-Zilgien, 1988). Does this variation represent the error resulting from a given neural output? It is unclear whether the same variation is apparent in the force and power production of legs. Results from our study of ghost crabs suggest considerable variation in force production for single legs (Blickhan and Full, 1987). Our work on insect leg force development showed somewhat less variation (Full and Tu, 1990, 1991; Full *et al.,* 1991). More important, what is most striking in all our kinetics studies is the regularity in the movement of the body or center of mass. It may be insufficient to determine the flow of information to individual legs without considering the movement produced by all the legs simultaneously. Analysis of an "effective" or "virtual" leg that represents the action of all legs would allow a link to be made between neural input and behavior. Defining leg function in the context of the motion of the whole body is essential to the understanding of locomotor control.

One might expect the whole body of many-legged animals to move like the body of some legged robots (Full *et al.,* 1989). Many multilegged vehicles or robots have been designed to be "wheel-like" (Fig. 2A). The more legs an animal has, the smoother the ride. Acceleration and deceleration of the body or center of mass are made negligible. The body of the robot is given a smooth ride to minimize energetic cost or for stabilization of sensors. Primarily vertically directed, ground reaction forces are observed. This can be achieved by using a telescoping leg, the equivalent of a bent knee. The smooth ride can also be attained by using a slider mechanism that is the equivalent of a pantograph leg design. Many models of multilegged locomotion also assume a quasi-static walking gait in which static equilibrium is maintained throughout a stride. More simply, the robot must be like a stool at all times with at least three legs supporting its body. This is especially

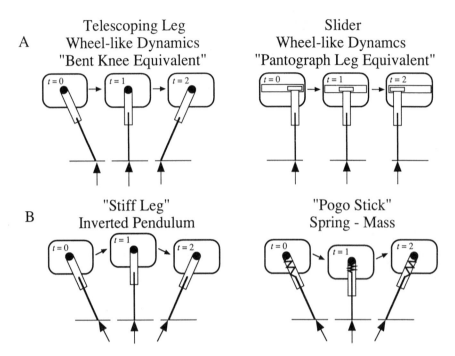

FIGURE 2. Leg design in terrestrial locomotion. (A) Many robots use wheel-like dynamics, where the body moves with little oscillation (horizontal arrows) and the ground reaction forces are directed vertically (vertical arrows), thereby reducing the acceleration and deceleration of the body. Wheel-like dynamics can be attained by using a telescoping joint, the equivalent of a bent knee, or by using a slide, the equivalent of a pantograph mechanism. Notice the large moments that are created around the hip of a telescoping leg. (B) In contrast to many robots, animals use legs that function as inverted pendulums and pogo sticks. Ground reaction forces show that the body is accelerated and decelerated as it oscillates up and down. In both mechanisms energy can be exchanged or stored and recovered. Reprinted from *The International Journal of Robotics Research* 9:2, "Three Uses for Springs in Legged Locomotion," by R. McN. Alexander by permission of the MIT Press, Cambridge, Massachusetts, © 1990 MIT Press.

true if the robot is close to the ground, because it doesn't take long to hit the substratum if the center of mass falls outside the base of support.

Surprisingly, multilegged arthropods are not restricted to wheel-like movement or balancing like stools. Locomotion of the whole body more closely resembles an inverted pendulum or spring–mass system than it does a wheel (Blickhan, 1989; Full, 1989, 1991; McGeer, 1990). During walking, two- and four-legged animals use an energy-conserving mechanism that is analogous to an inverted, swinging pendulum or an egg rolling end over end (Fig. 3) (Cavagna et al., 1976, 1977; Heglund et al., 1982). By using this pendulum-like mechanism, kinetic energy and gravitational potential energy fluctuate out of phase so as to allow recovery of

Chapter I Integration of Individual Leg Dynamics

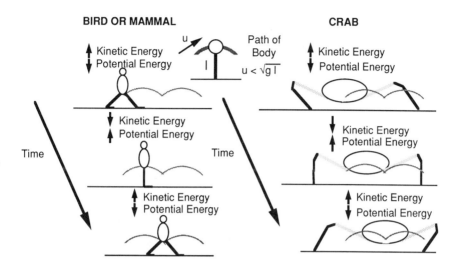

FIGURE 3. Inverted pendulum mechanism of walking for a biped and crab. One leg of a biped and four legs of a crab act to move the body through a series of arcs of radius, l. Potential energy and kinetic energy fluctuate out of phase such that kinetic energy can be recovered as potential energy and vice versa. The maximum speed (u) predicted must be less than the square root of the product of acceleration due to gravity (g) and radius. After Blickhan and Full (1987).

energy as the animal's center of mass rises and falls during a stride. Vaulting over a relatively stiffened leg in humans can conserve up to 70% of the energy that must otherwise be provided by muscles and tendons. Blickhan and Full (1987) have shown that eight-legged ghost crabs do not move with a constant velocity of the center of mass and are not wheel-like. Crabs can use a pendulum-like mechanism during walking (Fig. 3). Energy recovery in these arthropods can reach 55%.

Cockroaches don't appear to use this walking mechanism, even at lower speeds (Full and Tu, 1990, 1991). The percent recovery averages about 6–15% and is not a function of speed. At all intermediate speeds, cockroaches use a regular, symmetric alternating tripod gait. The right front, left middle, and right hind legs all move simultaneously in a step, while each moves out of phase with its contralateral pair (Fig. 4). In this gait, cockroaches do not move with a constant velocity of the center of mass and are also not wheel-like. Distinct maxima and minima in the whole ground reaction forces are apparent. Each vertical force peak is correlated with a step of an alternating set of legs. As the animal's body comes down on a tripod, it decelerates in the horizontal direction. Its vertical force increases above body weight. As the body lifts up, it is accelerated and the vertical force decreases below body weight. This pattern is repeated for the next step of the tripod. In contrast to a pendulum-like walk-

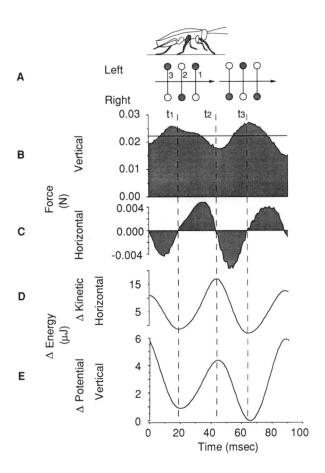

FIGURE 4. Gait, ground reaction force, and energy of the center of mass during one stride (i.e., one complete leg cycle) of a 2.3 g cockroach traveling at 0.25 m s^{-1}. Tracings represent the following: (A) The gait used is an alternating tripod in which the right front, left middle, and right hind legs all moved simultaneously, while each moved out of phase with its contralateral pair. In the gait diagram filled circles represent legs on the ground (retracting or in stance phase), whereas open circles show legs in protraction (swing phase). (B) Vertical and (C) horizontal forces obtained from a force platform. (D) Horizontal kinetic energy and (E) gravitational potential energy fluctuations of the center of mass. Horizontal line in B represents the animal's weight.

ing gait, potential energy and kinetic energy fluctuate in phase during cockroach locomotion.

In 1978 Alexander and Jayes made a link between the vertical force pattern and the gait an animal was using. They found that vertical force patterns could be described by a modified Fourier series and a single measure called the shape

Chapter I Integration of Individual Leg Dynamics

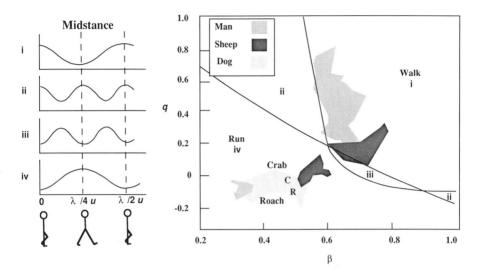

FIGURE 5. Gait characterization by shape and duty factor. The shape of the vertical force pattern (Fv) can be described at any time T by a modified Fourier series where $F\text{v} = a\,[cos(\pi t/T) - q\,cos(3\pi t/T)]$ from the time a foot is set down $(-T/2)$ to when it is lifted $(+T/2)$. For crabs and cockroaches, the shape factor describing the force pattern ($q = -0.1$ to 0) and duty factor (the fraction of a stride duration a foot is on the ground, $\beta = 0.5$) corresponds to a run or trot when compared with two- and four-legged vertebrates. From Alexander and Jayes (1978).

factor, q. If the shape factor is plotted as a function of duty factor (β), the fraction of a stride the leg is on the ground, different gaits or styles of locomotion can be separated (Fig. 5). For example, if an animal was walking, at midstance the position of the center of mass would be at its lowest point, whereas if the animal was running, the position of the center of mass would be at the highest point. On the basis of q and β, walks and runs fall into distinct zones. If the shape factor for cockroaches and crabs (i.e., at higher speeds) is plotted as a function of duty factor, both arthropods fall clearly in the area of a run (Fig. 5). Cockroaches and crabs can use a running gait even though at these trotting speeds they don't show aerial phases. In both species, potential energy and kinetic energy fluctuate in phase as they would if the body were bouncing with springs. McMahon *et al.* (1987) have shown that an aerial phase is not a requirement for the definition of a bouncing or running gait. Humans running with bent legs, like Groucho Marx, show this pattern and don't have aerial phases.

Despite the differences in morphology, at intermediate and high speeds, two-, four-, six-, and eight-legged animals produce ground force patterns that are fundamentally similar. All can run or bounce. Running humans, trotting dogs, cockroaches, and sideways-running crabs can move their bodies by producing alternating propulsive forces. Two legs in a trotting quadrupedal mammal, three legs in an insect, and four legs in a crab can act as one leg does in a biped during ground contact. The center of mass of the animal undergoes repeated acceleration and deceleration with each step, even when traveling at a constant average velocity (Blickhan and Full, 1987; Cavagna *et al.*, 1977; Full and Tu, 1990, 1991; Heglund *et al.*, 1982). Horizontal kinetic energy and gravitational potential energy of the center of mass fluctuate in phase.

The dynamics of crabs and cockroaches suggest that arthropods can use springs and bounce during running like mammals. Equivalent gaits may exist among pedestrians that differ greatly in morphology. Further evidence of this equivalence or similarity comes from examining the relationship between stride frequency and running speed. In quadrupedal mammals, stride frequency increases linearly with speed during trotting (Heglund *et al.*, 1974; Heglund and Taylor, 1988). Stride frequency becomes independent of speed as they switch to a gallop. Galloping quadrupeds move faster by increasing stride length. Blickhan and Full (1987) found a similar pattern in ghost crabs. At the fastest galloping speeds, ghost crabs use fewer legs, leap, and have aerial phase (Burrows and Hoyle, 1973). Full and Tu (1990) discovered a similar relationship in cockroaches. As speed increases, stride frequency eventually attains a maximum. After finding this pattern in the cockroach, we wondered whether equivalent gait transitions in two-, four-, six-, and eight-legged animals could be identified. We scaled the maximum sustainable stride frequency and the speed at which it was attained in crabs and cockroaches with the data already available for mammals (Heglund *et al.*, 1974; Heglund and Taylor, 1988). Surprisingly, when the effect of size is removed, legged animals attain a similar maximum sustainable stride frequency at a similar speed (Full, 1989, 1991). For example, a crab and a mouse of the same mass change gait at the same stride frequency (9 Hz) and speed (0.9 m/s; Blickhan and Full, 1987). These data suggest that legged animals may consist of multijointed springs.

IV. Quasi-Static Versus Dynamic Stability

Our results from the study of six- and eight-legged runners (Blickhan and Full, 1987; Full and Tu, 1990, 1991; Full *et al.*, 1991) provide strong evidence that dynamic effects cannot be ignored in multilegged runners that are maneuverable, as has been done in the design of multilegged robots (e.g., Bressonov and Umnov, 1973; Song, 1984; Song and Waldron, 1989).

Most six-legged insects use an alternating tripod gait at most speeds such that at least three legs are on the ground at any time during locomotion. The legs of arthro-

Chapter I Integration of Individual Leg Dynamics

TABLE I. Comparison of Legged Locomotion in Animals and Vehicles

	No. of legs	Hip height, h (m)	Speed u (m/s)	Frequency, f (Hz)	Froude no., $u^2/(gh)$
Crab walking	8	0.035	0.4	3.2	0.4
Man walking	2	0.9	1.6	1	0.3
Dog walking	4	0.5	1.3	1.6	0.4
Crab trotting	8	0.035	0.9	6.2	2.4
Cockroach trotting	6	0.004	0.3	13	1.7
Man jogging	2	0.9	3.3	1.6	1.2
Dog trotting	4	0.5	2.7	2.2	1.5
Turtle	4	0.07	0.1	0.6	0.02
3-D hopper	1	0.6	2.2	1.5	0.9
Kenkaku 1	2	0.6	0.8	1.1	0.1
PV II	4	0.4	0.02	—	0.0001
NCTU quadruped	4	0.6	0.01	0.02	0.00002
Quadruped trotting	4	0.6	2.2	1.2	0.9
Quadruped bounding	4	0.6	2.9	2.4	1.5
Sutherland hexapod	6	0.2	0.1	—	0.005
ASV	6	1.8	3.6	—	0.7
ReCUS	8	3.5	0.07	0.03	0.0001

After Alexander (1989).

pods generally radiate outward, providing a wide base of support. It has been suggested that the morphology of the limbs provides stability against such disturbances as wind and uneven terrain. Alexander (1982) has suggested that an insect such as a cockroach is so close to the ground that the animal must always have three legs in contact with the surface or it would fall to the ground before taking the next step. Hughes (1952) stated that the six-legged condition is the "end-product of evolution" because the animal can always be statically stable. Several robots have been designed with a quasi-static gait criterion that seems very similar to insect locomotion. The center of mass moves smoothly and is contained within a triangle or quadrilateral of support. However, these walking machines move very slowly, so a quasi-static condition is required (Alexander, 1989; Song and Waldron, 1989). Speeds and frequencies used by these robots are low compared to those of animals. Froude numbers (i.e., the ratio of inertial to gravitational forces) indicate that most robots do not move in a dynamically similar fashion to animals (Table I). Most legged robots are dynamically more similar to turtles. Raibert's amazing robots at MIT are an obvious exception because they rely on dynamic stability (Raibert and Sutherland, 1983).

Contrary to the hypothesis that static stability is one of the most important design criteria in arthropods and in the design of robots, we have found that dynamic stability is crucial even in small, multilegged animals (Ting et al., 1990). First, we have shown that crabs and cockroaches employ a running or bouncing gait that is dynamically similar to trotting in quadrupeds and to running in bipeds (Blickhan and Full, 1987; Full, 1989; Full and Tu, 1990). Second, at high speeds ghost crabs propel themselves with two legs on the trailing side of the body as they leap into the air (Blickhan and Full, 1987; Burrows and Hoyle, 1973). The American cockroach can run quadrupedally and bipedally at high speeds (Full and Tu, 1991). These gaits demand dynamic stability. Third, cockroaches with ablated middle legs run with a duty factor of less than 0.75 without falling (Pham and Full, 1989). Fourth, the stability margin (i.e., the minimum distance from the center of mass to the base of support) decreases linearly with speed and becomes negative at the lowest speeds (i.e., statically unstable, Fig. 6) (Ting et al., 1990). Moreover, the position of the center of mass moves posteriorly with speed. Stability margin is actually related to speed and momentum because if the animal attempts to stop instantaneously, it will keep moving forward, essentially tossing itself into the support triangle. Fifth, ground reaction forces create moments about the center of mass that cause pitching and rolling of the body. The resultant force or center of pressure is not directed through the center of mass. If the animal was stopped (i.e.,

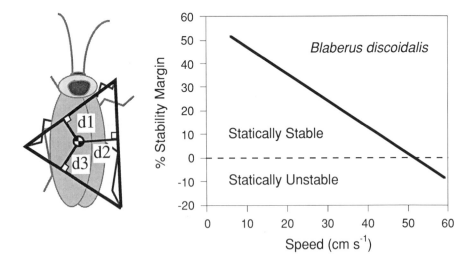

FIGURE 6. Percent stability margin as a function of speed in cockroaches. Stability margin is the minimum distance (e.g., d1) from the center of mass to the edge of the triangle of support. Percent stability margin is the stability margin normalized to the maximum possible stability margin. Static instability (less than zero percent stability margin) occurs when the center of mass falls outside the base of support.

was examined statically), the resultant force vector would create a moment that could cause the animal to flip over.

Dynamic stability appears to be an important consideration in the analysis of gaits, even in polypedal animals. In fact, at high speeds, the gait in the large tropical cockroach can be best explained using a dynamic, spring–mass model of running and hopping (Blickhan, 1989; McMahon and Cheng, 1990; Raibert *et al.*, 1986; Ting *et al.*, 1990) (Fig. 7). At low speeds, the cockroach uses a running gait, but due to limitation in vertical displacement the stride frequency is not matched to the natural frequency of the spring–mass model, causing increased angular rotation of the body. The best model at low speeds is one in which there are periods when two springs (i.e., two tripods or all six legs) are on the ground at the same time.

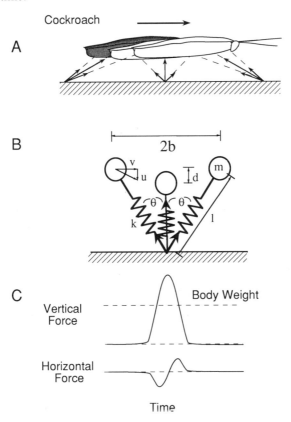

FIGURE 7. Spring–mass model for cockroaches. Three legs of a cockroach, one leg of a biped, two of a trotting quadruped, and four of a trotting crab can be represented by one spring. The dynamics of each of these runners can be described by a spring attached to a mass, the body.

V. Variation in Individual Leg Dynamics—Similarity in "Effective" Leg Function

Whole-body mechanics in two-, four-, six-, and eight-legged runners is dynamic, not wheel-like, and can be remarkably similar, despite variation in body form or morphology (Full, 1989, 1991; Full *et al.*, 1989). Pedestrians that vary in leg number and design can generate similar ground reaction force patterns. Similarities between mammalian and arthropod whole-body mechanics suggest that rigid constraints may exist on the possible mechanisms by which a leg or legs can function during walking and running.

Trotting quadrupedal mammals, such as dogs, bounce by producing nearly the same force pattern with each leg, just as do humans (Alexander, 1977). In fact, successful trotting quadrupedal robots have been designed so that the kinetics of each leg are the same, differing only in relative phase (Raibert *et al.*, 1986). Individual leg force patterns have the same shape as whole-body force patterns. We hypothesized that similarities in leg function were less likely to be characteristic of six-, and eight-legged runners, even though two-, four-, six-, and eight-legged animals all show a common whole-body, ground- reaction force pattern. It was not obvious how a common pattern could result in runners that have impressive differences in leg morphology.

To explain how diverse leg designs could result in common whole-body dynamics, we used a miniature force platform to measure the ground reaction forces produced by individual legs of a cockroach (Full *et al.*, 1991). Hexapedal runners are not like quadrupeds with an additional set of legs. At a constant average velocity, each leg pair of the cockroach is characterized by a unique ground reaction force pattern. The first leg decelerates the center of mass in the horizontal direction, whereas the third leg is used to accelerate the body. The second leg does both, much like legs in bipedal runners and quadrupedal trotters. Vertical force peaks for each leg are equal in magnitude and significant lateral forces are present.

The orientation of leg ground- reaction forces in the cockroach has several consequences (Fig. 8). Foremost, it rejects the hypothesis that legs result in wheel-like motion of the body. Arthropod legs do not necessarily function according to the design of many existing robot legs, (Fig. 8) where ground reaction forces are primarily vertical and little or no fluctuation of body position is observed (Fig. 2A) (Alexander, 1990). Arthropods legs do not function like sliders or operate like pantograph mechanisms. Instead, legs appear to function as inverted pendulums and springs that result in oscillations of the body (Fig. 2B). The center of mass undergoes accelerations and decelerations that are not necessarily unwanted. Legs in crabs that function as inverted pendulums allow as much as 50% of the energy to be exchanged and recovered (Blickhan and Full, 1987). The energy recovered need not be supplied

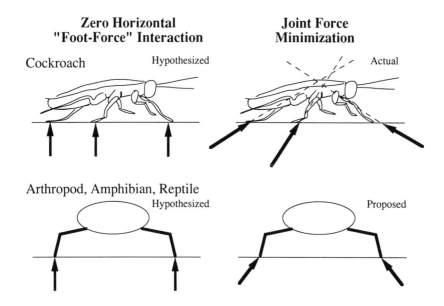

FIGURE 8. Zero horizontal foot-force interaction versus joint force or torque minimization. Zero foot-force interaction, a criterion in the design of many multilegged robots, minimizes accelerations and decelerations of the body but can result in large moments around the "hip" joints. Lateral, anterior, and posterior placement of the limbs results in large moments if zero foot interaction applies. Ground reaction forces in cockroaches are directed more in line with the hip (coxal joint) and tend to minimize torque as they may in other arthropods, amphibians, and reptiles.

by muscles or actuators. Legs in cockroaches may function more like springs (Full *et al.,* 1991). Energy can be absorbed, stored on landing as elastic strain energy, and returned upon takeoff. Both pendulum and spring-like leg functions result in forces that accelerate and decelerate the body, but both can recover or exchange energy.

An equally important consequence stemming from our study on cockroach leg function is related to muscle force production. In the cockroach, peak ground reaction forces are oriented toward the coxal joints (or "hip" equivalent), which articulate with the body (Fig. 8). This arrangement tends to minimize joint moments and muscle forces (Full *et al.,* 1991). Legs of animals do not generate vertically directed ground reaction forces that result in large torques about the "hip" as do some legged robots. They also do not operate under the horizontal, zero-foot force criterion used in the design of legged robots (Waldron, 1986). Legs or "feet" push against one another. Yet, production of horizontal and lateral ground forces that account for most of the mechanical energy generated during locomotion can actually reduce total muscle force by directing the ground reaction forces through the leg joints.

Locomotion with a sprawled posture as seen in small mammals, amphibians, reptiles, and arthropods does not necessarily result in large moments around joints or large muscle forces. This is consistent with the finding that the minimum metabolic cost of locomotion in species that differ in posture can be similar (Full, 1991).

Our investigation of cockroach leg function (Full *et al.*, 1991) has shown a strut/lever model to be insufficient. Whole-body dynamics common to two-, four-, six-, and eight-legged runners can be produced by different numbers of legs that vary in orientation with respect to the body, generate unique ground reaction force patterns, but combine to function as one leg of a biped.

VI. Conclusions

We assert that a dynamic analysis of arthropod locomotion is required to develop general principles of motor control. The "plant" must be studied in close association with the "controller." Static hypotheses of the "plant" are insufficient to explain the link between morphology and performance of terrestrial locomotion. Many-legged animals are not like wheels, cars, or most robots. They accelerate and decelerate their bodies. Dynamic stability is an important consideration in the analysis of polypedal locomotion. Locomotion in many pedestrians can be modeled by an inverted pendulum or spring–mass system, and this may explain why equivalent gaits may exist in species that differ so in morphology.

In general, animals do not function like many of the multilegged robots that have been built thus far. To design the most versatile legged robot possible:

1. Dynamic capacities must be integrated with the more thoroughly studied quasi-static capacities that are notable in many legged animals that have the advantage of high stability, a wide base of support, and a low center of mass (i.e., design a six- or eight-legged robot that has spring-like dynamics).
2. Isolated leg dynamics must be integrated with whole-body dynamics.
3. Compensation for environmental perturbations must be studied by integrating neural control with the dynamics of the musculoskeletal system.

Acknowledgments

Many thanks to T. McKenna for inspiring this manuscript by organizing the meeting at Woods Hole and to R. Kram for comments on the manuscript. Supported by ONR grant NOOO14-92-J-1250 and NSF grant DCB 89-0458689.

References

ALEXANDER, R. MCN. (1982). *Locomotion of Animals.* Blackie, Glasgow.
ALEXANDER, R. MCN. (1989). *Physiol. Rev.* **69,** 1199.
ALEXANDER, R. MCN. (1990). *Int. J. Robotics Res.* **9,** 53.
ALEXANDER, R. MCN. (1977). In Alexander, R. McN., and Goldspink, T. (eds.), *Mechanics and Energetics of Animal Locomotion* (pp. 168–203). Wiley, New York.
ALEXANDER, R. MCN., and JAYES, A.S. (1978). *J. Zool. Lond.,* the Zoological Society of London, **185,** 27.
BASSLER, U. (1987). *Biol. Cyber.* **55,** 397.
BLICKHAN, R., and BARTH, F. G. (1985). *J. Comp. Physiol.* **157,** 115.
BLICKHAN, R. (1989). *J. Biomech.* **22,** 1217.
BLICKHAN, R., and FULL, R. J. (1987). *J. Exp. Biol.* **130,** 155.
BRESSONOV, M. G., and UMNOV, N. V. (1973). *Proceedings of a Symposium on the Theory and Practice of Robots and Manipulators.* (pp. 87–97). Elsevier, New York.
BURROWS, M., and HOYLE, G. (1973). *J. Exp. Biol.* **58,** 327.
CAVAGNA, G. A., HEGLUND, N. C., and TAYLOR, C. R. (1977). *Am. J. Physiol.* **233,** R243.
CAVAGNA, G. A., THYS, H. and ZAMBONI, A. (1976). *J. Physiol.* **262,** 639.
CLARAC, F. (1981). In Herreid, C. F., and Fourtner, C. R. (eds.), *Locomotion and Energetics in Arthropods* (pp. 31–72). Plenum, New York.
CLARAC, F., and CRUSE, H. (1982). *Biol Cybern.* **43,** 109.
CRUSE, H. (1976). *J. Comp. Physiol.* **112,** 235.
CRUSE, H. (1985). In Gewecke, M., and Wendler, G. (eds.), *Insect Locomotion.* Paul Parey, Berlin.
CRUSE, H. (1990). *Trends Neurosci.* **13,** 15.
CRUSE, H., CLARAC, F., and CHASSERAT, C. (1983). *Biol. Cybern.* **47,** 87.
DELCOMYN, F. (1981). In *Locomotion and Energetics in Arthropods.* New York, Plenum.
DELCOMYN, F. (1984). In *Comprehensive Insect Physiology, Biochemistry and Pharmacology.* Pergamon, New York.
DELCOMYN, F. (1985). In GEWECKE, M., and WENDLER, G. (eds.), *Insect Locomotion.* (pp. 1–18). Paul Parey, Berlin.
DELCOMYN, F. (1991). *J. Exp. Biol.* **156,** 503.
DELCOMYN, F., and COCATRE-ZILGIEN, J. H. (1988). *Biol. Cybern.* **59,** 379. Full, R. J. (1989). In Wieser, W. and Gnaiger, E. (eds.), *Energy Transformation in Cells and Animals.* (pp. 175–182). Thieme, Stuttgart.
FULL, R. J. (1991). In Blake, R. W. (ed.), *Concepts of Efficiency, Economy and Related Concepts in Comparative Animal Physiology.* (pp. 97–131). Cambridge University Press, New York.
FULL, R. J., BLICKHAN, R., and TING, L. H. (1991). *J. Exp. Biol.* **158,** 369.
FULL, R. J., and TU, M. S. (1990). *J. Exp. Biol.* **148,** 129.
FULL, R. J., and TU, M. S. (1991). *J. Exp. Biol.* **156,** 215.
FULL, R. J., TU, M. S., and TING, L. (1989). *Proc. Am. Soc. Mech. Eng. DSC* **17,** 35.
GRAHAM, D. (1985). *Adv. Insect Physiol.* **18,** 31.
HARRIS, J., and GHIRADELLA, H. (1980). *J. Exp. Biol.* **85,** 263.
HEGLUND, N. C., TAYLOR, C. R., and MCMAHON, T. A. (1974). *Science.* **186,** 1112.
HEGLUND, N. C., and TAYLOR, C. R. (1988). *J. Exp. Biol.* **138,** 301.
HEGLUND, N. C., CAVAGNA, G. A., and TAYLOR, C. R. (1982). *J. Exp. Biol.* **79,** 41.

HIROSE, S. (1984). *Int. J. Robotics Res.* **2,** 113.
HUGHES, G. M. (1952). *J. Exp. Biol.* **29,** 267.
KLARNER, D., and BARNES, J. P. (1986). *J. Exp. Biol.* **122,** 161.
MCMAHON, T. A., VALIANT, G., and FREDERICK, E. C. (1987). *J. Appl. Physiol.* **62,** 2326.
MCMAHON, T. A., and CHENG, G. C. (1990). *J. Biomech.* **23,** 65.
MCGEER, T. (1990). *Proc. R. Soc. Lond.* **240,** 107.
PEARSON, K. G. (1985). In Barnes, W. J. P., and Galdden, M. H. (eds.), *Feedback and Motor Control in Invertebrates and Vertebrates.* (pp 307–315). Croom Helm, London.
PHAM, D. T., and FULL, R. J. (1989). *Am. Zool.* **29,** 128A.
RAIBERT, M. H., CHEPPONIS, M., and BROWN, B., JR. (1986). *IEEE J. of Robotics and Autom.* **RA-2,** 70.
RAIBERT, M. H. and SUTHERLAND, I. E. (1983). *Sci. Am.* **248,** 44.
SONG, S. M. (1984). Ph.D. dissertation, Dept. of Electrical Engineering, Ohio State University.
SONG, S., and WALDRON, K. (1989). *Machines That Walk: The Adaptive Suspension Vehicle.* MIT Press, Cambridge, MA.
TING, L., FULL, R. J., BLICKHAN, R. and TU, M. S. (1990). *Am. Zool.* **30,** 135A.
WALDRON, K. J. (1986). *IEEE J. Robotics Autom.* **RA-2,** 214.
ZAJAC, F. E., and GORDON, M. E. (1989). *Exerc. Sport Sci. Rev.* **17,** 187.
ZILL, S. N. (1985). In Barnes, W. J. P., and Galdden, M. H. (eds.), *Feedback and Motor Control in Invertebrates and Vertebrates.* (pp. 187–208). Croom Helm, London.

Chapter II

The Walking of Cockroaches—Deceptive Simplicity

FRED DELCOMYN

Department of Entomology and Neuroscience Program
University of Illinois
Urbana, Illinois

I. Introduction

The past few years have seen a sharp increase of interest in the design of legged robotic vehicles. This has been reflected in the appearance of books (Donner, 1987; Song and Waldron, 1989) and the convening of conferences (Meyer and Wilson, 1991) on the subject. There have even been articles in the popular press (Freedman, 1991).

Growing with this interest has been an awareness of the features of walking of the six-legged biological machines that we call insects (e.g., Donner, 1987). Walking insects are exceptionally adaptable in the way they move and use their legs. They can walk on irregular surfaces like leaves and branches almost as well as they can on flat, level ground. They can walk forward and backward, and sometimes even sideways, and do so right side up or upside down. Insects can also walk after injury or damage to their legs or even after suffering the complete loss of one or more legs.

Many of these features would be useful in a legged robotic vehicle. For this reason, it has been suggested that study of the insect system can contribute to the design of such a machine. However, few engineers are sufficiently familiar with walking insects to appreciate all that they have to offer. My purpose in this chapter is to alleviate this unfamiliarity and thereby foster communication between biologists and roboticists. I will fulfill this purpose in three parts. First, I will describe the normal walking of one particular insect, *Periplaneta americana,* the American cockroach, and the motor patterns that underlie this walking. Second, I will describe the responses of the cockroach locomotor system to perturbations and point out the implications of these responses for its organization. Finally, I will outline the implications for the design of a robotic legged vehicle of what we know about cockroach (and other insect) walking.

II. Normal Walking

A. Behavior

Most insects exhibit a single, stereotyped gait (a pattern of leg movements) during walking, independently of how fast or how slowly they may be moving (Delcomyn, 1981). That this is not the perception of most engineers is due largely to a single, influential article by the late D. M. Wilson (1966). Before the advent of motion picture photography, it was impossible to be sure of the sequence of leg movements of any but the most slowly moving insects, and visual inspection seemed to reveal a plethora of different gaits. Wilson's brilliant contribution was to show that by following only a few simple rules, one could in theory generate all the gaits that had been described (Fig. 1).

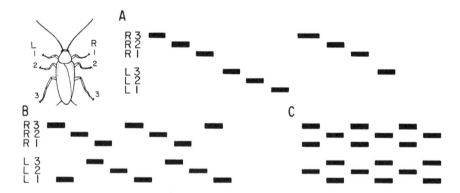

FIGURE 1. Diagrams showing Wilson's (1966) scheme for generating different insect gaits. See text for assumptions. Each leg of the insect is designated by a letter and a number. A black bar in the stepping diagram represents the swing phase of the stepping cycle; the blank space between bars represents the stance phase. (A) Slow walking. Each leg steps once before any leg steps a second time. (B) Stance phase has been shortened so that some legs swing forward at the same time as others. (C) Fast walking. Overlap between the swing phases of different legs has generated the alternating tripod gait, in which front, rear, and contralateral middle legs step together. Note that *cockroaches rarely show patterns other than that shown in (C)*. From Delcomyn (1981).

Chapter II The Walking of Cockroaches

Wilson used three rules: (1) The duration of the forward swing of each leg was constant; (2) the two legs on opposite sides of the body in a single segment moved in "antiphase," meaning essentially in opposite directions at any given time; and (3) legs on one side of the body moved in a metachronal wave from rear to front, with a constant lag from the beginning of movement of one leg to the beginning of movement of another. By following these rules, Wilson was able to generate a variety of gaits merely by varying the duration of the stance period of the leg cycle, as shown in Fig. 1.

The main feature of Wilson's model is that gait is a function of the frequency of leg movement. This is because the period of a cycle of movement is determined by the durations of the swing phase (which is held constant) and the stance phase (which varies to generate the different gaits) of leg movement. Because speed of forward progression in an insect is a function of the frequency with which the insect moves its legs, the model also predicts that gait will vary with speed of progression. The "typical" insect gait, in which the front, rear, and contralateral (opposite) middle legs move together and alternate in their movements with the remaining three legs (hence called the alternating tripod or triangle gait), is generated in this scheme at relatively high speeds of walking.

Wilson knew that each of these rules was violated in one case or another, but he used them anyway, presumably because he thought the violations were exceptions. In the single, careful cinematographic study of insect locomotion that existed at that time (Hughes, 1952), the rules did seem to hold. However, subsequent study has shown that for most insects, not only is the alternating tripod gait common (Fig. 2)

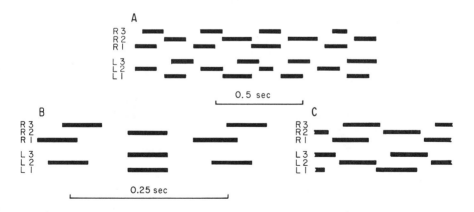

FIGURE 2. Diagrams of stepping in three different insects, each showing the alternating tripod gait. Conventions as in Fig. 1. (A) A grasshopper (*Romalea*). (B) A beetle (*Phosphuga*). (C) A primitive insect, the silverfish (*Petrobius*). There is no time scale for (C); stepping is probably about 3–5 steps per second. From Delcomyn (1981).

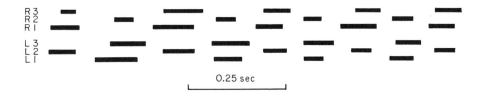

FIGURE 3. Stepping diagram for a slowly walking cockroach. Conventions as in Fig. 1. The insect uses the alternating tripod gait even at this slow walking speed, about 4 steps per second. From Delcomyn (1981).

but also this gait is used over most of the range of walking speeds exhibited by insects (reviews: Delcomyn, 1981, 1985). Certainly for the American cockroach the alternating tripod is the only gait to be seen at stepping rates in the range 3–24 steps per second (Delcomyn, 1971) (Fig. 3). Even at stepping rates as low as 1.5 steps per second (see later), it is *rare* for the American cockroach to depart from this gait. The strong commitment to the alternating tripod gait can also be seen in adjustments an insect makes to adopt the gait after short pauses in its walk (Fig. 4).

B. Motor Patterns Underlying Walking

During walking, the back-and-forth movements of the legs are caused mainly by the reciprocal action of flexor and extensor muscles in the legs. It is hence to be expected that the pattern of activity in these muscles, that is, the timing of muscle action in one leg relative to that in another, will show the same tripod pattern as the legs themselves do. In all animals, there is a minute electrical potential associated with muscle contraction. It is therefore possible to determine the specific times

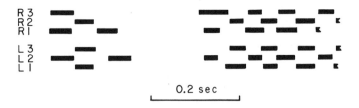

FIGURE 4. Stepping diagram for a rapidly walking cockroach. Conventions as in Fig. 1. Note the spontaneous pause in the middle of the sequence. The right rear leg (R3) did not swing forward with the other members of its tripod just before the pause, and consequently it shows an unusually long swing phase when the insect starts to walk again. This brings the leg back into synchrony with the other members of its tripod. From Delcomyn (1981).

Chapter II The Walking of Cockroaches

when muscles are active by inserting fine wires into the muscles and recording the electrical activity associated with their contraction (Delcomyn, 1989). Such recordings, taken from an extensor muscle (i.e., from a muscle active during the stance phase of the stepping cycle) of each leg during walking, show the tripod pattern, whether the animal is moving slowly (Fig. 5A) or rapidly (Fig. 5B).

The timing of muscle activity, which determines the timing of leg movements, is crucial to the generation of a particular gait. If two legs that normally move synchronously do not do so, the normal gait will be disrupted. Timing between two cyclic events can be described and quantified by calculating the phase of the events (phase = lag/period). If the events are synchronous, the phase is about 0.0 (or 1.0, which is the same); if the events are 180° out of synchrony with each other, the phase is 0.5. Phase is often used as a descriptive parameter in insect walking. In the alternating tripod gait, the phase of any leg relative to any adjacent leg (a leg just in front of it, just behind it, or across the body from it) is about 0.5, since adjacent legs are always in different functional tripods.

Calculation of the phases of muscle activity for pairs of legs reveals the consistency of the alternating tripod gait, as well as other features of walking in cockroaches. Figure 6 shows summary plots of phases from all seven unique pairs of adjacent legs. The solid line drawn through each scattergram represents the 10th-order regression polynomial calculated for each distribution. Several features of these distributions stand out. First, it is clear that the insects use an alternating tripod gait even at very long step periods (600–700 ms, which is 1.6–1.4 steps per

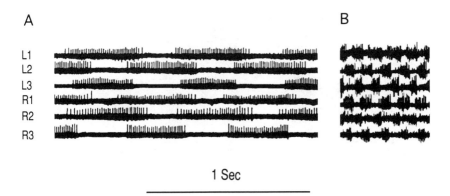

FIGURE 5. Records of the electrical activity in extensor muscles in the coxa of each leg during walking in *Periplaneta americana*. Extensor muscle activity is approximately coincident with the stance phase of walking. The two traces are shown to the same scale. (A) Slow walking (about 2½ steps per second). (B) Fast walking (about 10 steps per second) of a different insect. Note that in both sequences the insect uses the alternating tripod pattern of leg movements.

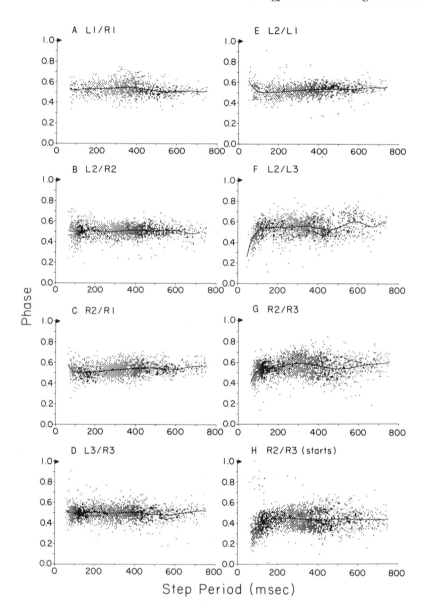

FIGURE 6. Summary plots of the distributions of phase of bursts for every pair of adjacent legs in 1000–2400 steps from 15–39 insects. Phases are calculated on periods measured from burst end to burst end, except for H. Note that phases do not drift significantly at slow speeds of walking (periods of greater than 200 ms), indicating that insects maintain the alternating tripod gait even at that speed. The scatter of the points is due largely to slight differences in mean phase for different individual insects (see Fig. 7). From Delcomyn (1989).

Chapter II The Walking of Cockroaches

second), as shown by an average phase for muscle activity in adjacent legs of about 0.5. The deviation from phases of 0.5 for short periods in adjacent ipsilateral (same side) legs (which is not seen in phases calculated from actual leg movements) is probably due to slight differences in the durations of the swing and stance phases of front, middle, and rear legs, which would show up as shifts in phase as period shortened. Note that the departure from a phase of 0.5 is not present in phases between contralateral legs in a single segment, such as L1/R1.

A second feature of the motor performance of walking cockroaches that is apparent in these plots is the variability of individual phases. Variability, a common feature of biological systems, can be quite large. That it can be significant is illustrated in Fig. 7, which shows the mean and 95% confidence intervals for the phase of motor bursts in one rear leg relative to bursts in the other for 23 different

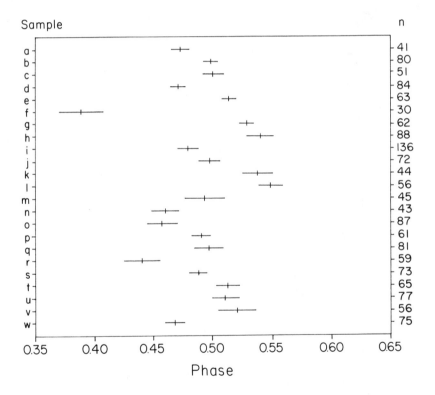

FIGURE 7. Plot showing the mean phases and 95% confidence intervals of bursts in a left rear leg relative to the period of bursts in a right rear leg for 23 intact, freely walking cockroaches. Note the large dispersion between the means compared to the confidence intervals for some insects. From Delcomyn and Cocatre-Zilgien (1988).

animals. Statistical analysis of this type of distribution has shown that insects cannot all be treated as members of a single statistical population (Delcomyn and Cocatre-Zilgien, 1988).

C. The Neural Basis of the Walking Pattern

A physiologist considers behavior to be the consequence of properly selected and timed muscle action. Muscles are activated by impulses from the central nervous system, so from a physiological point of view, the central issue in understanding a behavior is to understand how the nervous system is able to produce the pattern of signals to muscles that brings about the behavior.

For rhythmic activities like walking, it has been shown that there are usually two factors involved in the production of a coordinated pattern. These are an intrinsic ability of the central nervous system to generate a rhythmic, patterned output and, at the same time, sensory feedback from moving appendages that helps to provide signals that stabilize and help time the pattern.

It has been clearly established that nervous systems are capable of generating rhythmic patterns of motor output quite similar to those recorded from intact animals. These patterns are generated by networks of neurons located within the central nervous system and referred to as central pattern generators (CPGs). CPGs can generate a rhythmic output in the complete absence of any sensory feedback, and they seem to underlie all types of rhythmic behavior (Delcomyn, 1980). In insects, the most direct evidence that pattern generators help produce the rhythmic movements of the legs during walking is the demonstration by Pearson and Fourtner (1976) that rhythmic motor patterns resembling those of walking could be generated in a deafferented cockroach (one in which all sensory feedback from the legs had been eliminated).

Although central pattern generators seem to underlie the production of all rhythmic behavior, the contribution of the CPG to the final motor output is not equally strong in every case. For example, cockroaches seem to rely heavily on the centrally generated pattern and ignore sensory input when they walk rapidly (Delcomyn, 1991), whereas the more slowly moving stick insects seem to rely little on the central pattern. Some evidence even suggests that in these insects the central pattern is so weak that it is insufficient to produce well-coordinated motor patterns on its own (Bässler and Wegner, 1983).

In such cases, the other factor that influences the pattern of motor output, sensory feedback from the moving legs, plays a strong role. In stick insects, for example, experiments have demonstrated the strong influence of many different types of sensory input from the legs on leg coordination (see summaries by Graham, 1985;

Bässler, 1987). Even in animals in which CPGs play a dominant role, sensory feedback may be important in fine-tuning movements or in allowing the motor output to remain stable after perturbation. I shall discuss below evidence of the role of sensory feedback in coordinating walking movements in cockroaches.

III. Perturbation of Walking

Examination of the motor patterns and behavior of intact, freely moving insects gives one an idea of how tightly the movements of the legs are regulated by the control system for locomotion. However, it reveals little about how the control system itself functions. One popular and successful technique used to study the organization of the locomotor control system is to perturb the walking system in some way. The rationale is that perturbing walking changes the sensory feedback provided to the central nervous system. If that feedback is important in coordinating leg movements, changing it ought to be reflected in some change in the movements. The particular changes in movements can then be used to form inferences about specific neural pathways or mechanisms that are important in coordination.

Interference with the walking system can be done directly, by destroying or otherwise manipulating the sense organs in the legs, or indirectly, by interfering with or otherwise disturbing normal leg movements. The direct approach can yield dramatic changes in leg movements. For example, surgically altering the attachment of a sense organ in the leg of a locust or a stick insect so that it signals leg extension rather the normal leg flexion results in the leg being held up in an abnormal "salute" position while the rest of the legs continue to walk (Bässler, 1979; Graham and Bässler, 1981). On the other hand, destroying one or more sense organs in the legs often has only minor effects on walking or its coordination (e.g., Usherwood et al., 1968; Wong and Pearson, 1976). This may be because there is considerable redundancy in sensory feedback (i.e., many different sense organs provide essentially similar information) (Hustert, 1983; Bräunig and Hustert, 1985).

Many of the indirect techniques have the advantage that they emulate events or situations that an insect may encounter in nature. Three types of perturbation experiments have been done. The first involves allowing the insect to walk on a different surface or with a different orientation than usual. Both rough terrain (Pearson and Franklin, 1984) and slopes (Spirito and Mushrush, 1979) cause some small changes in stepping. A second type of experiment is to interfere with leg movements. These experiments cause a more dramatic alteration in stepping, one that is dependent on when in the stepping cycle an obstruction is encountered (Dean and Wendler, 1982). The third method is removal of a leg, which emulates natural leg autotomy that may result from accident or attack by a predator. This type of experiment also has dramatic effects on locomotion. I will discuss it in more detail in the next section.

A. *Freely Walking Insects*

1. Behavioral Effects. It is not surprising that the behavioral effects of some leg amputations are dramatic. Recall that in the alternating tripod gait, one side of the body is being supported by only a single middle leg while the other side is being supported by the front and rear legs together. Hence, removing the two middle legs will leave an insect without support on first one then the other side of the body during each step. Unless some adjustment is made in the gait, it would seem that under these conditions the insect would fall over each time it attempted to walk.

But freely walking insects do not fall over when they walk under these conditions. For over a hundred years, it has been recognized that after amputation of the middle legs insects do switch gaits, such that the front and rear legs step alternately instead of in synchrony with one another. Because this switched gait is similar to a gait of four-legged animals, it has been called the tetrapod gait. For many decades, it was thought that all insects used the tetrapod gait at all times after middle leg amputation. It was not until the application of motion picture analysis to the walking of insects that researchers discovered that at least some insects could use other gaits after amputation, as well.

Examination of high-speed motion pictures of American cockroaches showed that during slow walking after middle leg amputation, the front and rear legs did move alternately (Delcomyn, 1971) in the well-known tetrapod gait (Fig. 8A). However, when these insects walked rapidly, the front and rear legs reverted to a near-normal relationship in which they moved nearly synchronously (Fig. 8B).

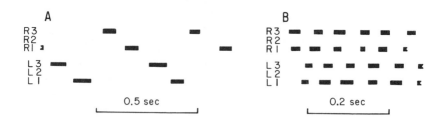

FIGURE 8. Stepping diagram showing the effects of double middle leg amputation on the walking of a cockroach. Conventions as in Fig. 1. Note the difference in time scale in the two diagrams. During slow walking (A, about 2 steps per second), the insect adopts a tetrapod gait (see text). During fast walking (B, about 17 steps per second), it uses the alternating tripod gait without the two middle legs. From Delcomyn (1981).

It is worth emphasizing that most insects can only walk relatively slowly. Insofar as such slow-moving insects have been studied, they always exhibit the tetrapod gait after middle leg amputation. It is only in the case of a very swift insect like *P. americana* that the timing of the front and rear legs can approach normal in the amputee insect. This is presumably due in part to the high frequency of leg movement in this insect at high speeds of walking—as much as 24 steps per second (Delcomyn, 1971). At even 20 steps per second, each leg is off the walking surface for only 25 ms (about half the cycle time of 50 ms), which is a short enough period that the body has time to fall only about 3 mm. The result is a rather lurching forward progression that looks awkward as the body tips from side to side, but that is nevertheless effective in getting the insect rapidly from one place to another.

2. Physiological Effects. The timing of motor bursts in the legs of walking cockroaches missing one or more legs can be predicted from a knowledge of the leg movements that these bursts produce. However, amputation has effects on the motor pattern that are not apparent from a study of the behavior only. I will describe one of these, multiple bursting, before describing the physiological basis of the gait switch that occurs after amputation.

Amputations are normally done at the trochanter, the small leg segment that is fused with the femur and articulates with the coxa. The coxa is the first segment of the leg and the one from which recordings of muscle activity are normally made. Hence, it is possible to record the activity of coxal muscles in the stump of an amputated leg. Because this activity is an expression of the locomotor motor pattern (rather than, for example, an attempt by the animal to find a foothold; Delcomyn, 1988), study of the pattern can lead to further insights into the organization of the motor control system.

An unexpected feature of the motor pattern in the stump of an amputated leg was the appearance of multiple bursts. These are bursts that occur at two or three times the frequency of bursts in the neighboring, intact, legs. Careful observation of the stump while recordings were being made showed that when double bursting occurred, the stump actually moved back and forth two or three times for every single cycle of movement of an adjacent leg (Delcomyn, 1988). Not every insect showed multiple bursting after amputation, and among those that did, multiple bursts appeared only during some sequences of walking, another expression of variability in the locomotor system.

Multiple bursts have preferred timing relationships to single bursts in neighboring legs. Two typical examples are shown in Fig. 9. An interesting feature of these plots is that the timing between the muscle bursts in the middle leg stump and those in the front leg (Fig. 9A) exhibits less variability than does the timing between bursts in the stump and the intact rear leg (Fig. 9B). The basis for this difference is not known; it could, however, represent differences in coupling strength between pattern generators regulating movements of the different legs.

The biggest behavioral effect of a middle leg amputation is a shift in timing between the front and rear legs ipsilateral to the amputated leg during slow walking. When the behavioral experiments were done (Delcomyn, 1971), it

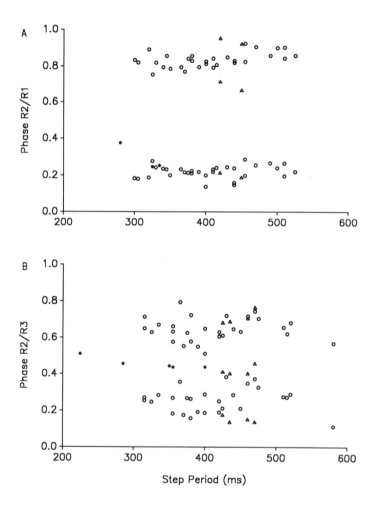

FIGURE 9. Plots of the phase between motor bursts in a right middle leg stump and (A) bursts in the leg in front of it and (B) bursts in the leg behind it. Note that phase is independent of the step (burst) period. Key: filled circles, single bursts in the stump; open circles, double bursts (see text); triangles, triple bursts.

Chapter II The Walking of Cockroaches

was thought that the switch from alternating tripod to tetrapod gait was relatively abrupt. However, examination of the phases of front to rear leg bursts after middle leg amputation showed clearly that the timing of the bursts from these legs, and hence the timing of the leg movements that the bursts drive, is a continuous function of burst period (which is the duration of a full cycle of leg movement) (Fig. 10). This means that rather than being abrupt, the shift in timing is a continuous function of the speed at which the insect walks. This finding suggests that the gait change is brought about by some parameter that changes gradually as speed of walking changes, rather than one that acts like an on–off switch.

Examination of the timing of motor bursts in the stump of an amputated leg provides a further clue to the source of the signal that causes the gait shift. When bursts in the stump are 1: 1 to bursts in the intact legs (i.e., when there is no multiple bursting), the timing of the stump bursts is different relative to bursts in the leg in front of the stump than it is relative to bursts in the leg behind the stump. There is a strong speed-dependent shift in phase between motor bursts in the stump and the front leg (Fig. 11A) and absence of any such phase shift between stump bursts and bursts in the rear leg (Fig. 11B)

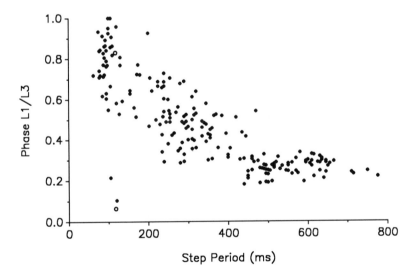

FIGURE 10. Plot of the phase between bursts in a left front leg relative to bursts in the ipsilateral rear leg in a freely walking cockroach in which the left middle leg had been amputated. The timing between bursts in these legs shows a strong period-dependent distribution. In intact insects, the average phase between bursts in these legs is 0.048.

(Delcomyn, 1991). The timing shift is not dependent on amputation of the middle leg specifically, because a similar shift can be seen between bursts in the stump of an amputated rear leg and the middle leg in front of it (Delcomyn, 1991). In contrast, there is no significant shift of timing between motor activity in a stump and the intact leg on the opposite side of the body, for either middle or rear leg amputations (Delcomyn, 1991), although timing between bursts in such contralateral leg pairs is quite variable.

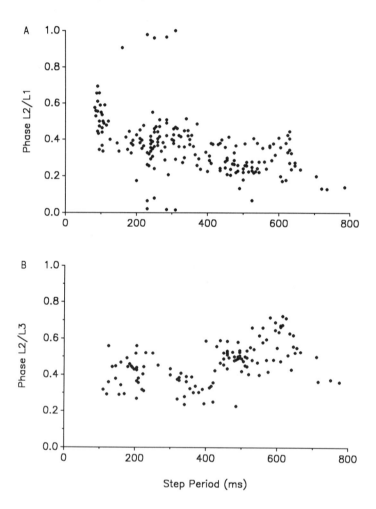

FIGURE 11. Plots of the phase between motor bursts in a left middle leg stump and (A) bursts in the leg in front of it and (B) bursts in the leg behind it. These plots show the phases of single bursts. Figure 9 showed the phases of multiple bursts. Note the tendency for the L2/L1 phase to decline as step period increases (A) and the absence of such a trend in the L2/L3 phase (B).

One can draw several inferences from these differences in the effects of an amputation on the timing of motor bursts in the stump relative to other legs. First, mechanisms that ensure proper coupling between contralateral leg pairs are probably different from mechanisms that ensure proper coupling between ipsilateral ones. This statement is based on the observation that amputating a leg has a strong effect on timing between some ipsilateral legs but none on contralateral ones. Second, information from sense organs in the legs is critical to proper timing between legs, but only during slow walking. The lack of importance of sensory feedback during fast walking is suggested by the disappearance of all of the special features of amputee motor patterns at high speeds (Delcomyn, 1991), as well as other data (Zill, 1985; Zill, this volume). During slow walking, however, it is clear that the changed sensory condition causes the dramatic shift in gait. And finally, particular information from specific leg sense organs about leg movements seems to travel only in specific directions in the central nervous system. This statement is based on the observation that the effects of amputation of a leg are ipsilaterally asymmetrical; that is, the effect on timing is different relative to the leg in front of the amputated leg than it is relative to the leg behind it.

Although the experiments described so far indicate that sensory information is important, it does not allow us to say what the source of the observed ipsilateral timing shift is. There are two possibilities. On the one hand, it could be due to the loss of sensory input from the missing leg. Not only are the sense organs in the missing part of the leg also missing, but also the campaniform sensilla on the trochanter, which are thought to be important in sensing the cuticular stress associated with load bearing, must have a sharply reduced output because the stump is too short to touch the walking surface. On the other hand, there are also differences in the strength and timing of sensory input from the intact legs after an amputation compared to before, because of the changed load distribution as a consequence of the amputation. Of course, these effects are not mutually exclusive.

B. Suspended Insects

In order to gain a thorough understanding of the locomotor system, it is important to understand the relative contributions of these two effects of an amputation. They can be evaluated as described below by suspending an insect and allowing it to walk on a surface such as graphite-covered glass that is too slippery for the animal to grip. This arrangement causes a significant reduction in feedback from cuticular stress receptors during walking, because the reduction of force on the legs due to slippage results in reduced stimulation of these receptors.

In spite of this reduction in feedback, intact, suspended cockroaches show patterns of motor activity similar to those seen in intact, freely walking cockroaches (Fig. 12). That is, bursts in adjacent legs show phases of about 0.5 relative to one another, independent of the frequency of leg movement of the insect, yielding the alternating tripod gait. There are sometimes minor differences in the exact timing between motor bursts in different legs, as well as differences in the scatter of the timing, but overall these are not statistically significant.

Amputating a leg of a suspended insect allows the effects of loss of sensory input to be revealed with minimum contamination from the effects of a changed sensory input from the intact legs, because the suspended insect experiences no significant change in load as a consequence of amputation. Hence, if the effect of amputation in a suspended insect is the same as the effect of amputation in a freely walking one, then the loss of sensory input from the missing leg is the cause of the changed timing between motor bursts and the change in gait. On the other hand, if the effects of amputation in a suspended insect are different from those in a freely walking one, then the changed load on the intact legs in the free cockroach must be the main cause of the gait change that follows amputation.

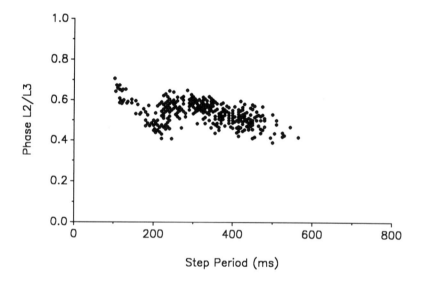

FIGURE 12. Phase between bursts in a left middle leg and those in the leg behind it in a suspended, intact insect walking on a slippery glass plate. Most of the phases fall in the range 0.4 to 0.6, as do phases for free walking insects (see Fig. 6). There is a slight disturbance of phase due to the walking conditions, but there is no significant change of phase at low speeds of walking.

Chapter II The Walking of Cockroaches   37

What happens after amputation? In most cases, amputation of a middle leg does cause strong changes on the timing of motor activity in some of the ipsilateral legs. First, the walking speed–dependent phase shift between bursts in the front and rear legs on the same side as the amputated one that is so obvious in freely walking amputee insects is apparent in suspended amputee insects as well. In most insects, there is also a phase shift between motor bursts in the stump and those in one of the intact ipsilateral legs. However, whereas in the freely walking middle leg amputee

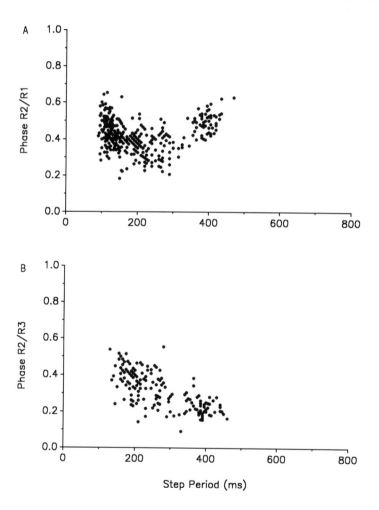

FIGURE 13. Phase between bursts in the stump of a right middle leg and (A) bursts in the leg in front of it and (B) bursts in the leg behind it, in a suspended insect. See Fig. 11. In spite of the increased scatter in the distribution, the plots show a tendency for the R2/R3 phase to decline as step period increases and the lack of such a trend in the phase of R2/R1.

the phase shift is relative to bursts in the leg in front of the amputated one, in suspended cockroaches the phase shift is relative to bursts in the leg *behind* the stump (Fig. 13B). Timing between bursts in the stump and in the front leg is more or less independent of the period of the bursts (Fig. 13A).

These results indicate that in freely walking insects, the altered gait and altered motor burst timing that follow loss of a middle leg are due largely to the altered load on the intact legs rather than the loss of sensory information from the missing leg. In suspended insects, on the other hand, the effects of middle leg amputation are due mainly to loss of information from the missing leg. This must be so because there is no significant change in load on the legs of the suspended insect as a consequence of amputation.

These experiments lead to two conclusions about mechanisms of coordination in cockroaches. First, there must be strong feedback from a leg onto the pattern generator for that leg itself, since the changed loading on the intact legs can cause a significant change in coordination. Second, it is also the case that sensory feedback from at least the middle legs has an influence on the timing of muscle activity in the leg behind it, since there is an effect of amputation in suspended insects (in which there is no changed loading). An influence of sensory feedback on muscle activity in a leg in front may exist as well, but the amputation experiments provide no support for it.

IV. Implications for Robotics

One objective of neurobiologists studying the neural mechanisms by which rhythmic behavior is generated is to understand how specific neural interactions can lead to a coordinated motor performance. One objective of engineers designing a legged vehicle is to develop a control system that will allow coordinated movement. It is thus obvious that specific control mechanisms revealed in the biological system may be of some use to the engineer, either by providing a direct method for solution of a specific control problem or by giving the engineer an idea for a practical solution even if it is not similar to the way in which the biological system handles the problem.

A. *Summary of the Insect System*

From the point of view of engineering design, three aspects of the control system for insect walking may be relevant to the construction of a robotic vehicle. First, and perhaps trivially, the biological system uses specific feedback mechanisms to ensure coordination. That is, input from specific sense organs seems to be used in certain pathways to help coordinate the movements of specific legs. In other words, not every piece of sensory information is distributed to every element of the system. Information flows in particular directions to specific destinations only.

Second, in fast-moving insects like cockroaches, the control mechanism is sensitive to the speed of progression of the animal. It is apparent that during fast walking, cockroaches rely largely on central mechanisms of coordination. That is, the intrinsic pattern of motor activity laid down by the interconnected CPGs is fed directly to the muscles with little influence from feedback from sense organs in the legs. If, on the other hand, the insect is walking slowly, feedback from phasically stimulated sense organs in the legs plays a dominant role in coordination, as evidenced by the dramatic disruption of the walking pattern when sense organ output is disturbed.

The dependence of a coordinated pattern on sensory feedback during slow walking leads to a third characteristic, that of flexibility or adaptability of response. If normal feedback is interfered with, such as by loss of a leg, the locomotor control system adapts automatically to the loss. This adaptation ensures that the insect will still be able to make reasonable forward progress.

B. Possibilities for Control

These biological characteristics suggest two elements of design that might be useful for engineers to consider. First, the optimal design of the control system for slow progression may be different from that for fast progression. It may be important for engineers to consider carefully the performance parameters desired for their machines and to design the control system to optimize that performance, rather than trying to build a machine that will do everything equally well. Alternatively, it may be necessary to use different control mechanisms for different parts of the performance range of the device. Control could be switched from one mechanism to another based on the desired speed range, or control could be influenced in a graded manner in a way similar to the system used by insects.

Second, it may be important for engineers to think in terms of performance after damage without repair. In automobiles, it is expected that damage will be repaired, so there are few provisions for continued operation after damage has occurred. In space vehicles, performance after damage or other failure is more important. There the design approach is redundancy, so that if one control mechanism fails, another will be available to take its place.

Insects use a third approach. If a part of the body is damaged, the control system automatically adapts to allow the remaining parts to continue to function reasonably normally. This kind of design might be especially useful in robots used in planetary exploration because the cost of duplicate control mechanisms might be prohibitive; a system that could function in spite of damage would have obvious benefits.

C. Conclusions

Whether insect systems are used as a model for emulation or as a source of inspiration (Beer, this volume) in the design of robotic legged vehicles, neither biologists nor engineers should expect every detail of the insect locomotor control system to provide useful information. It is important to bear in mind that insects have been shaped by millions of years of evolution. Evolutionary pressures have not been applied exclusively toward the development of an efficient system for the control of walking, since other functions must also be considered. Variability of response, for example, a concept largely foreign to mechanical devices, may be an important component in allowing insects to produce appropriate and adaptive responses to unforeseen situations.

In the end, any attempt at interaction between biologists and engineers will stand or fall on open lines of communication. If the biologist is willing to entertain the perhaps naive biological questions of the engineer and the engineer is willing to put up with the biologist's perhaps simplistic ideas about control, interactions between the two may lead to real progress in both realms.

Acknowledgment

The original research described in this article was supported by grants from the National Institutes of Health and the Whitehall Foundation.

References

BÄSSLER, U. (1979). *Physiol. Entomol.* **4,** 193.
BÄSSLER, U. (1987). *Biol. Cybern.* **55,** 397.
BÄSSLER, U., and WEGNER, U. (1983). *J. Exp. Biol.* **105,** 127.
BRÄUNIG, P., and HUSTERT, R. (1985). *J. Comp. Physiol. A* **157,** 73.
DEAN, J., and WENDLER, G. (1982). *J. Comp. Physiol.* **148,** 195.
DELCOMYN, F. (1971). *J. Exp. Biol.* **54,** 443; *ibid.* **54,** 453.
DELCOMYN, F. (1980). *Science* **210,** 492.
DELCOMYN, F. (1981). In C. F. Herreid and C. R. Fourtner (eds.) *Locomotion and Energetics in Arthropods,* p. 103. Plenum, New York.
DELCOMYN, F. (1985). In G. A. Kerkut and L. I. Gilbert (eds.), *Comprehensive Insect Physiology Biochemistry and Pharmacology,* Vol. 5, p. 439, Pergamon Press, Oxford.

DELCOMYN, F. (1988). *J. Exp. Biol.* **140,** 465.
DELCOMYN, F. (1989). *Biol. Cybern.* **60,** 373
DELCOMYN, F. (1991). *J. Exp. Biol.* **156,** 483; *ibid.* **156,** 503.
DELCOMYN, F., and COCATRE-ZILGIEN, J. H. (1988). *Biol. Cybern.* **59,** 379.
DONNER, M. D. (1987). *Real-Time Control of Walking,* Birkhäuser, Boston.
FREEDMAN, D. H. (1991). *Discover* **12**(3), 42.
GRAHAM, D. (1985). *Adv. Insect Physiol.* **18,** 31.
GRAHAM, D., and BÄSSLER, U. (1981). *J. Exp. Biol.* **91,** 179.
HUGHES, G. M. (1952). *J. Exp. Biol.* **29,** 267.
HUSTERT, R. (1983). *J. Comp. Physiol.* **150,** 77.
MEYER, J. A., and WILSON, S. W. (1991) *(eds.),* (1991). *From Animals to Animats. Proceedings of the First International Conference on Simulation of Adaptive Behavior.* MIT Press, Cambridge.
PEARSON, K. G., and FOURTNER, C. R. (1976). *J. Neurophysiol.* **38,** 33.
PEARSON, K. G., and FRANKLIN, R. (1984). *Int. J. Robotics Res.* **3,** 101.
SONG, S. -M., and WALDRON, K. J. (1989). *Machines That Walk: The Adaptive Suspension Vehicle,* MIT Press, Cambridge.
SPIRITO, C. P., and MUSHRUSH, D. L. (1979). *J. Exp. Biol.* **78,** 233.
USHERWOOD, P. N. R., RUNION, H. I., and CAMPBELL, J. I. (1968). *J. Exp. Biol.* **48,** 305.
WILSON, D. M. (1966). *Annu. Rev. Entomol.* **11,** 103.
WONG, R. K. S., and PEARSON, K. G. (1976). *J. Exp. Biol.* **64,** 233.
ZILL, S. N. (1985). In W. J. P. Barnes and M. H. Gladden, (eds.) *Feedback and Motor Control in Invertebrates and Vertebrates,* p. 187. Croom Helm, London.

Chapter III Load Compensatory Reactions in Insects: Swaying and Stepping Strategies in Posture and Locomotion

SASHA N. ZILL
Department of Anatomy and Cell Biology
Marshall University School of Medicine
Huntington, West Virginia

I. Introduction: Adapting Posture and Locomotion in Unstable Environments

Most of the world isn't flat, and much of it represents extremely unstable and uneven surfaces to stand and walk on. As noted by Marc Raibert, "only about half the earth's landmass is accessible to existing wheeled and tracked vehicles, whereas a much larger fraction can be reached by animals on foot" (Raibert, 1986, p. 1). These obvious facts are compelling reasons for studying the principles and mechanisms underlying the control of animal posture and locomotion in order to construct legged robots. In order to stand on and traverse irregular substrates, both animals and legged robots must utilize information from sensors that detect variations in the environment. Furthermore, this information must be integrated with data provided to the standing or walking control system about the position of the legs and the forces they exert on the substrate to maintain stable or dynamic support and avoid falling.

In this chapter I will examine how sensory inputs are utilized to adapt posture and legged locomotion in one of the most diverse and successful animal groups, insects. Although the nervous systems of insects have been used as models for the design of several systems to control legged robots, it has not been clear to some investigators whether, in theory, these animals actually have the ability to adjust posture and walking to perturbations and irregular terrains. For example, in considering postural control using an inverted pendulum model, Marc Donner reasoned, based on their size and reaction time, that insects may not possess mechanisms for maintaining balance in walking (Donner, 1987). He noted that "if there were a simple reflex capable of handling balance it might be able to respond fast enough" but "if balance required any more complex decision making...involving multiple legs or sensors...it couldn't be performed by insects" (*ibid.,* p. 100).

In the following sections, I will review recent literature that demonstrates that insects are able to respond to perturbations occurring in posture and during walking. Furthermore, as animals, insects are of interest in studying postural control because they do not possess sense organs that are specialized for detection of gravitational forces or body acceleration, such as the vestibular apparatus of vertebrates or the crustacean statocyst. Instead, insects apparently use information from sensory receptors, known as *proprioceptors,* that are located in the legs to calculate gravitational vectors (Markl, 1971; Wendler, 1971). These types of sense organs monitor variables such as joint angles and forces exerted on the legs. These receptors also provide the nervous system with ongoing data about changes in the environment and with inputs reflecting parameters of the animal's own walking movements. The nervous system uses these inputs in a process known as *load compensation* to generate appropriate muscle contractions and movements to adjust for irregularities in the walking substrate and to compensate for perturbations.

Before I discuss the load compensatory responses of insects, however, I will review several paradigms that have been used to examine characteristics of responses to mechanical perturbations in posture and locomotion of vertebrates, particularly humans, and the results that have been obtained. The literature on vertebrates is considerably more extensive than that on invertebrates and provides a conceptual framework and gives insights into basic strategies that may be common to adaptive behaviors of a number of terrestrial animals. In addition, many of the techniques used in those studies have been adapted to examine the types of load compensatory reactions demonstrated by insects.

The goal of this chapter is to delineate strategies used by insects in incorporating information from leg sense organs to generate different types of adaptive responses. It is hoped that some of these strategies may be found useful in the design of artificial walking control systems.

II. Characteristics of Load Compensatory Reactions in Vertebrates in Posture and Locomotion

A. Swaying Strategies in Postural Control

Mechanical perturbations of standing subjects have been widely used in a number of paradigms to characterize the types of responses and patterns of muscle activities that occur in postural reactions of vertebrates. In the paradigm developed by Nashner (Nashner, 1976, 1977; Nashner et al., 1979; Nashner and McCollum, 1985), human subjects stand on a platform that is suddenly rotated or moved horizontally or vertically (Fig. 1A). Humans react to substrate movements by using discrete reactions and motor strategies. Platform displacements of small or intermediate magnitude produce responses that consist of swaying strategies, in which the foot remains fixed on the platform and changes occur in the angles of leg joints. For example, when the platform is moved horizontally backward for a short distance, subjects sway forward at their ankle joints and these changes in joint angle are countered by contractions of extensor muscles on the back of the legs (Horak and Nashner, 1986) (Fig. 1B). These muscle contractions are automatic responses, in that they occur at latencies that are considerably shorter than voluntary reaction times. When somewhat larger displacements are applied, a hip strategy is used. The changes in hip joint angles that occur in those reactions are also resisted by short-latency, stereotypical contractions of hip muscles. Similar types of reactions have also been demonstrated and widely studied in other vertebrates, such as cats (MacPherson, 1988).

B. Stepping Strategies

Another paradigm that has been used to test dynamic postural reactions is the postural stress test (Wolfson et al., 1986). In this paradigm, a weight is attached by a pulley and wire to a padded belt around a subject's waist (Fig. 1C). Postural perturbations are generated by suddenly dropping the weight. As the person being tested stands facing away from the pulley, this paradigm measures a subject's ability to counter destabilizing forces that pull the center of gravity behind the base of postural support (Duncan et al., 1990). Normal human subjects respond to release of small weights by swaying strategies and changes in leg joint angles similar to those seen in the platform tests. However, larger weights cannot be effectively countered by simple changes in joint angles but instead evoke stepping strategies. In those responses, the foot is lifted from the substrate and repositioned and one or more steps are taken to produce a more stable basis of postural support (Wolfson et al., 1986). Such stepping strategies have also been demonstrated in studies in

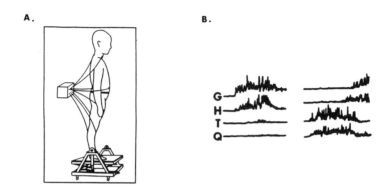

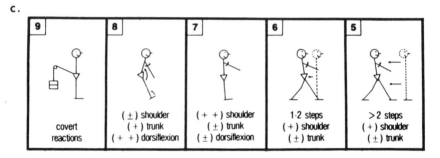

FIGURE 1. Methods used to test postural load compensatory reactions of humans. (A) In Nashner's paradigm, a subject stands on a movable platform and responses of muscles are recorded myographically with wires. (B) Muscle activities during compensatory responses. Left: suddenly moving the platform backward causes a subject's body to tilt forward, reflexively activating muscles on the back side of the leg (G, gastrocnemius; H, hamstrings). Right: displacement in the opposite direction generates contractions of muscles on the front of the leg (T, Tibialis anterior; Q, Quadriceps). Body tilt in these reactions is accompanied by changes in leg joint angles. C. In Wolfson's paradigm, a weight is attached to a subject's waist by a wire and pulley. Suddenly releasing small weights induces body tilting and swaying reactions. Large weights cause stepping reactions in which a leg is lifted and re-positioned. (A) reprinted from Woollacott and Shumway-Cook, *Phys. Ther.* **70**, 799, 1990, with the permission of the American Physical Therapy Association; (B) reprinted from Nashner, Woollacott and Tuma, 1979; (C) from Wolfson, L.I., Whipple, R., Amerman, P., and Kleinberg, A., Stressing the postural response: a quantitative method for testing balance, *J. Amer. Ger. Soc.* **34**, 845–850, 1986, © American Geriatric Society.)

which forces were applied to subjects that were standing in initially unstable postures, such as leaning far forward (Do et al., 1982). Although less well studied, stepping responses have also been reported to be elicited in Nashner's paradigm at high rates of platform displacement (Horak, 1987).

C. Load Compensatory Reactions during Walking

Responses to perturbations during walking have also been extensively studied in vertebrates. In the early studies of Nashner (Nashner, 1980), a movable platform was placed in the path of a walking human subject. When a foot was placed on the platform during the stance phase, small to intermediate displacements produced muscle contractions that consisted of swaying strategies similar to those used when a person was standing still. For example, moving the platform horizontally backward produced flexions at the ankle joint followed by activation of the extensor muscles. These studies implied that similar strategies were used to regulate joint angles in posture and in locomotion, the difference being that joint angles were fixed in standing but followed a prescribed trajectory during locomotion. These observations were confirmed and extended by the studies of Dietz and colleagues (Berger *et al.*, 1984). In those studies subjects walked on a treadmill and perturbations were produced by suddenly increasing or decreasing the treadmill speed. Again, swaying strategies were used in the leg in the stance phase and muscle contractions opposed changes in joint angles. However, when muscle activities were examined in the leg that was in the swing phase, a different strategy, qualitatively similar to a movement or stepping strategy, was revealed. Sudden changes in the speed of the treadmill induced movements of the leg that were generated by muscle contractions. Accelerations of the treadmill caused the leg in the swing phase to be rapidly brought forward to provide support in front of the body. In contrast, decelerations actually caused the leg in swing to reverse its direction of movement and it was placed down behind the body. In both these cases, the movements or "steps" taken by the leg in swing caused it to be repositioned to provide a dynamically stable base of support.

D. Resetting the Pattern of Locomotion

A last notable effect of many types of perturbations in vertebrates is that they can reset the rhythm of muscle contractions and movements during walking, depending on the specific phase of walking in which they are applied. When a cutaneous sensory nerve is electrically stimulated in the hind leg of a walking cat, for example, it produces a flexion of that leg to pull it away from the noxious stimulus (Gauthier and Rossignol, 1981). This stimulation also produces movements of the opposite leg whose direction depends on the phase of the step cycle. The opposite leg is flexed if that leg is in the swing phase but extended if the leg is in the stance phase. Those movements rapidly increase the opposite leg's role in postural support. However, when the stimulus is applied in the time of transition from swing to stance, the animal takes an additional quick step with the contralateral hind leg that subsequently resets the rhythm of walking. Similar resetting effects have been shown during stepping strategies of humans perturbed when walking on a treadmill (Berger *et al.*, 1984). These studies imply that while many perturbations can be effectively compensated by swaying or stepping strategies, some types of disturbances can be countered only by steps that reset the walking system as a whole.

E. Contributions of Different Sensory Modalities to Vertebrate Postural Control

Although these classes of motor responses have been well characterized in vertebrates, there is still considerable controversy about the neuronal mechanisms that underlie these compensatory responses and the types of sensory modalities that contribute to these reactions (Nashner and McCollum, 1985; Cohen and Keshner, 1988). The early studies by Nashner and colleagues suggested that proprioceptive inputs from limb mechanoreceptors were primarily responsible for generating compensatory reactions (Nashner *et al.*, 1979). For example, ankle muscles were activated in postural swaying tests at latencies similar to those seen in studies of long-latency stretch reflexes (knee jerk reflexes) in nonstanding subjects, suggesting that leg proprioceptors such as ankle muscle spindles or joint receptors mediated postural responses (Nashner, 1977). However, more recent studies have shown that neck, spinal, and abdominal muscles are also recruited at similar latencies, suggesting that other, non-proprioceptive pathways were also active (Keshner *et al.*, 1988). Thus, the current literature suggests that inputs from the visual, vestibular, and leg proprioceptive systems affect reactions to postural perturbations (see Nashner and McCollum, 1985 for review; Cohen and Keshner, 1988; Keshner and Cohen, 1988).

F. Summary of Characteristics of Load Compensatory Reactions in Vertebrates

1. In posture, small perturbations produce changes in joint angles that are followed by activation of muscles that resist those changes = swaying strategies.
2. Larger perturbations, produced by sudden destabilizing forces, elicit steps in which a leg is lifted from the substrate and repositioned = stepping strategies.
3. In walking, substrate displacements also produce changes in joint angles that are followed by activation of muscles to resist those perturbations during the stance phase.
4. In walking, some perturbations can produce additional steps and some can reset the rhythm of stepping, depending on the phase of the step cycle in which they occur.

Chapter III Load Compensatory Reactions in Insects

5. Many of the characteristics of load compensatory reactions resemble stretch reflexes (mediated by muscle spindles). However, a number of sensory modalities can potentially contribute to these reactions, including visual, vestibular, and proprioceptive systems.

G. Outline of Load Compensatory Reactions in Insects

I will now

1. Briefly characterize the response properties, reflex effects, and some of what is known of the central connectivity of insect proprioceptive sense organs, focusing on one type of joint angle receptor known as a chordotonal organ.
2. Review recent studies that have directly examined load compensatory reactions in insects. These studies have sought to characterize the motor strategies that are used by insects in load compensation and to determine some of the variables that are controlled in these responses and the types of sense organs that contribute to them.
3. Examine some preliminary data from experiments that have tested the responses of walking cockroaches to mechanical perturbations. I will also review evidence in the literature that supports the idea that the effects of proprioceptive sense organs depend on the speed of walking and that those effects are either limited or modified during rapid locomotion in cockroaches.

III. Functions of Proprioceptive Sense Organs in Insects: A Capsule Summary

Studies of the functions of different types of proprioceptive sense organs in generating and adapting posture and walking in insects have been greatly aided by the fact that many of these receptors are identifiable, in that the same sense organs can be found in different animals. This fact has permitted characterization of the response properties of individual proprioceptors and study of their central connectivity and reflex effects in motoneurons of leg muscles. In the following I will briefly review some of the types of proprioceptive sense organs that are found in the legs of insects. The reader is referred to much more extensive reviews of the subject (Bässler, 1983; Zill, 1990).

A. *Types of Proprioceptors*

1. Receptors that monitor joint angles. In insects at least four types of sense organs encode the angles of leg joints: hair plates, multipolar receptors, chordotonal organs, and strand receptors. Hair plates are groups of innervated hairs that are located on the surface of the exoskeleton adjacent to joints. The hairs are bent by joint flexion. Multipolar receptors, chordotonal organs, and strand receptors are found inside the exoskeleton and consist of neurons that either span joints or are linked to ligaments that extend between leg segments. The redundancy of information provided to the nervous system by these different types of receptors ensures that the full range of joint position and movements is unequivocally encoded.
2. Receptors that monitor muscle contractions and leg load. Campaniform sensilla are receptors that monitor the effects of muscle tensions and external loads by the mechanical strains they produce in the exoskeleton. These sense organs, which essentially function like strain gauges used in structural engineering, consist of sensory neurons that have a process (dendrite) embedded in the exoskeleton. Forces generated by the animal's weight and muscle contractions produce strains in the exoskeleton that generate afferent activity. The magnitude of contractions of some muscles is also encoded by multipolar receptors attached to muscle tendons.

B. *Reflex Effects and Potential Functions of Leg Proprioceptors*

Many types of receptors can control movements and muscle tensions by negative feedback. Stimulation of a number of proprioceptive sense organs has been shown to produce reflex effects in motoneurons to leg muscles that form negative feedback loops. These include many chordotonal organs, multipolar receptors, and some groups of campaniform sensilla. This feedback limits changes in joint angles or forces imposed on the leg. These reflexes could contribute to load compensatory reactions if perturbations produce postural changes in joint angles or alterations in leg load. Some of these reflexes can also function to limit the amplitude of leg movements during normal, unperturbed walking. The hair plate located on the trochanteral segment of the leg, for example, has been shown to aid in setting the anteriormost position of a leg at the end of the swing phase (Pearson *et al.,* 1976). Ablation of this receptor results in "overstepping," in which the swing phase is exaggerated.

Some sense organs can potentially amplify muscle contractions by positive feedback. The trochanteral campaniform sensilla can respond to strains that result from contractions of the trochanteral extensor muscle, which is active in the stance phase. Mechanical stimulation of these sense organs, in turn, excites extensor motoneu-

Chapter III Load Compensatory Reactions in Insects   51

rons. These groups of receptors have been postulated to function in a positive feedback loop to amplify the force exerted in the stance phase (Pearson, 1972).

Other receptors have multiple modes of reflex action, some of which may induce steps. In several insects, mechanical stimulation of the femoral chordotonal organ has been shown to elicit more than one type of reflex action in motoneurons to leg muscles. I will focus on this sense organ because these reflex effects may contribute to the types of strategies used in load compensation. The femoral chordotonal organ consists of a group of sensory neurons that encode the angle of the femorotibial joint of the leg and also the rate of change and, potentially, acceleration of joint angle (Fig. 2A) (Usherwood *et al.,* 1968; Zill, 1985a; Matheson, 1990). Selective mechanical stimulation of the organ is effected by pulling on its ligament, which spans the femorotibial joint, producing an afferent discharge that mimics an actual change in joint angle. In stick insects, Bässler first demonstrated that the reflex effects of this stimulus depended on the specific behavior of the animal (Bässler, 1976, 1988). In "inactive" animals, which were not moving nonstimulated legs, displacements of the ligament of the organ produced resistance reflexes that opposed apparent joint movements. For example, pulling on the ligament, producing an afferent discharge mimicking joint flexion, excited the extensor muscle. However in "active" animals, which were moving other legs, the same stimulus elicited a complex series of coordinated muscle contractions which Bässler termed the "active" reaction. In the front leg, this series consisted of an initial excitation of muscles of different leg joints that were normally active during the stance phase, followed by excitation of muscles that fire during the swing phase (Bässler, 1986). The frequency of occurrence of the active reaction also depended on the initial angle of the femorotibial joint, being most frequent in the extreme ranges of joint flexion. Bässler postulated that this sequence of contractions represented those that normally occur in walking in the transition between the stance phase and the swing phase and, essentially, that the stimulus evoked the initiation of a step.

Similar complexity in modes of reflex action has also been demonstrated for the femoral chordotonal organ of the locust hind leg. Mechanical stimulation of that sense organ has been demonstrated to produce two distinct modes of reflex activation in both restrained and freely standing animals (Field and Burrows, 1982; Zill, 1985b; Zill and Jepson-Innes, 1988, 1990). In the first mode, termed the resistance mode, displacement of the main ligament of the chordotonal organ elicits resistance reflexes in tibial muscles that are directional and oppose apparent joint movement (Fig. 2B). In the second mode, termed the flexion mode, stimuli indicating movements in any direction produce activation of excitatory axons in the flexor muscle (Fig. 2C). By applying selective mechanical stimulation of the chordotonal organ in freely standing animals (using a piezoelectric crystal attached to the leg), it was also demonstrated

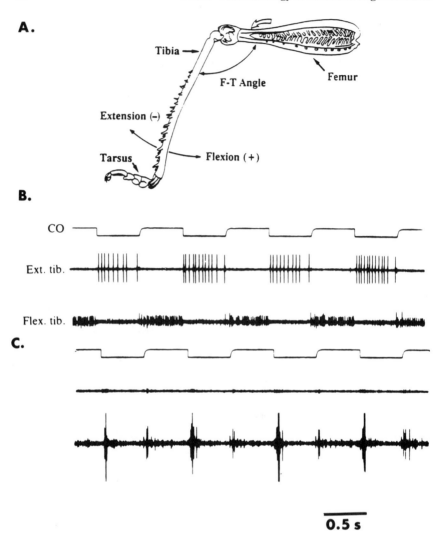

FIGURE 2. Location and reflex modes of action of a chordotonal organ of the locust hind leg. (A) The locust hind leg consists of segments that are linked to one another by joints. The femoral segment (femur) is attached to the more distal segment, the tibia, by a hinge joint that permits flexion and extension movements. The femoral chordotonal organ (located at the point indicated by the open arrow) is a sense organ that monitors the angle of the femorotibial joint. (B) and (C) Reflex modes of action of the chordotonal organ. The reflex effects of the chordotonal organ on motoneurons to leg muscles were tested by mechanically moving the ligament of the organ while monitoring the activities of the tibial extensor and flexor muscles. In the resistance mode, the chordotonal organ elicited bursts of activity in leg muscles to oppose apparent joint flexions (up on position trace) and extensions. (C) In the flexion mode, displacements in any direction produced bursts of activity in the tibial flexor muscle. (A) reprinted from Matheson, 1990; (B) reprinted from Zill, 1985b.

Chapter III   Load Compensatory Reactions in Insects

that the frequency of occurrence of the flexion mode depended on the initial joint angle, being greatest in extreme joint flexion (Zill and Jepson-Innes, 1990). Similar activation of flexor motoneurons in non-resistance-type patterns has also been found in responses to simple joint movements (Siegler, 1981). I will discuss the potential functions of these reflex modes in load compensation in another section.

However, it is of interest to note that Burrows and colleagues have extensively characterized the central connections of afferents that underlie these reflexes (Burrows, 1987, 1988; Burrows et al., 1988). Individual chordotonal afferents synapse on local circuit spiking and nonspiking interneurons and, in some cases, on motoneurons to leg muscles directly. Among the spiking interneurons, individual cells have been identified that respond directionally, as in the resistance mode of action, while others respond nondirectionally, as in the flexion mode (Fig. 3) (Burrows, 1988). Similar complexity in the processing of chordotonal inputs has also been demonstrated in the stick insect (Büschges,

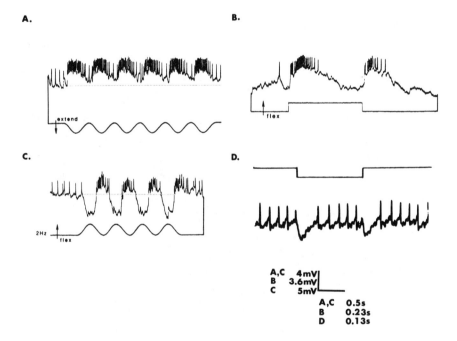

FIGURE 3. Interneurons of the central nervous system that receive chordotonal inputs. (A) Spiking local circuit interneurons have been identified that receive inputs from the chordotonal organ. Interneurons can fire directionally to displacements of the ligament (A and C) in the pattern of a resistance reflex or fire or be inhibited nondirectionally (B and D) in the flexion reflex mode. (A–C) reprinted from Burrows, 1988; (D) Zill, unpublished record.

1989, 1990; Bässler and Büschges, 1990). Thus, there is a considerable amount of parallel processing in this circuitry and a number of levels at which modulation and reflex switching can occur, to account for the plasticity of reflexes of the chordotonal organ.

IV. Postural Load Compensatory Reactions in Grasshoppers

Although postural load compensatory responses have been extensively studied in vertebrates, until recently these reactions have only rarely been examined in invertebrates. (Barnes *et al.*, 1972) first demonstrated that freely moving crabs showed compensatory discharges in leg muscles to resist forces imposed on a leg (via an attached solenoid) during walking. Similar adjustments have also been shown to occur in lobsters walking on a treadmill (Chasserat and Clarac, 1983). To test whether insects show similar types of responses in simply maintaining stable postures, we placed grasshoppers in a chamber that was mounted on a swivel joint and repetitively swayed by a motor (a rough adaptation of Nashner's paradigm) (Fig. 4A) (Zill and Frazier, 1990; Zill *et al.*, 1992). Tests were performed when animals stood on a screen placed on the wall of the cage with their heads pointed up. When sinusoidal movements were imposed, grasshoppers were repeatedly forced toward and away from the side of the chamber on which they stood. Myographic activities were recorded from a number of leg muscles, as well as the position of the wall of the cage relative to the horizontal plane. In addition, in some experiments we monitored the animal's behavior and measured the angle of the femorotibial joint of the hind leg by attaching light-emitting diodes to the leg and videotaping animals.

The first major finding of these studies was that grasshoppers were readily able to maintain stable postures when standing on a moving surface and fell off in less than 1% of the tests. This ability was not affected by covering the animals' eyes and ocelli, depriving them of visual inputs. Records of muscle activities and videotapes of animals' reactions showed that two different strategies were used during displacement tests, similar to the swaying and stepping strategies seen in vertebrates. In swaying reactions the foot (tarsus) of the animal was held in place on the side of the cage, while in stepping responses the leg as a whole was lifted and moved to a new position. In the hind leg, this position was most often one of full flexion of the femorotibial joint. The characteristics of each of these types of reactions are discussed below.

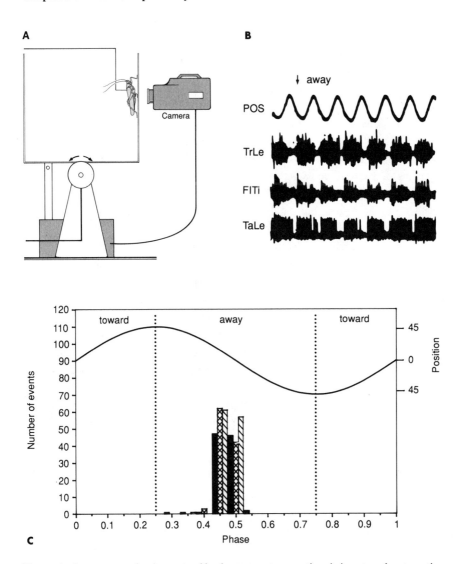

FIGURE 4. Apparatus used to test postural load compensatory reactions in insects and motor activities during swaying reactions. (A) Load compensatory reactions of grasshoppers were tested by placing them in a cage that was repeatedly swayed back and forth. These movements caused animals to be pushed toward and away from the side of the cage. Muscle activities were recorded myographically and the angle of the femorotibial joint of the hind leg was measured by videotaping animals. (B) Swaying reactions. Animals showed regular bursting during repeated movements of the cage (POS indicates the position of the cage) in muscles that pulled the animal toward the wall of the chamber. (C) A histogram of the time of onset of muscle bursting during displacement tests. The onset of bursting is tightly coupled in all leg muscles and was initiated during the phase when the cage was being pulled away from the animal. (A–C) reprinted from Zill, *et al.*, 1992.

A. Swaying Strategies

Swaying strategies predominated in the middle leg and were often seen in the hind leg. In swaying responses, regular repetitive bursting occurred in groups of leg muscles (flexors and levators) that was phase linked to the repeated movements of the chamber (Fig. 4B). Plots of the time of onset of muscle activity showed that bursting in these muscles occurred nearly synchronously during the phase of movement when the wall of the cage was being pulled away from the animal (Fig. 4C). Despite the fact that these muscles are located at different joints, they all function to decrease joint angles and thus would act to pull the animal toward the surface that was being moved away from it. Activities of the antagonist muscles at these joints did not normally occur during tests but could be induced by placing weights on the animal.

Measurements of the angle of the hind leg femorotibial joint during swaying reactions showed that it also systematically changed in phase with the movement of the cage, increasing during the time when the wall was being moved away from the animal and decreasing during the opposite phase. Experiments in which both muscle activities were recorded and the femorotibial joint angle was measured gave insight into one of the variables that was being controlled during swaying reactions. Figure 5 shows histograms representing the onset of muscle bursting and the mean change in joint angle recorded simultaneously over a number of cycles of movement. It can be seen that the following events occur during the phase when the wall is being moved away from the animal: (1) the joint angle begins to increase at the start of the phase; (2) then muscle bursting is initiated about halfway through that phase; and (3) the onset of muscle activity is followed by a dampening of the rate of change of joint angle. Thus, one factor that is apparently controlled during swaying reactions is the rate of change of joint angle.

B. Stepping Strategies and Rules of Use

The second type of response that grasshoppers showed in the hind legs in tests of compensatory reactions consisted of stepping strategies in which the leg was lifted from the substrate and moved, most often to full flexion of the femorotibial joint (Zill et al., 1992). Stepping responses did not appear to be artifacts or startle reactions as they did not occur in any particular sequence in a series of tests and, therefore, showed no evidence of habituation. However, the frequency of occurrence of stepping reactions was correlated with the initial position of the joint at the onset of a displacement test (Fig. 6). Stepping responses occurred most frequently when the femorotibial joint was in the extreme ranges of joint extension or flexion, but not already in a fully flexed position. In contrast, animals used swaying strategies in the mid-range of joint angles. That finding is of considerable interest because

Chapter III Load Compensatory Reactions in Insects

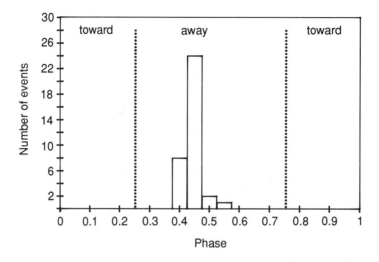

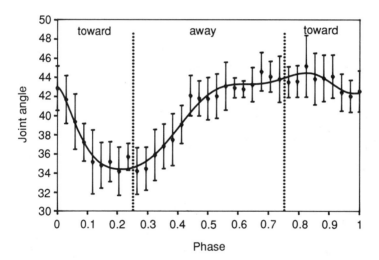

FIGURE 5. Changes in joint angles during swaying reactions. These histograms plot the onset of muscle bursting (top) and the changes occurring in the femorotibial joint angle (bottom) from an experiment in which both variables were measured simultaneously. The onset of muscle activity is followed by a significant dampening in the rate of change of joint angle. (Reprinted from Zill, et al., 1992.)

the mechanical advantages of the tibial extensor and flexor muscles are smallest when the leg is either highly extended or flexed and are much greater closer to the midrange. Thus, grasshoppers appear to use relatively simple rules in compensatory responses of the hind leg:

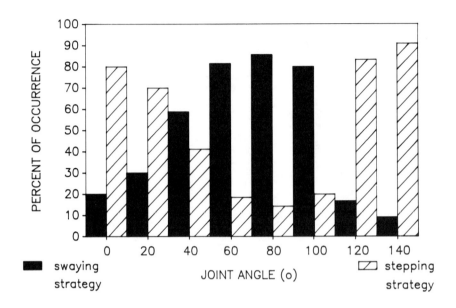

FIGURE 6. Occurrence of swaying and stepping (flexion) strategies at different joint angles. This is a histogram of the frequency of occurrence of swaying and stepping reactions at different initial angles of the femorotibial joint. Swaying reactions occurred most frequently in the midrange of joint angle while stepping reactions occurred in the extremes. Thus, the animal moves the leg when perturbations occur in positions of postural instability. (Reprinted from Zill, et al., 1992.)

1. They use swaying strategies to hold the leg in place at joint angles that are stable and in which muscle tensions can be effectively developed to counter perturbations.
2. They use stepping strategies to move the leg out of ranges of joint angle that are unstable and in which mechanical advantages of muscles are low.

If these rules are valid, why is the hind leg moved most often to full flexion instead of the midrange? The reason for that discrepancy may lie in the particular features of the hind leg that are used in generating the force for jumping, for which the greatly enlarged hind leg is specialized. Prior to a jump, the tibial segment of the leg is moved to full flexion of the femorotibial joint (Heitler and Burrows, 1977). In that position, the joint has an anatomical locking mechanism, due to a bifurcation of the end of the flexor muscle tendon which is pulled over a thickened mass of cuticle known as the "lump" (Heitler, 1974). In jumping, this mechanism permits muscle tensions to be built up until they are sufficient to propel the grasshopper off the substrate by merely inhibiting tension in the flexor muscle. In load compensatory reactions, animals apparently utilize this feature to essentially immobilize the femorotibial joint, rather than attempt to adjust muscle tensions to external perturbations. I should also note that in 13% of tests of load compensation the joint of the hind leg was moved to the midrange, as were all movement reactions of the middle legs, which possess no such locking mechanism (S. N. Zill, unpublished observations).

C. *Contribution of Chordotonal Organ Reflexes to Load Compensatory Reactions*

There are similarities between the types of responses obtained in tests of load compensation in grasshoppers and the modes of reflex action of the femoral chordotonal organ that suggest that chordotonal inputs can strongly contribute to these responses (Field and Burrows, 1982; Zill, 1985b). Swaying reactions resemble resistance reflexes in that they (1) are accompanied by systematic changes in joint angles; (2) are directional, in that they occur in fixed phases in the cycles of displacement; and (3) occur most frequently in the midrange of joint angles. Similarly, stepping reactions resemble the flexion reflex mode in that they (1) are nondirectional and generate flexor bursts to forces producing joint flexions or extensions and (2) occur in the extreme ranges of joint angle. However, there are also significant differences between reactions to substrate displacements and chordotonal reflexes: (1) resistance reflexes usually elicit reciprocal bursts in antagonist muscles to changes in joint angles, whereas activities in antagonist muscles did not occur in swaying responses in the absence of added load; and (2) flexion reflexes of the chordotonal organ are only phasic, unlike stepping reactions, in which flexor firing was prolonged. It should first be noted that a number of studies have shown that the extensor muscle of the leg can have a considerable amount of resting tension, in the absence of motoneuron activity (Hoyle, 1978). That tension could allow the muscle to function as a passive shock absorber in load compensation. It should also be noted that other leg sense organs such as the tarsal spurs (Burrows and Pflüger, 1986) and isolated hairs on the leg (Pflüger, 1980) excite flexor motoneurons and these sensory receptors could potentially contribute to the prolonged flexor discharges. Thus, in sum, the known reflex activities and connectivity of the chordotonal organ suggest that it could strongly contribute to elements of responses in swaying and stepping reactions. Clearly, however, additional sensory receptors and other central nervous system pathways must be activated during these global reactions.

V. Preliminary Studies of Compensatory Reactions during Cockroach Walking

A. *Evidence That Sensory Inputs Are Limited or Modified during Rapid Walking*

Previous studies strongly suggest that the effects of leg sense organs must be either limited or modified at different rates of walking, particularly during very rapid walking. The initial evidence for this conclusion is as follows: (1) In slow walking, motoneuron activity immediately precedes the development of muscle tensions and the generation of movements. During rapid running (at rates over 10 cycles per second) motoneuron activities occur far in advance of the movements they produce (Delcomyn and Usherwood, 1973). For example, extensor firing is initiated as early as halfway through the preceding flexion movement. These phase shifts may be due to the fact that there is an inherent delay between the onset of motoneuron activity and the development of muscle tensions sufficient to generate leg movements. However, they imply that the timing of reflex effects of leg sense organs that detect movements must be quite different in rapid walking. (2) Pearson has demonstrated that increased loading of a walking animal augments the intensity of motoneuron bursting in slow walking but has little effect in rapid walking (Pearson, 1972). (3) In studies in which the activities of tibial campaniform sensilla were recorded in freely walking cockroaches, we found that afferent activity was timed to provide effective feedback control of muscle tensions and leg loading during slow walking (Zill and Moran, 1981). However, at rates over 5–10 Hz, sensory firing was sufficiently delayed in phase relative to motoneuron activity that it could not modify that activity within a single step cycle. New evidence also supports this hypothesis. (1) Delcomyn (1991a, b) has extensively examined the effects of leg amputations on the patterns of coordination in cockroach walking. He has found that severing a limb substantially disrupts coordination in slow walking, presumably due to altered sensory inputs, but has very little effect on rapid running. (2) Lastly and most remarkably, Full and Tu (1991) have shown that cockroaches running very rapidly can actually have a partially airborne posture and can use quadrupedal and bipedal gaits (in which only the hind legs come in contact with the substrate). This finding also implies that inputs from sensory receptors of the middle and forelegs must be substantially altered in rapid running. All these studies imply that sensory inputs are either completely suppressed during rapid running or, more interestingly, are somehow altered by the central nervous system to effect more than one step cycle.

B. *Use of Swaying and Stepping Strategies in Cockroach Walking*

In order to test the abilities of cockroaches to show compensatory reactions to perturbations during walking, we have constructed an arena that can be suddenly

Chapter III Load Compensatory Reactions in Insects

displaced horizontally (Fig. 7A). The walking arena is mounted on wheels and is linked to a large solenoid. Voltages applied to the solenoid produce brief (40–60 ms in duration), small (1 cm) displacements of the entire walking surface. In preliminary experiments, we either implanted myogram leads in leg extensor and flexor muscles and recorded walking patterns or simply videotaped cockroaches that were walking freely with no attached wires (using a high-speed video system kindly provided by Roy Ritzmann). We have begun by analyzing data taken during sudden displacements perpendicular to the walking direction.

Preliminary experiments in which the angles of the femorotibial joints of the hindlegs were measured during walking at slow to moderate speeds have shown that changes in joint angles occur during perturbations that are qualitatively similar to the swaying and stepping strategies seen in studies of perturbations in posture. Figure 7B–D is a series of plots of the angles of the femorotibial joints during sequences in which perturbations were applied at different walking speeds. The joint angles increase during the stance phase and decrease in swing and, as the two hindlegs are used in an alternating pattern, the joint angles of the legs change in opposite phase with one another. When the arena is suddenly displaced during walking at a moderate rate (4.3 Hz), a swaying response is elicited in which the joint angle of the leg in stance transiently increases but the foot (tarsus) remains on the substrate (Fig. 7B). This increase (approximately 20°) is rapidly countered and is followed by a slightly faster swing phase, so that the overall pattern of stepping is unchanged. Thus, cockroaches have the ability to counter perturbations within a single step cycle at slow to moderate rates of walking using swaying strategies.

However, in some tests, we have found that displacements of the arena during certain phases of the step cycle can produce movement or stepping responses. In the example shown in Fig. 7C, the perturbation was applied late in the stance phase of an animal walking at 5.0 Hz. Instead of an increase in joint angle, the leg was rapidly lifted from the walking surface, the joint was flexed, and a small step was taken (note the decrease in joint angle following the bar). The leg was then placed down in support and retracted in an additional stance phase. After a small additional step, the overall rhythm of walking was unchanged. Thus, in addition to swaying strategies, animals use some stepping responses to effectively counter perturbations during walking.

In preliminary studies, we have also applied displacements of the arena when cockroaches were walking very rapidly. Figure 7D is an example of a perturbation applied to an animal that was running at 12.5 Hz. The effects of arena displacement are much more dramatic than those seen during slow walking. There are subsequent large (50°), rapid changes in the angles of the femorotibial joints of both legs, which are suddenly retracted in stance, and the pattern of leg movements is disrupted. In addition, the rhythm of walking is completely reset.

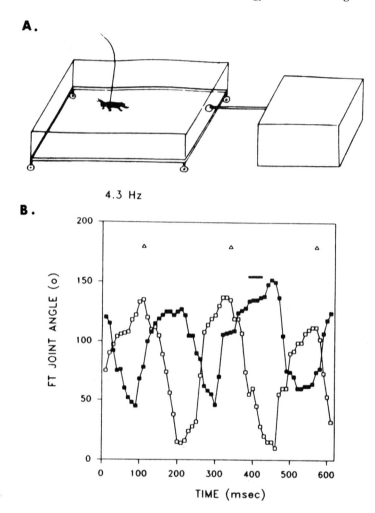

FIGURE 7. Load compensatory reactions during walking. (A) Techniques. We have constructed a walking arena that is supported by wheels and can be suddenly displaced horizontally by a solenoid. Load compensatory reactions are monitored in freely walking cockroaches myographically. (B–D) These graphs show the angles of the hind leg femorotibial joints when perturbations were applied during walking (during the time indicated by the bar). Displacements produced swaying (B) and stepping (C) reactions at slow to moderate speeds of walking. Perturbations greatly disrupted walking during rapid running (D) and the walking rhythm was reset. (A–D) Zill, unpublished.

Thus, our initial findings are that perturbations can be effectively countered using stepping or swaying strategies at slow to moderate rates of walking but that displacements are considerably more disruptive in very rapid walking. These results

Chapter III Load Compensatory Reactions in Insects

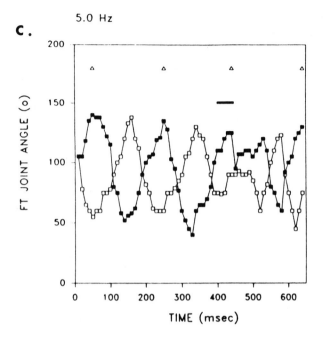

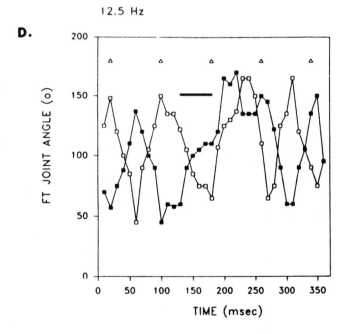

FIGURE 7 continued.

support the idea that sensory inputs are less effective in generating compensatory reactions in a single step cycle at high rates of locomotion. However, we have not yet resolved whether cockroaches use proprioceptive information to maintain dynamic equilibrium in rapid walking by resetting the rhythm of stepping.

VI. The Need for Modulation of Proprioceptive Reflexes

From the studies of the effects of perturbations of grasshoppers in posture and from preliminary studies of substrate displacements of walking cockroaches, it would seem that one variable that is controlled during load compensatory reactions of insects is joint angle and/or its rate of change. Furthermore, inputs from joint angle receptors, such as chordotonal organs, can elicit reflexes that could strongly contribute to these responses. However, several important factors are relevant to modeling of this system and have implications about how the central nervous systems of insects process proprioceptive inputs. The first fact is that there is no evidence in these systems that these sense organs are under any direct, mechanical efferent control, as are vertebrate muscle spindles and crustacean myochordotonal organs. Those sense organs are coupled to specialized muscle fibers under control of the nervous system whose contraction can change afferent sensitivities and firing patterns. Given that fact, it is apparent that resistance reflexes of joint angle receptors cannot simply be constantly active, because they would produce reflex muscle contractions that would oppose any voluntary or walking movement made by the animal. However, the studies that have been reviewed in this chapter have shown that these animals are able to make rapid adjustments to perturbations that produce unexpected deviations in joint angles during walking. Those findings imply two possibilities that are not mutually exclusive: (1) proprioceptive inputs are modulated or adapted by the central nervous system during voluntary movements and in walking, and (2) there are mechanisms in the nervous system that compare the anticipated changes in joint angles with those that are actually occurring, as signaled by proprioceptive inputs, and those mechanisms adjust motor activity accordingly. There is now a considerable amount of evidence to support the first of these possibilities, and proprioceptive reflexes have been shown to be modulated at a number of levels. Reflexes can be changed by (1) neurohormonal control of afferent sensitivity, (2) presynaptic inhibition of primary afferents in the central nervous system, (3) interactions between spiking and nonspiking interneurons and among groups of nonspiking interneurons (Burrows, 1988), and (4) modulation of reflexively evoked neural activity at the neuromuscular junction. Any of these mechanisms, used independently or in combination, could adjust the effectiveness or gain of propriocep-

Chapter III Load Compensatory Reactions in Insects

tive reflexes during movements generated by the animal itself. These mechanisms also could be utilized to alter the effects of proprioceptive inputs at different rates of walking, as is implied by the studies of the effects of perturbations during running in cockroaches (Delcomyn, 1991a,b; Delcomyn, Chapter II, this volume). However, it is still unclear what elements actually comprise mechanisms that are used to compare anticipated movements or muscle contractions with those actually occurring during walking. This important problem is an exciting area for experimentation on the circuitry of the nervous system which processes proprioceptive inputs. Its solution also will be aided by further studies of the types of variables that are controlled by insects in responding to postural perturbations and in traversing irregular terrains.

VII. Conclusions

In this chapter, I have presented evidence to support the following conclusions:

Insects can generate compensatory reactions in posture and walking in response to mechanical perturbations. In posture, these reactions consist of the organized contractions of groups of limb muscles.

Load compensatory reactions can occur as two different types of responses reflecting different strategies: (1) resistance or swaying strategies in which the foot remains on the ground, and (2) movement or stepping strategies in which a leg is lifted and moved to a new position.

Muscle contractions can be initiated to regulate the rate of change in joint angles in posture and in walking, although other variables may also be controlled. Inherent muscle tensions can also play a role in load compensatory responses.

The reflex effects of a number of identified leg proprioceptive sense organs can contribute to load compensation, particularly joint angle receptors known as chordotonal organs.

I have also presented some preliminary data suggesting that swaying and stepping strategies may be used by cockroaches to compensate for perturbations during slow to moderate speeds of walking. However, these data indicate that cockroaches may be less able to compensate adaptively for perturbations in single step cycles during rapid running. The latter preliminary finding also supports the idea that the mechanisms that can adapt locomotion to environmental perturbations are probably different in slow and very rapid walking.

From an evolutionary standpoint, the general ability of insects to exhibit compensatory responses probably confers several advantages on them. First, these animals must be able to stand on and traverse many surfaces, such as grasses or plants, that can, in natural circumstances, be quite unstable. Further, the ability to adjust motor activities to changes in external load is efficient in that it avoids falls or gross movements of the animal as a whole. In evolving these mechanisms, insects have also probably been able to adapt to a much greater variety of habitats and environments.

From the standpoint of a control system, there are several notable characteristics about the types of information and processing that insects utilize in generating compensatory reactions. First, the system apparently relies strongly on inputs from leg sensors to make reflex adjustments at a local level, within the leg itself. Also, the types of motor strategies that are used are determined, in part, by parameters that are monitored in the leg. For example, swaying or stepping strategies are used to regulate posture preferentially when the leg is in different ranges of joint angles. Within a single leg, however, these reactions can involve the patterned and coordinated contractions of diverse muscles of a number of limb joints, so that the control mechanisms are more complex than simple, single connections between sensors and motor outputs. This plasticity and divergence of control is also reflected in what is known of the connectivity of proprioceptive sense organs in insects, which shows considerable parallel processing in the central nervous system and potential modulation at a number of levels. However, the mechanisms that underlie leg control may ultimately be functionally determinable experimentally, as the system utilizes information from a relatively limited number of types of identifiable sense organs.

A considerable amount of information has yet to be experimentally determined in this system. In particular, it is necessary to establish whether the mechanisms used to respond to environmental perturbations are similar to those utilized in adjusting walking movements over uneven terrains, a subject that has been investigated in only a few studies in insects (Pearson and Franklin, 1984; Cruse *et al.,* 1989). It should be noted that, for insects, the absence of a separate gravity orientation system, like a vestibular apparatus, requires that adjustments be made using simple inputs from leg sensors when walking on an undetermined, irregular substrate that are quite similar to those presented by a surface that is moving. It is also necessary to know more about how proprioceptors of other modalities influence motor outputs, such as those responding directly to forces acting on the legs. The evidence that we have to date suggests that both these problems may be resolved into a limited number of types of strategies that can be utilized by insects and in the design of legged robots to similar advantage.

Acknowledgment

I am extremely grateful to Stephen Fish, Karen Thompson, and Fred Delcomyn for their helpful comments on this chapter. Some of the work reported here resulted from collaborations with Faith Frazier and Greg Larsen. This work was supported by grants from the Whitehall Foundation and by NIH NINDS grant NS22682.

References

BARNES, W. J., SPIRITO, C. P., and EVOY, W. H. (1972). *LZ. Vergl. Physiol.* **76**, 16.
BÄSSLER, U. (1976). *Biol. Cybern.* **24**, 47.
BÄSSLER, U. (1983). *Neural Basis of Elementary Behavior in Stick Insects.* Springer-Verlag, New York.
BÄSSLER, U. (1986). *J. Comp. Physiol.* **158**, 351.
BÄSSLER, U. (1988). *J. Exp. Biol.* **136**, 125.
BÄSSLER, U., and BÜSCHGES, A. (1990). *Biol. Cybern.* **62**, 529.
BERGER, W., DIETZ, V., and QUINTERN, J. (1984). *J. Physiol.* **357**, 109.
BURROWS, M. (1987). *J. Neurosci.* **7**, 1064.
BURROWS, M. (1988). *J. Comp. Physiol. A* **164**, 207.
BURROWS, M., and PFLÜGER, H. J. (1986). *J. Neurosci.* **6**, 2764.
BURROWS, M. LAURENT, G. J., and FIELD, L. H. (1988). *J. Neurosci.* **8**, 3085.
BÜSCHGES, A. (1989). *J. Exp. Biol.* **144**, 81.
BÜSCHGES, A. (1990). *J. Exp. Biol.* **151**, 133.
CHASSERAT, C., and CLARAC, F. (1983). *J. Exp. Biol.* **107**, 219.
COHEN, H., and KESHNER, E. A. (1988). *Am. J. Occup. Ther.* **43**, 331.
CRUSE, H., RIEMENSCHNEIDER, D., and STAMMER, W. (1989). *Biol. Cybern.* **61**, 71.
DELCOMYN, F. (1991a). *J. Exp. Biol.* **156**, 483.
DELCOMYN, F. (1991b). *J. Exp. Biol.* **156**, 503.
DELCOMYN, F., and USHERWOOD, P. N. R. (1973). *J. Exp. Biol.* **59**, 629.
DO, M. C., BRENIERE, Y., and BRENGUIER, P. (1982). *J. Biomech.* **15**, 933.
DONNER, M. D. (1987). *Real Time Control of Walking.* Progress in Computer Science, vol. 7.
DUNCAN, P. W., STUDENSKI, S., CHANDLER, J., BLOOMFELD, R., and LAPOINTE, L. K. (1990). *Phys. Ther.* **70**, 88.
FIELD, L. H., and BURROWS, M. (1982). *J. Exp. Biol.* **101**, 265.
FULL, R. J., and TU, M. S. (1991). *J. Exp. Biol.* **156**, 215.
GAUTHIER, L., and ROSSIGNOL, S. (1981). *Brain Res.* **207**. 303.
HEITLER, W. J. (1974). *J. Comp. Physiol.* **89**, 93.
HEITLER, W. J., and BURROWS, M. (1977). *J. Exp. Biol.* **66**, 203.
HORAK, F. B. (1987). *Phys. Ther.* **67**, 1881.
HORAK, F. B., and NASHNER, L. M. (1986). *J. Neurophysiol.* **55**, 1369.
HOYLE, G. (1978). *J. Exp. Biol.* **73**, 173.
KESHNER, E. A. and COHEN, H. (1988). *Am. J. Occup. Ther.* **43**, 320.
KESHNER, E. A., WOOLLACOTT, M. H., and DEBU, B. (1988). *Exp. Brain Res.* **71**, 455.
MACPHERSON, J. M.(1988). *J. Neurophysiol.* **60**, 204.
MARKL, H. (1971). In Gordon, S. A., and Cohen, M. J. (eds.), *Gravity and the Organism* (p. 185). University of Chicago Press, Chicago.
MATHESON, T. (1990). *J. Comp. Physiol. A.* **166**, 915.
NASHNER, L. M. (1976). *Exp. Brain Res.* **26**, 59.

NASHNER, L. M. (1977). *Exp. Brain Res.* **30,** 13.
NASHNER, L. M. (1980). *J. Neurophysiol.* **44,** 650.
NASHNER, L. M., and MCCOLLUM, G. (1985). *Behav. Brain Sci.* **8,** 135.
NASHNER, L. M., WOOLLACOTT, M., and TUMA, G. (1979). *Exp. Brain Res.* **36,** 464.
PEARSON, K. G. (1972). *J. Exp. Biol.* **56,** 173.
PEARSON, K. G., and FRANKLIN, R. (1984). *Int. J. Robotics Res.* **3,** 101.
PEARSON, K. G., WONG, R. K. S., and FOURTNER, C. R. (1976). *J. Exp. Biol.* **64,** 251.
PFLÜGER, H. J. (1980). *J. Exp. Biol.* **87,** 163.
RAIBERT, M. H. (1986). *Legged Robots That Balance.* MIT Press, Cambridge, MA.
SIEGLER, M. (1981). *J. Neurophysiol.* **46,** 310.
USHERWOOD, P. N. R., RUNION, H. I., and CAMPBELL, J. I. (1968). *J. Exp. Biol.* **48,** 305.
WENDLER, G. (1971). In Gordon, S. A., and Cohen, M. J. (eds.), *Gravity and the Organism* (p. 195). University of Chicago Press, Chicago.
WOLFSON, L. I., WHIPPLE, R., AMERMAN, P., and KLEINBERG, A. (1986). *J. Am. Geriat. Soc.* **34,** 845.
ZILL, S. N. (1985a). *J. Exp. Biol.* **116,** 345.
ZILL, S. N. (1985b). *J. Exp. Biol.* **116,** 463.
ZILL, S. N. (1990). In Huber, I., Masler, E. P., and Rao, B. R. (eds.), *Cockroaches as Models for Neurobiology: Applications in Biomedical Research* (vol. 2, p. 247). CRC Press, Boca Raton, FL.
ZILL, S. N., and FRAZIER, S. F. (1990). *Brain Res.* **535,** 1.
ZILL, S. N., and JEPSON-INNES, K. A. (1988). *J. Comp. Physiol. A.* **164,** 43.
ZILL, S. N., and JEPSON-INNES, K. A. (1990). *Brain Res.* **523,** 211.
ZILL, S. N., and MORAN, D. T. (1981). *J. Exp. Biol.* **94,** 57.
ZILL, S. N., FRAZIER, S. F., LANKENAU, J., and JEPSON-INNES, K. A. (1992). *J. Comp. Physiol.* **170,** 761.

Chapter IV   Integration by Spiking and Nonspiking Local Neurons in the Locust Central Nervous System. Importance of Cellular and Synaptic Properties for Network Function

GILLES LAURENT
Biology Division, Computation and Neural Systems Program
California Institute of Technology
Pasadena, California

The study of sensory–motor integration in animals encompasses several levels of analysis. One can, for example, try to determine some of the principles that underlie the internal representation of a sensory stimulus. One can also try to assess the neural mechanisms by which a coordinated sequence of muscle contractions is generated. This complex problem is sometimes known as the "implementation problem" (Georgopoulos, 1990). One must in the end, however, try to understand how sensory and motor representations are put together so that a given sensory signal evokes an appropriate corrective motor response. Alignment of sensory and motor maps in the superior colliculus of vertebrates, for example, appears to be one such mechanism, contributing to some sensory–motor transformations (Sparks *et al.*, 1990).

The study of sensory–motor processing in vertebrate and invertebrate nervous systems also revealed that such complex information transformations involve both long-range and short-range interneuronal interactions. Long-range interactions in the rat whisker somatosensory system, for example, could be those between thalamus and vibrissa cortex or those between different barrels within layer IV of the somatosensory cortex (Simons *et al.*, 1989). Local or short-range interactions would, on the contrary, be those that take place between spiny stellate and aspinous or sparsely spined nonpyramidal neurons in one barrel. Very few preparations, however, be they from vertebrate or invertebrate animals, have allowed the study of such local information processing in as much detail as in the thoracic nervous system of insects (Burrows,

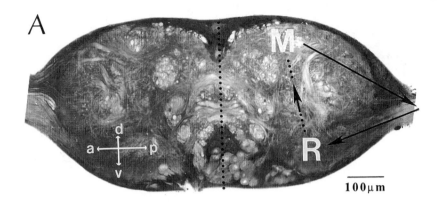

B THE LOCAL CIRCUITS

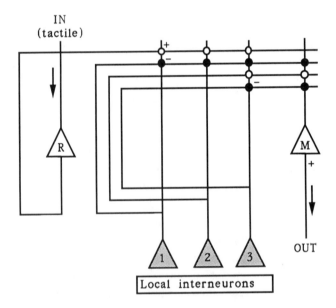

FIGURE 1. The local circuits. (A) Semithin (1 μm) tranverse section of a locust mesothoracic ganglion stained with ethylgallate. Mechanosensory afferents from the legs project to the ventral and intermediate regions of the corresponding thoracic neuropil. The hair afferent terminals form a topographical representation of the leg in the ventral neuropil (Newland, 1991). The anteroposterior and ventrodorsal (a–p, v–d) axes of the leg are respectively represented by the horizontal/transverse and vertical axes in the neuropil. The afferent

Chapter IV Integration by Spiking and Nonspiking Local Neurons

1989). Indeed, the local circuits responsible for sensory–motor integration in the locust thoracic ganglia are known in considerable detail and summarized in Fig. 1. A relatively small number of spiking (1 and 2, Fig. 1B) and nonspiking (3, Fig. 1B) local interneurons receive afferent information from many thousands of mechanoreceptors on or in the legs (R, Fig. 1), and, from it, form an appropriate reflex motor output (M, Fig. 1) (Burrows and Siegler, 1982; Siegler and Burrows, 1986; Laurent and Burrows, 1988; Laurent and Hustert, 1988). I will present some of the recent findings on the integrative and synaptic properties of the two main types of local neuron (nonspiking and spiking) that form these circuits. I will emphasize the aspects of single cell processing that may play an important role at the level of the network.

I. Nonspiking Interneurons: Compartmentalization and Electrical Geometry

A series of recent experiments has shed some new light on the integrative properties of this class of premotor interneurons (Laurent and Burrows, 1989a,b; Laurent, 1990, 1991a). In the following I will summarize the main results, which are related to synaptic or voltage-dependent membrane nonlinearities.

A. *Local Conductance Increase and Shunting of Synaptic Potentials*

Nonspiking interneurons receive converging sensory inputs from many sources. These inputs can originate from the leg whose motor neurons the interneuron controls (through afferents: Burrows *et al.,* 1988; Laurent and Burrows, 1988; or through spiking local interneurons: Burrows, 1987b) but also from a leg in an adjacent segment (through intersegmental interneurons: Laurent and Burrows, 1989a,b).

FIGURE 1. (cont.) (or receptor, R) signals are then conveyed to the motor neurons (M), which arborize mainly in the dorsal and lateral regions of the neuropil. The pathways underlying this transfer are summarized in B. (B) Schematic diagram of the local sensory-to-motor pathways. Connections are indicated by open (excitatory) or filled (inhibitory) dots. In this figure, each neuron (whose soma is here symbolized by a triangle) represents a *population* of neurons. R, receptor; M, motor neuron; 1, spiking local interneurons in the midline population (Siegler and Burrows, 1984); 2, spiking local interneurons in the anteromedial population (Nagayama, 1989); 3, nonspiking local interneurons (Siegler and Burrows, 1979). All the indicated connections have been demonstrated, except for that among spiking local interneurons in the midline population (1), which is as yet only suspected to exist. Hair (tactile) afferents are only exceptionally known to make direct connections with motor neurons (Laurent and Hustert, 1988). Hair sensory inputs to motor neurons thus generally come exclusively from the intercalated interneuronal layers. This is different with proprioceptors, for these afferents do connect with motor neurons directly (Burrows, 1987a).

Furthermore, the synaptic contacts made onto a nonspiking interneuron by some of these presynaptic neurons are spatially restricted. For example, the volume of neuropil occupied by the terminals of a single hair afferent from the leg is very much smaller than that occupied by a postsynaptic nonspiking neuron (Newland, 1991; Laurent and Burrows, 1988). The contacts underlying any connection between them must therefore occur on only a few neighboring neurites. Laurent and Burrows (1989b) thus supposed that certain postsynaptic potentials (PSPs) evoked in a nonspiking neuron might act locally, not to increase or decrease transmitter release at neighboring synapses but rather to shunt coincident inputs by a simple conductance increase mechanism, similar to that hypothesized for directional selectivity in the visual system (Torre and Poggio, 1978). Laurent and Burrows (1989b) therefore predicted that if such interactions existed, they should be specific.

A nonspiking interneuron was selected which received inputs from three different sources; one was a set of local (homosegmental) mechanosensory afferents on a hind leg, and the others were two different intersegmental sources. These were provided by intersegmental interneurons with different receptive fields on an ipsilateral middle leg. It was observed that the local afferent-evoked excitatory PSPs (EPSPs) in the nonspiking interneuron were significantly reduced in amplitude if they occurred at the same time as one of the two intersegmental inputs but not if they coincided with the other. In addition, it was found that *some, but not all,* of the intersegmental inputs to a nonspiking neuron are accompanied by a detectable conductance increase at the recording site (Laurent and Burrows, 1989b). Since all these inputs appeared to be mediated by chemical synapses (Laurent and Burrows, 1989a), the simplest explanation for these combined observations was that a postsynaptic conductance increase was detected only when the synaptic sites were close to the recording electrode. (All the intracellular recordings from nonspiking neurons were made from a neurite, rather than the soma.) It was concluded from these two sets of experiments that some synaptic currents in a nonspiking neuron could indeed be shunted by certain, but not all, coincident inputs. This suggests (1) very precise "wiring" of the networks at the level of a single neuron (to allow for the "appropriate" shunting interactions) and (2) a possible mechanism for selective gain control of local sensory–motor loops (see later).

B. *Voltage-Dependent Membrane Properties*

The use of discontinuous current-clamp techniques *in situ* made it possible to inject relatively large currents (>1 nA) through a high-resistance microelectrode and yet make accurate measurements of membrane voltage at the recording site with the same microelectrode. In addition, the use of the discontinuous voltage-

Chapter IV Integration by Spiking and Nonspiking Local Neurons

clamp technique allowed the characterization of several voltage- and ligand-gated currents (Laurent, 1990, 1991a; Laurent and Sivaramakrishnan, 1992), which may be determinant for temporal and spatial integration by nonspiking interneurons.

The first conspicuous feature emerging from these experiments was that the neuropilar membrane of the nonspiking interneurons expresses at least two voltage-activated outward currents: one is a fast-activating, rapidly inactivating A-type K^+ current, turned on by depolarization around −60 mV, and with a voltage of half-inactivation of about −60 mV (slope factor ≈ 8 mV) (Laurent, 1991a). This current is thus not totally inactivated at a resting potential of −58 mV. The other is a delayed-rectifier K^+ current with very slow inactivation kinetics. The second conspicuous feature was that some, but not all, recordings revealed slow regenerative potentials triggered by depolarization. These were not blocked by tetrodotoxin (TTX), suggesting that they were not carried by Na^+. An interesting possibility is that the currents underlying these slow action potentials were only locally expressed and thus revealed only when the recording site was one such locus of expression. Alternatively, it is possible that these currents were activated only in this voltage range, when under the influence of specific endogenous neuromodulators, present at the time of the recordings in only few animals. These experiments thus revealed that the membrane of nonspiking neurons, although unable to generate conventional action potentials, expresses several voltage-dependent conductances and behaves in a highly nonlinear fashion (Laurent, 1990).

C. Consequences for Temporal Integration

The development of a preparation in which "fictive" locomotor patterns could be evoked reliably (S. Ryckebusch and G. Laurent, in preparation) led to the finding that nonspiking neurons have a large voltage operating range (G. Laurent, H. Zhang, and M. Burrows, in preparation). During a bout of rhythmical activity lasting for 1 to 2 s, the membrane potential of a nonspiking interneuron can travel from a resting potential of about −55 mV to a most depolarized level of about −40 mV or a most hyperpolarized level of about −75 mV. These values overlap those at which the outward currents described earlier are activated and inactivated. This leads to interesting membrane nonlinearities. Indeed, if short test current pulses are injected in a nonspiking interneuron, dramatic changes in input resistance and membrane time constant can be observed when the membrane is successively held at the different voltages of this operating range (Fig. 2A). In some cases, an approximately 10-fold decrease in membrane time constant is observed over 30 mV (Laurent, 1990). The immediate consequence of this observation is that synaptic

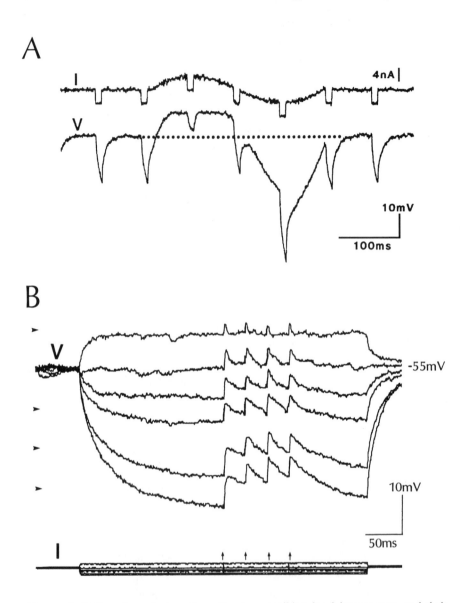

FIGURE 2. Voltage-dependent nonlinearities in nonspiking local interneurons, and their importance for temporal integration. (A) Short negative current pulses are superimposed on a sine wave current injection (I) in discontinuous current clamp. The short voltage deviations in the interneuron (V) are an indication of its input resistance and of its membrane time constant. Notice the dramatic drop in input resistance and time constant as the membrane is depolarized by some 10 mV only from resting potential. Note also that there is an increase in input resistance as the membrane is hyperpolarized, indicating that some of the

Chapter IV Integration by Spiking and Nonspiking Local Neurons

potentials evoked in a nonspiking interneuron decay slowly at negative potentials, but up to 10 times faster at more positive potentials. This dramatic effect is shown in Fig. 2B for mechanosensory afferent-evoked EPSPs. It should be noted that the time course of the underlying synaptic currents is not strongly voltage-dependent (personal observation, not shown). The modulation of the time course of the synaptic potentials is therefore mostly explained by the voltage-dependent changes in membrane conductance. A similar effect is observed with hyperpolarizing inhibitory PSPs (IPSPs), which, rather than increase in amplitude as the postsynaptic membrane is depolarized (as they would be expected to, because of the increase in their driving force), decrease in amplitude as well as duration because of an increased postsynaptic input conductance (Laurent, 1990). If K^+ channel blockers are applied, both the outward rectification and the effects of membrane polarization on the time course and amplitude of PSPs are reduced (Laurent, 1991a). The consequences of these properties are most clearly seen when evoking a train of EPSPs at about 30 Hz (Fig. 2B). If these EPSPs are evoked when the nonspiking neuron is held at −85 to −70 mV, they summate because their decay is slow; if they are evoked when the interneuron is held at resting potential or at a slightly more depolarized potential, they fail to interact and temporal summation does not occur (Fig. 2B). Similar observations were made with IPSPs (Laurent, 1990). These experiments were performed at steady state, i.e., while the membrane potential of an interneuron had reached a more or less constant value. If, however, PSPs were evoked during, say, sinusoidal current injection into the nonspiking neuron, interesting dynamic modulation of synaptic potential amplitude and duration was observed. Because the A current inactivates rapidly, the membrane conductance increase is, for a given instantaneous value of membrane potential, greater during a depolarizing phase than during a hyperpolarizing one (Laurent, 1990). This difference might thus be used to modulate the "efficacy" of synaptic inputs during one phase of the locomotor cycle but not during the other.

FIGURE 2. (cont.) voltage-activated conductances may contribute to the resting potential (see Laurent, 1990). Reprinted with permission from Oxford University Press, Laurent, G., *Journal of Neuroscience,* **10,** 2268–2280 (1990). (B) This outward rectification, (due to the activation of at least two K^+ voltage-gated currents (Laurent, 1991a), causes significant changes in the temporal integrative properties of the interneurons. Excitatory postsynaptic potentials (EPSPs) evoked by stimulating a group of mechanoreceptors on the leg (small vertical arrows) decay slowly when the interneuron is hyperpolarized but decay rapidly and thus fail to summate when the interneuron is at resting potential or more depolarized. The arrowheads indicate 10-mV increments in membrane potential (V). Current injected: 1, 0, −0.5, −1, −1.5, −2 nA. Discontinuous current clamp, from a neuropilar recording site.

D. Consequences for the Network

Since the nonspiking interneurons are premotor neurons embedded in sensory–motor pathways, we must attempt to consider the possible advantages conferred on the networks by these intrinsic cellular properties.

1. A Synaptic Gain Control Mechanism. As demonstrated previously, nonspiking interneurons release neurotransmitter tonically (Burrows and Siegler, 1978) and small deviations in membrane potential such as single EPSPs are sufficient to increase transmitter release significantly (Burrows, 1979a; Laurent and Burrows, 1989b). Since the amplitude and time course of PSPs both depend critically on the postsynaptic membrane potential at the time when they are evoked (see earlier), two counteracting phenomena come into play. As a nonspiking interneuron is depolarized, the gain of an input synapse is decreased (by virtue of the decreased input resistance), whereas the gain of an output synapse is increased (by virtue of the increased dynamic gain of the synaptic transfer; Laurent, 1990). This could lead to stabilization of the gain of a sensory-to-motor-neuron disynaptic pathway. The extent to which these two phenomena actually counteract each other is so far unknown because the synaptic transfer at these nonspiking synapses has not yet been studied quantitatively. Experiments to solve this problem are therefore needed (G. Laurent, H. Zhang, and M. Burrows, in preparation).

2. Global Organization of Movement vs. Local Gain Control. Burrows and Siegler's original experiments (1976, 1978) showed that single nonspiking interneurons can control several pools of motorneurons, so that the depolarization of a single interneuron often leads to a coordinated set of movements about several joints of the same leg. In addition, because nonspiking interneurons are interconnected by lateral inhibitory synapses, activation of one pool of motor neurons by one nonspiking interneuron is often accompanied by inhibition of the antagonistic pool (Burrows, 1979b). Such global control of a large population of motor neurons might sometimes require fine adjustments that a single nonspiking neuron, with its distributed outputs, would appear to be unable to provide. If, however, we consider that nonspiking interneurons could, under certain conditions, become functionally compartmentalized, then such "global action" and "local gain control" would become compatible. This functional compartmentalization would have to meet the following requirements:

1. That output synapses to different pools of motor neurons be spatially and electrically segregated, at least some of the time.

Chapter IV   Integration by Spiking and Nonspiking Local Neurons

2. That specific synaptic inputs to a nonspiking interneuron be located in register with (close to) the "correct" pool of output synapses, thus allowing specific local input–output relationships.

Unfortunately, the extent to which these two criteria are met is thus far unknown. There is, however, some indication that synaptic interactions on nonspiking interneurons are specific. Indeed, the experiments summarized in Section I.A suggest that local dendritic shunts could be used to affect integration of synaptic potentials in some neurites only of a nonspiking interneuron. In addition, the experiments summarized in Section I.B indicate that the temporal integrative properties of nonspiking interneurons are voltage-dependent. The spatial properties of nonspiking interneurons should similarly be modulated in a voltage-dependent fashion, because, for a passive cable at least, the space constant is proportional to the square root of the membrane resistance (Jack *et al.,* 1983). The electrical distance between synaptic sites could therefore theoretically be increased when the neuron is depolarized (and thus in a region of high synaptic transfer gain) and decreased when the neuron is hyperpolarized (i.e., in a region of lower synaptic transfer gain).

II. Spiking Local Interneurons: Synaptic Development and Specificity

Experiments on the population of midline spiking local interneurons (Burrows and Siegler, 1984; Siegler and Burrows, 1984) have led to several new findings pertaining to the study of synaptic development and physiology.

A.   *Embryonic Development: Branching Patterns and Synaptogenesis*

Using intracellular labeling techniques, Shepherd and Laurent (1992) showed that all spiking local interneurons in the midline population of each thoracic hemiganglion appear to be produced by the same unique neuroblast, NB4-1. The interneurons are produced in the latter stages of the neuroblast lineage and could be characterized as members of this interneuron population only after 55% of embryonic development. Their morphological developmental could be divided into three main phases. First, a phase of directed outgrowth (55–70% of embryonic development), during which the primary neurite crosses the midline and establishes the basic scaffolding of neurites of the mature local interneuron. Second, a period of sustained and rapid growth (70–80%), during which many side branches are produced from the primary and secondary neurites. At this stage, ventral and dorsal branches could not yet be distinguished from each other by their morphological appearance (fine and branched ventrally vs. varicose and sparse dorsally). Third, a period of maturation (80–90%), during which some of the neurites produced previously appear to be pruned, most significantly in the dorsal field of branches. It is only then that specific members of the population known from intracellular fills in adults could be recognized as individuals, from morphological criteria (Shepherd and Laurent, 1992). Although the underlying basis for this rearrangement of arborizations is so far unknown, it will be critical to establish whether the observed "pruning" is linked to activity-dependent factors, such as the arrival of specific signals from incoming mechanosensory afferents, and can be correlated with synapse elimination or redistribution.

A first attempt at answering this question was made by studying the development of synapses on the spiking local interneurons at the ultrastructural level, in parallel with that of their membrane excitability (Leitch *et al.*, in press). At the time when spiking local interneurons first develop their initial scaffold of neurites, only filopodial and punctate contacts are seen. The interneurons fail to generate action potentials upon intracellular current injection, although some outward rectification is observed, indicating a possible early expression of K^+ channels (Leitch *et al.*, in press). During the period of rapid growth (70–80%), asymmetrical contacts and mature synapses are seen, with presynaptic vesicular profiles. Synaptic noise is now detected from intracellular recordings and action potentials can be generated. At this time, output synapses predominate on the entire neurons. The proportion of outputs to inputs on ventral branches, for example, is 7.5 : 1. In adults, this ratio is 1 : 2, making the ventral field an "input" region (Watson and Burrows, 1985). Finally, during the period of maturation (80–90%), the proportion of mature synapses increases and the ratios of output to input synapses in both the dorsal and ventral fields of branches progressively tend toward the values found in adults. At 85–90%, the ratio of dorsal outputs to inputs is 8.5 : 1 (6.5 : 1 in adults; Watson and Burrows, 1985), and that of ventral outputs to inputs is reduced to 2 : 1 (1 : 2 in adults). It is therefore possible that the period of embryonic development between 75 and 90% is one when synaptic contacts are first made and later refined, since some branches are seen to disappear (Shepherd and Laurent, 1992). At this stage, some of the known targets of the midline spiking local interneurons (motorneurons: Burrows and Siegler, 1982; intersegmental interneurons: Laurent, 1987) have already been produced by their parent neuroblasts, which correlates well with the observation that output synapse formation is an early event (Leitch *et al.*, in press). Since the only known inputs to the spiking local neurons are leg mechanosensory afferents, we must now determine whether the emergence of sensory receptors and the growth of their axon terminals into the central nervous system correlate well with (and maybe cause) the synaptic maturation process observed in the spiking local interneurons.

Chapter IV Integration by Spiking and Nonspiking Local Neurons

B. Release Properties at the Synapses Made by a Single Interneuron on Different Central Targets

We know from several studies that a single spiking local interneuron can make divergent connections to several target neurons. These may be several individuals in the same population (nonspiking interneurons: Burrows, 1987b; intersegmental interneurons: Laurent, 1987) or individuals in different populations (nonspiking interneurons and motorneurons: Laurent and Sivaramakrishnan, 1992). Experiments were conducted here to determine whether the properties of release at the synapses made by a single spiking local interneuron onto different targets are similar or not. The finding that differences do exist would indicate that refined adjustment of transmitter release properties take place *at the level of single functional synapses* and would thus suggest that specific regulatory interactions between pre- and postsynaptic neurons take place during embryonic development or during postembryonic life. Long-term plasticity in those sensory-motor networks would thus become an important issue.

Simultaneous recordings were thus made from connected spiking and nonspiking interneurons, and large numbers of IPSPs were evoked at a constant frequency of 2–7 Hz. The statistical properties of release at each synapse were then studied. It was found that the distribution of PSP amplitudes could be described by a simple binomial model (Redman, 1990), implying uniformity of p (the probability of release at a single site) for each synapse. The mean quantal amplitude was 290 ± 110 µV, and the mean quantal content m of the IPSPs (average number of released quanta) was 6.2 ± 3. The mean values of binomial n (average size of the releasable pool) and p were 13.1 ± 3 and 0.45 ± 0.16, respectively (Laurent and Sivaramakrishnan, 1992).

Monte Carlo numerical simulation experiments were performed to test whether the IPSP amplitude distribution histograms might be misleadingly indicative of quantal release. Indeed, the generally small sample size of each physiological experiment (400–800 IPSPs) and the arbitrary binning procedure could have introduced false structure in the histograms and thus influenced our interpretations and conclusions. These simulation experiments suggested that the structure of the IPSP amplitude distribution histograms was indeed indicative of quantal release and therefore probably not an artifact or a consequence of the inherent limitations of the physiological experiments and binning procedures (Laurent and Sivaramakrishnan, 1992).

Finally, experiments were carried out in which a single spiking local interneuron was impaled and several of its target nonspiking interneurons were successively sampled within a few minutes of each other. Quantal analysis was then performed with the different IPSPs evoked, under identical conditions, by the same presynaptic interneuron, and the quantal parameters were compared between the synapses. It was generally found that the binomial parameters n and p and their product m (or mean quantal content) differed between the synapses made by a single spiking interneuron onto different target neurons. Figure 3 summarizes the results obtained from one such experiment, showing the best-fitting distributions for the two synapses made by a spiking interneuron with two nonspiking interneurons. For the synapse with the first nonspiking interneuron, n and p were 15 and 0.69, respectively, giving a mean quantal content of about 10. For the synapse with the second interneuron, n and p were 10 and 0.2, respectively, giving a mean quantal content of 2 (Fig. 3, and see Laurent and Sivaramakrishnan, 1992). These experiments showed that quantal contents can indeed vary for the many synapses made centrally by one interneuron and suggested that this variability may arise partly from differences in release probabilities between the sites associated with different synapses.

Theoreticians often assume that the individual "synapses" made by a presynaptic "unit" can be modified individually, without requiring that the efficacy of other synapses made by the same unit also be altered. Insofar as these plastic modifications are due to presynaptic changes, the occurrence of synapse-specific changes would require that different release sites of a given neuron have different release properties. The work reported above helps establish that such a scheme is indeed biologically

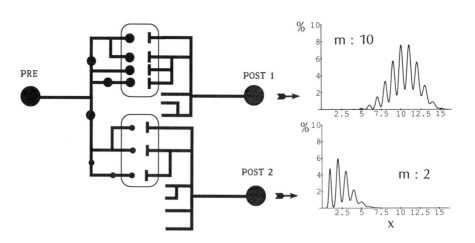

FIGURE 3. The properties of transmitter release differ between the different synapses made by single spiking local interneurons (pre) onto several nonspiking local interneurons (post 1 and 2). The best-fitting binomial distributions for two synaptic connections made by a single spiking local interneuron on two different nonspiking interneurons are plotted on the right of the figure, with the mean quantal contents (m) of each IPSP. These are normalized frequency distribution histograms, where x represents the number of quantal components in the IPSP (see Laurent and Sivaramakrishnan, 1992).

plausible. In addition, the finding that quantal analysis could be performed at all with the synapses between spiking and nonspiking local interneurons indicates that the postsynaptic recording electrode had similar "electrical access" to the different release sites forming a synapse. Had this not been the case, quantal variance would probably have been too large for individual peaks to be distinguished in the amplitude distribution histograms (Laurent and Sivaramakrishnan, 1992). Since nonspiking neurons are not thought to be electrically compact (Laurent, 1990), this result suggests that the release sites forming one functional synapse between one spiking and one nonspiking local interneuron were at similar electrical distances from the postsynaptic electrode and were therefore possibly clustered, rather than distributed over the whole arbor. Such clustering of input synapses onto the nonspiking local interneurons would be compatible with the scheme presented in Section I.D.1, whereby local compartmentalization of nonspiking interneurons would allow specific control over *some, but not all* of the input-output transfers operated by single interneurons.

III. Design Principles of Local Circuits

If we are to design synthetic analogs (e.g., numerical models) and also hardware devices working in real time with locomotor "behaviors" (with apologies to ethologists) similar to those of insects, we must attempt to draw design principles of local circuit organization from the study of biological networks. I will present several such potentially relevant principles. They emerge from the study of mechanosensory processing in locusts and relate to local and intersegmental information processing, i.e., that which takes place within a single hemisegment or between adjacent segments.

1. *Tactile and proprioceptive signals are sparsely distributed.* Afferent signals provided to the central nervous system (CNS) by a small subset of adjacent hairs on a leg, for example, are processed in parallel by several (but a small proportion only) of the local interneurons (Burrows and Siegler, 1985; Siegler and Burrows, 1983; Burrows, 1992; Laurent and Burrows, 1988). These interneurons are generally thought to have overlapping but not identical receptive fields. The organization of these receptive fields, however, appears to be highly specific. Burrows (1992) showed that the effectiveness of the different hairs within the excitatory region of a receptive field of a spiking local interneuron varies in a nonrandom fashion as one moves along the length or the width of the receptive field. In addition, a single hair afferent that makes excitatory connections with two different interneurons may have different effects on each of them. The patterns of connections between the component neurons of the networks are therefore both generally sparse and specific.

2. *The networks lack internal feedback pathways.* Although I hesitate to make this a definite statement, it remains that some degree of "hierarchy" is apparent in the local circuits and that no reciprocal connections between the known populations of neurons have yet been found. Nonspiking interneurons, for example, do not appear to make output connections with any of the spiking local or intersegmental interneurons that are known to be presynaptic to them (Burrows, 1987b; Laurent and Burrows, 1989a,b). Lateral connections, however, do exist (Fig. 1B). They have been demonstrated for nonspiking interneurons (Burrows, 1979b) and are suspected (but not known yet) to exist amongst midline spiking interneurons. Only inhibitory lateral interactions have been described so far. They often seem to enhance the definition of a receptive field (spiking local interneurons), or enhance a motor output, by preventing the coactivation of antagonistic pools (nonspiking interneurons).

3. *A functional understanding of the receptive field of an interneuron may require knowledge of its "projective field."* Often, the fine structure of the receptive field of an interneuron makes what one would call functional sense. Two inhibitory regions flanking an excitatory center, for example, probably sharpen the edges of the excitatory region. In such cases, knowing the output connections of this interneuron does not necessarily add much to our intuitive understanding of the makeup of its receptive field. In many instances, however, such simple rationalizations are not possible. The tactile receptive field of another interneuron, for example, might possess excitatory and inhibitory regions that are not adjacent. The excitatory region might be on the tarsus and the inhibitory one on the femur, such that their coactivation by an object during normal behavior would appear unlikely. Why then, are these receptive fields so organized? A possible explanation came from the study of intersegmental interneurons and their projections. Laurent and Burrows (1989a) found that a given nonspiking local interneuron can receive converging inputs from several intersegmental interneurons. If one of these intersegmental interneurons is excitatory and the other one inhibitory, one might expect that their coactivation by the same sensory stimulus should be avoided, for their converging effects would be conflicting. One simple way of preventing their simultaneous activation would be to design their receptive fields in such a way that, when one is excited, the other is probably inhibited. Such a pattern was indeed demonstrated (Laurent and Burrows, 1989a). The inhibitory receptive field of one intersegmental interneuron in such a converging pair therefore appeared to underlie a more precise coding of the input only when it was realized that the two interneurons converged on a same target. This result suggested that the receptive field properties of an interneuron

Chapter IV Integration by Spiking and Nonspiking Local Neurons

may be partially determined by its "projective" field. This reasoning, of course, is an *a posteriori* rationalization of a set of complex experimental observations. Theoretical approaches as well as experimental manipulations are thus now needed to test and explore these possibilities in a systematic way.

4. *Global organization of movement by single premotor units does not preclude fine tuning.* First, the finding that the "gain" of the synapses made by a neuron with its many targets is not identical at all synapses (see Section 2.B) indicates that fine tuning of interneuronal connections can occur at the level of single functional synapses. It suggests that the relevant computational unit of these circuits is probably smaller that the individual neuron. This result also suggests the existence of long-term plastic adjustments in those circuits. Second, it is proposed that, even if a single nonspiking interneuron can control several sets of motor neurons that act at several joints of a leg, fine and independent control of each output pool may be allowed by functional compartmentalization of the nonspiking interneuron itself. The cellular mechanisms allowing such compartmentalization would include shunting synaptic interactions and voltage-dependent changes in the electrical geometry of single local interneurons.

5. *The coexistence of spiking and nonspiking integrative modes in the same networks must be important.* Although this statement is more a declaration of faith than a demonstrated fact, I believe it is important to explore the following question: Why do premotor interneurons in these locomotor networks not use spikes for intra- and intercellular communication? The study of nonspiking integration in the thoracic nervous system of insects reveals that, although nonspiking interneurons receive direct afferent inputs, they are nevertheless not the primary "integrators" of leg mechanosensory signals. Rather, this role is played by other populations of local neurons, which all appear to use action potentials. Therefore, the property of graded synaptic transmission and absence of conventional action potential–generating currents is tightly associated with premotor control and not a randomly or widely distributed property found in all the "layers" of the sensory–motor networks. The question posed above can thus be extended to: Why do the primary integrators of the mechanosensory signals not use analog signaling? Speed of inter- and intracellular communication, signal-to-noise ratios, variable gain control, and smooth control of motor output are some of the issues that will occupy us for the years to come. As we tackle these challenging problems, however, we must also realize that all is not known about the networks yet. For example, although nonspiking interneurons certainly play an important role in controlling the motor output, they are clearly not alone. More basic groundwork is therefore still needed.

Acknowledgments

Parts of the work summarized here were, over the last 5 years, funded by grants and/or fellowships from the Royal Society, the SERC (UK), the Hasselblad Foundation, the National Institute of Health, the Human Frontier Science program, the Office of Naval Research, and the Searle, Alfred P. Sloan, and McKnight Foundations. Many thanks are due to my collaborators Malcolm Burrows, Beulah Leitch, David Shepherd, and Anand Sivaramakrishnan, with whom parts of the work presented here were carried out, and to Harald Wolf for his critical comments on the manuscript.

References

BURROWS, M. (1979a). *Science* **204**, 81.
BURROWS, M. (1979b). *J. Neurophysiol.* **42**, 1108.
BURROWS, M. (1987a). *J. Neurosci.* **7**, 1064.
BURROWS, M. (1987b). *J. Neurosci.* **7**, 3282.
BURROWS, M. (1989). *J. Exp. Biol.* **146**, 209.
BURROWS, M. (1992). *J. Neurosci.* **12**, 1477.
BURROWS, M., and SIEGLER, M. V. S. (1976). *Nature* **262**, 222.
BURROWS, M., and SIEGLER, M. V. S. (1978). *J. Physiol. (Lond.)* **285**, 231.
BURROWS, M., and SIEGLER, M. V. S. (1982). *Science* **217**, 650.
BURROWS, M., and SIEGLER, M. V. S. (1984). *J. Comp. Neurol.* **224**, 483.
BURROWS, M., and SIEGLER, M. V. S. (1985). *J. Neurophysiol.* **53**, 1147.
BURROWS, M. (1988). LAURENT, G., and FIELD, L. W. *J. Neurosci.* **8**, 3085.
GEORGOPOULOS, A. P. (1990). "Neural Coding of the Direction of Reaching and a Comparison with Saccadic Eye Movement," *Cold Spring Harbor Symp. Quant. Biol.* **55**, 849.
JACK, J. J., (1983). NOBLE, D., and TSIEN, R. W, (1983). *Electric Current Flow in Excitable Cells* (p. 518). Oxford University Press, England.
LAURENT, G. (1987). *J. Neurosci.* **7**, 2977.
LAURENT, G. (1990). *J. Neurosci.* **10**, 2268.
LAURENT, G. (1991a). *J. Neurosci.* **11**, 1713.
LAURENT, G. (1991b). In (D. Armstrong and B. M. H. Bush (eds.), *Mechanisms of Arthropod and Vertebrate Locomotion* (p. 11). Manchester University Press, New York.
LAURENT, G., and BURROWS, M. (1988). *J. Comp. Physiol.* **162A**, 563.
LAURENT, G., and BURROWS, M. (1989a). *J. Neurosci.* **9**, 3019.
LAURENT, G., and BURROWS, M. (1989b). *J. Neurosci.* **9**, 3030.
LAURENT, G., and HUSTERT, R. (1988). *J. Neurosci.* **8**, 4349.
LAURENT, G., and SIVARAMAKRISHNAN, A. (1992). *J. Neurosci.* **12**, 2370.
LEITCH, B., LAURENT, G., and SHEPHERD, D. (1992). The Embryonic Development of Synapses on Spiking Local Interneurones in Locust. *J. Comp. Neurol.*, in press.
NAGAYAMA, T. (1989). *J. Comp. Neurol.* **283**, 189.

NEWLAND, P. L. (1991). *J. Comp. Neurol.* **312,** 493.
REDMAN, S. (1990). *Physiol. Rev.* **70,** 165.
SHEPHERD, D., and LAURENT, G. (1992). *J. Comp. Neurol.,* **319,** 438.
SIEGLER, M. V. S., and BURROWS, M. (1979). *J. Comp. Neurol.* **183,** 121.
SIEGLER, M. V. S., and BURROWS, M. (1983). *J. Neurophysiol.* **50,** 1281.
SIEGLER, M. V. S., and BURROWS, M. (1984). *J. Comp. Neurol.* **224,** 463.
SIEGLER, M. V. S., and BURROWS, M. (1986). *J. Neurosci.* **6,** 507.
SIMONS, D. J., (1989). CARVELL, G. E., and LAND, P. W.(1989). In J. S. Lund (ed.), *Sensory Processing in the Mammalian Brain* (p. 67). Oxford University Press, Oxford, New York.
SPARKS, D. L., LEE, C., and ROHRER, W. H. (1990). Population Coding of the Direction, Amplitude, and Velocity of a Saccadic Eye Movement by Neurons in the Superior Colliculus, *Cold Spring Harbor Symp. Quant. Biol.* **55,** 849.
TORRE, V., and POGGIO, T. (1978). *Proc. R. Soc. Lond. (Biol.)* **202,** 409.
WATSON, A. H. D., and BURROWS, M. (1985). *J. Comp. Neurol.* **240,** *219.*

PART II

NEUROETHOLOGY II: CONTROL OF ORIENTATION

Chapter V

Multisensory Processing for Movement: Antennal and Cercal Mediation of Escape Turning in the Cockroach

CHRISTOPHER M. COMER AND JOHN P. DOWD

Neurosciences Group
Department of Biological Sciences
University of Illinois at Chicago
Chicago, Illinois

I. The Sensory Guidance of Orienting Movements

The study of animal orientation deals not with the mechanisms of animal locomotion *per se,* but rather with "the directions in which they walk or swim, and the reasons why particular directions are selected" (Fraenkel and Gunn, 1961). Virtually all animal reactions to sensory inputs are spatially directed in this sense. However, although it has long been appreciated that most motor behavior is oriented either toward or away from specific sensory cues, the neural circuitry underlying such sensory guidance has only recently been analyzed in detail.

In vertebrate animals, orienting movements generally depend on information encoded by sensory interneurons and premotor interneurons within "computational maps" (Knudsen *et al.*, 1987). The best-analyzed examples are orienting movements of the eyes and/or head toward visual cues (e.g., Lee *et al.*, 1988) and visually triggered reaching movements of the arm (e.g., Georgopoulos *et al.*, 1986). For both of these types of movements there is at present a reasonable solution to what Georgopoulos has termed the "specification problem," i.e., determining the neural code by which direction of the movements is specified by the brain (Georgopoulos, 1990). In both instances, stimulus location and angle of movement are uniquely specified by impulse activity spatially distributed across a *large* interneuronal ensemble.

The evasive behavior of the cockroach, *Periplaneta americana,* provides an example of a hexapod orienting response for which the neural processing underlying directional movement can be dissected at the level of individually identifiable nerve cells. Initially, we were interested in how the direction of cockroach escape responses is specified by interneurons within the cercal wind-sensory system. There is reason to believe that the solution to the specification problem in this system also hinges on ensemble coding, but involving an extremely *small* neuronal population (Dowd and Comer, 1988; Camhi and Levy, 1989). However, we have recently found that escape can be initiated by another sensory modality and therefore additional interneurons. This provides opportunities to compare the control of directional movement by two different ensembles within the same organism.

II. Cockroach Evasive Behavior as an Orienting Movement

Orthopteroid insects such as cockroaches and crickets possess a pair of sensory organs (cerci) extending caudally from the abdomen. Some of the receptors on each cercus are filiform hairs that deflect in response to wind. Roeder (1948) analyzed how cercal hair afferents activated several large-diameter, rapidly conducting axons within the central nervous system (CNS) (the ventral nerve cord). Roeder basically proposed that the escape response of the cockroach was mediated by these elements as part of a three-neuron reflex arc: cercal afferents activated abdominal giant axons, giant axons transmitted signals rapidly to the thoracic ganglia, and there leg motor neurons were activated to initiate running. The timing of signal passage through these cells was found to be more than rapid enough to account for the relatively short behavioral latency of escape. The time from wind arrival to first leg movement averaged 54 ms in Roeder's experiments (Roeder, 1959).

However, the circuit for wind-triggered escape is not quite as simple as it appeared in Roeder's model. First, responses to abrupt wind stimuli do not consist simply of running. Animals typically pivot *away* from a wind puff and then run (Fig. 1). This directionality was first shown by Camhi and colleagues (Camhi and Tom, 1978, Camhi *et al.,* 1978). Second, at about the same time, the physiology of the giant axons was characterized and it became clear that more than just one or two of them are required to orient the turning component of escape (see later). Third, the giant axon-to-motor circuitry is polysynaptic and involves several types of thoracic interneurons (Ritzmann and Pollack, 1990; Ritzmann, this volume).

Chapter V Multisensory Processing for Movement

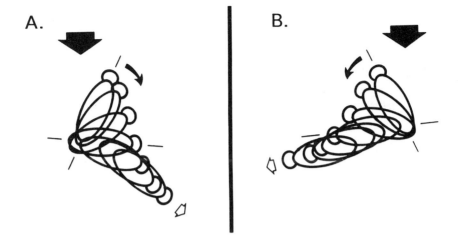

FIGURE 1. Evasive response to wind begins with a turn oriented away from the wind source. (A and B) Adult male cockroaches responding to standard wind puffs generated by a wind machine. Ellipse and circle represent body and head of cockroach on successive video frames. Large filled arrow shows wind direction; open arrow shows direction of subsequent run. (A) normal animal responds to left frontal wind with right turn. (B) Another animal (with lesion described in text) responds to right frontal wind with a left turn.

III. Interneurons and Wind-Mediated Evasive Turning

A. *Central Representation of Wind-Sensory Information*

Each giant axon arises from a cell body in the last abdominal ganglion. Seven pairs of giants are individually identifiable within the abdominal portion of the nerve cord, and they have come to be called giant interneurons (GIs). Furthermore, on each side of the cord, GI axons are clustered into two subgroups: four ventral axons (vGIs), which are the largest axons in the abdominal cord, and three smaller dorsal axons (dGIs). Each GI has been characterized by intracellular recording and staining. Each is wind sensitive, and five of the seven are directionally selective in their response to wind—i.e., they fire more impulses, and at shorter latencies, when wind arrives from a preferred direction (Westin *et al.*, 1977). Presumably, then, the relative activity of different GIs might be a code for wind direction.

Although it has not yet been possible to record from all 14 GIs simultaneously, the reliable variation in axon size and location makes it possible to ask how wind-evoked neural activity varies across the components of the GI population. This has been done using simultaneous extracellular multiunit recordings from each cord connective and signal sorting algorithms that can distinguish spikes from vGIs, dGIs and non-GIs (Smith *et al.*, 1988, 1991). Figure 2 shows a summary of relative wind-evoked neural activity (i.e., number of wind-evoked impulses) within the GIs. Note that all neural components show activity for all wind angles, and there is a pronounced pattern across the population: at all angles (except straight ahead) the GIs—and especially the vGIs—are more active ipsilateral to a wind puff and substantially less active on the contralateral side.

B. Models for Integrating Wind-Sensory Information

This laterality difference in GI activity as a function of wind angle is consistent with Westin's single-neuron data, as well as information on connectivity between each cercus and the GIs. In fact, the known connectivity has also served as the basis for lesion studies designed to "imbalance" wind-evoked activity in the GIs. Figure 3 compares the behavioral effects of two such lesions (Comer and Dowd, 1987). Animals normally turn *away* from winds on either side; i.e., about 80% of all turns are contraversive. This is seen in Fig. 3A for the responses of a group of control (sham-lesioned) animals. Figure 3 (bottom) shows that animals with one cercus removed would have a substantial reduction in wind-evoked GI activity ipsilateral to the lesion but only a slight reduction contralaterally. Animals with this lesion showed a reliable turning bias: a tendency to turn toward winds on the side of the lesion (note elevation of ipsiversive turns in response to wind from the left in Fig. 3B). Complete "imbalance" was produced by surgically severing one connective of the cord (Fig. 3C), and in these animals virtually all responses to winds on the side of the lesion were ipsiversive. In short, the rule appears to be: turn away from the side of the CNS with more wind-evoked GI activity.

Interpreting lesion studies of this sort depends on the presumption that the shifts in behavior result from alterations specifically in the GI pathway and not changes in other cells that might also follow from cercal ablation or cutting an entire connective. There is evidence for this presumption from selective deletion of individual GI axons by intracellular injection of proteolytic enzymes (pronase) (Comer, 1985, and see later).

The normal physiology and the lesion studies just cited suggest that the way wind information is displayed bilaterally within the CNS is related to an animal's direction of turning. To investigate this in some detail, we constructed computational models of the integration of wind-evoked GI activity (Dowd and Comer, 1988). The models were then tested by computer simulation. While the basic directional choice an animal makes (left vs. right) can be modeled by integrating across all GIs and assigning varying weights to each (Heetderks and Batruni, 1982), a relatively simple comparison of GI activity on the two sides of the CNS can simulate turning

Chapter V   Multisensory Processing for Movement

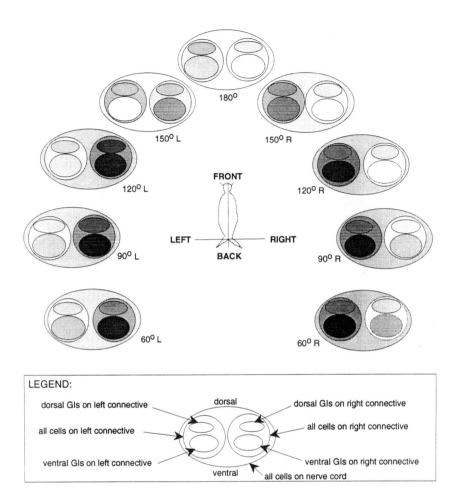

FIGURE 2. Profile of wind-evoked neural activity in interneurons that make up the ascending GI pathway. Legend at bottom is a schematic drawing representing a cross section through the abdominal portion of the nerve cord and shows the groups that could be separated from extracellular multiunit recordings. (Actual anatomical pattern of GI axons can be seen in Fig. 5A). Wind was presented from each of nine different angles in the horizontal plane as indicated around schematic cockroach at center. 0° is directly behind animal; 180° is directly in front. Areas in nerve cord schematic at each angle are shaded in proportion to average number of action potentials recorded within 50 ms of a standard wind puff for that cell group. All white represents 100% of a group's maximum activity, black represents 0%. Data averaged from five animals. Shading indicates greater relative numbers of impulses on left side for left winds and on right side for winds from the right. Reprinted from "Investigation of a Multicellular Neural Code for Directed Movement," by Smith, Wheeler, Dowd, and Comer, in the *Proc. Ann. Int. Conf. IEEE Engineering in Medicine and Biology Society.* © 1991 IEEE.

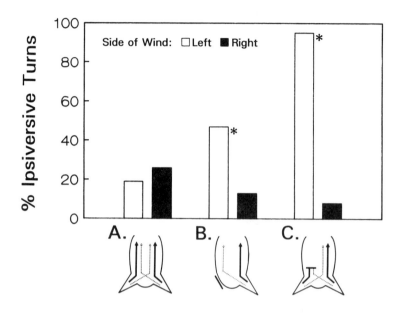

FIGURE 3. Direction of wind-evoked escape turning in animals with unilateral lesions within the cercal-to-GI system. Lesions were made on the left or the right side of the animals, but data are normalized to represent all lesions as left side. Histograms summarize percentage of turning responses that were ipsiversive (toward wind) when wind puff came from the left (side of lesion) or the right (opposite side) as shown by white and black bars. Condition of each group is shown schematically at bottom. Arrows show information flow from each cercus to GIs on each side of nerve cord. (Thick lines) GIs receive substantial input from ipsilateral cercus; (dotted lines) GIs receive less input from contralateral cercus; (A) Surgical controls; (B) one cercus removed; (C) one connective of nerve cord cut. Asterisk indicates a statistically significant difference in % ipsiversive turns on the two sides (2×2 chi-square test). N = five animals per group. Adapted from Comer and Dowd (1987, Figs. 1 and 8).

behavior with a pattern that matches behavior seen in real cockroaches (e.g., Comer and Dowd, 1987). This pattern includes (1) a choice of left or right turn; (2) a grading of turn amplitude—large turns for rostral winds, smaller turns for caudal winds; and (3) variation in turn angles and also direction (i.e., left or right). In the following description we will concentrate on choice of (and variation in) turn direction. For discussion of turn amplitude, consult Dowd and Comer (1988).

General features of the models we analyzed can be summarized. Wind puffs from various angles were the inputs. Outputs were turn angles computed according to algorithms that assigned a direction and angular amplitude to each response. On every simulated behavioral trial the computer randomly generated a wind stimulus angle and then calculated [based on Westin's data on spike numbers and timing

(Westin et al., 1977)] a probabilistic estimate of the number of impulses in each GI. (Impulse number could vary, but there was no variation in the temporal pattern of GI activity over trials, i.e., relative timing of spike trains either within or between GIs.) We proposed two simple algorithms for processing GI spike trains (Fig. 4A). First, a comparison of impulse activity on the two sides of the cord could determine turn direction. We assumed no differential weighting of the cells. Thoracic integrators simply counted spikes without regard for which GI the spike came from. Second, a summing of GI activity on one side of the cord was proportionate to turn amplitude. These two algorithms were first applied to the integration of information in the set of vGIs. Using just this information, simulated cockroaches were capable of producing a pattern of oriented turning with a degree of variation similar to empirical data and under a wide variety of conditions.

In our simulations, variation in the specific turn angle for a given stimulus angle, including turns toward the wind, could be produced only when the physiological

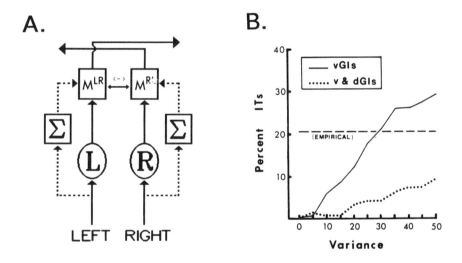

FIGURE 4. (A) Simple model for the GI control of the orientation of wind-mediated evasive turning. Arrows indicate direction of information flow. L and R, integrators that count GI impulses; M, motor center; LR and RL, left or right GIs initiating a right or left turn, respectively; sigmas, counters that set amplitude (specific angle) of turn—not dealt with here. (B) Influence of different numbers of GIs on operation of the type of model shown in A. ITs, Ipsiversive turns observed in simulated turning behavior; variance, degree of variation in GI impulse number (from the mean number) that was allowed over trials. Each curve was constructed by taking the average of three simulations (250 trials/simulation) at 11 intervals covering a range of cellular variance from 0 to 50%. (Dashed line) %ITs seen in our behavioral studies of real cockroaches. After Dowd and Comer (1988, Figs. 2 and 8).

response of the GIs was allowed to vary over trials. Behavioral variance was a function of cellular variance. When models included the integration of both ventral and dorsal GI activity, normal behavior (i.e., typical levels of behavioral variance) could not be produced at any reasonable level of cellular variance. The simulated cockroach became far more accurate than its biological counterpart (Fig. 4B). *The empirically observed pattern of turning in normal animals was best explained by assuming only vGI mediation of turning.*

Could a model that accurately simulated the behavior of intact animals also predict lesion effects? We found a surprisingly high degree of agreement between the effects of cutting one connective (Fig. 3C) and simulated unilateral lesions of all vGIs. In both cases overall response rate to wind dropped (to 73% empirically and to 72% in the simulations) and turning direction was shifted (an increase in ipsiversive turns on the side of the lesion: to 95% empirically, 100% in simulations). The effects of more discrete lesions will be described later.

Comparators have been invoked before to explain animal orienting movements (e.g., Huber *et al.,* 1984). However, the details of how a neural comparator might operate have not been worked out. It is therefore worth noting what our studies have revealed about the properties expected of a comparator that could orient cockroach escape turning. Our models were of two types: an "absolute" and a "relative" comparator. In the absolute comparator model there was no sharing of information between left and right GIs, and thoracic integrators were concerned solely with the pattern of activity arriving via ipsilateral cells. Whichever side had a sufficient level of activity (a given number of impulses occurring within a specified time interval) "won the race" and directed the animal to turn contraversively. In such a scheme, *the relative level of activity on the two sides is unimportant on any given trial.*

In relative comparator models, GI activity on one side was summed by a thoracic integrator on that side and either inhibited or excited the integrator on the other side. In the case of inhibitory interactions, for example, thoracic integrators are effectively computing, on each trial, the difference in activity on the two sides of the nerve cord. Relative comparator models with inhibition work no better than absolute comparators at simulating the behavior of normal or lesioned cockroaches (Dowd and Comer, 1988). This is significant because it says something fundamental about the representational character of information in the windsensory system: *There do not have to be any cells in the thorax that compute relative amounts of wind-evoked GI activity on the two sides of the nerve cord—even if the animal behaves, over many trials, such that directions of turn are predicted by relative activity levels of left vs. right GIs.* If there were excitatory interactions between the two sides, differences between GIs in each connective must be somewhat obscured. Nonetheless, normal patterns of behavior can still be simulated by such a model. Presumably such an arrangement would maximize rapid stimulus detection but must limit accurate stimulus localization. Physiological studies of thoracic interneurons that show convergent GI input (Westin et al., 1988; Ritzmann and Pollack, 1988) indicate that they sum inputs from one or both sides, but cells comparing left and right sides have not been described.

The simulations of normal behavior and of large unilateral lesions could be generated using either absolute or relative comparators. Absolute comparators are clearly the simplest type that can explain the integration of GI information and are consistent with the known physiology of thoracic cells.

C. *Specific Lesions of Wind Interneurons*

Further refinement of our understanding of the integration process must look to the level of specific cells. We have done this by analyzing the behavioral consequences of killing GIs individually and in various combinations. To do this, one or several axons are impaled with microelectrodes and pressure injected with pronase. The nerve cord is otherwise left intact, and escape behavior is subsequently monitored. The identity of cells killed by injection is determined from postmortem histology (e.g., Fig. 5A). Injected axons are known to degenerate within 48 h (Comer, 1985).

In initial studies (with nymphal cockroaches) it was found that when GIs were deleted unilaterally, it was possible to produce shifts in turning behavior similar to those produced with surgical lesions. The specific effect depended on which GIs were ablated and how many were ablated (Comer, 1985). More complete studies have now been conducted on adults and the results parallel those seen in nymphs (Comer and Stubblefield, 1988, and in preparation). For example, unilateral removal of GI-2 (an omnidirectional cell) had no apparent effect on direction of turning, but removal of GI-1 or GI-3 (both directionally selective cells) produced noticeable increases in ipsiversive turning in response to winds on the side of the lesion (Fig. 5B). Even greater shifts in turning were produced when multiple GIs were simultaneously deleted from one side of the cord (nymphs: Comer, 1985; adults: Comer and Stubblefield, 1988). In Table I, note that there was a trend that as one, two, or three GIs were deleted unilaterally, the bias in turning increased correspondingly. It is clear that *shifts in direction of wind-evoked turning can be produced by unilateral deletions exclusively within the GI population.*

An interesting experiment that is possible with lesion techniques is to delete or silence multiple cells, but to do so symmetrically. Comparator models (of both types described earlier) predict that such lesions would reduce the probability of responding to wind but should not result in any turning bias. We tested this predic-

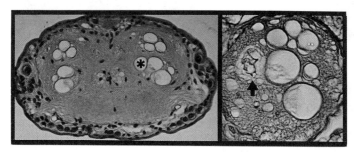

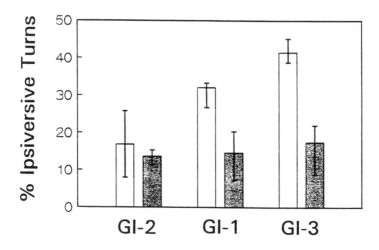

FIGURE 5. Selective unilateral deletion of GIs alters direction of wind-evoked turning. (A) Evidence for killing GI with pronase. (Left) Cross section through abdominal ganglion taken 1 week after injecting left GI-1 with pronase (since left GI-1 is absent, right GI-1 is marked with asterisk for comparison). Top of picture is dorsal; axons lateral to GI-1 = GI-2, axons ventral to GI-2 = GI-3. For scale, diameter of the ventral GI axons is in range 25–40 µm. Adapted from Comer (1985, Fig. 2). (Right) Section through abdominal connective shortly after injecting GI-1 on right in another animal. Note that crenulated profile of degenerating GI axon is visible (arrow). (B) Turning behavior of adult cockroaches after unilateral GI deletions. Lesions were made on the left or right side of the animals, but data are normalized to represent all lesions as left side. Histograms summarize mean percentage of ipsiversive turns (ITs) in response to winds from left (side of lesion, open bars) or from right (opposite side, stippled bars). Error bars indicate range of %ITs seen in individual animals. $N = 3$ animals, and at least 200 trials per group.

Chapter V Multisensory Processing for Movement

TABLE I. Percentage of Ipsiversive Turns (ITs) Following Various Unilateral GI Lesions Made by Pronase Injection

Lesion (left)	N (n)	%ITs left	%ITs right
GI-1	3 (270)	32.1	14.7
GI-3	3 (262)	41.5	17.5
GIs 1,3	2 (191)	53.1	15.6
GIs 1,2,3	3 (373)	61.9	12.9

Animals were lesioned on left or right side of nerve cord, but all data are normalized to represent lesion on left side. 1,3 = GIs 1 and 3 both removed on the same side of the cord; 1,2,3 = GIs 1, 2, and 3 all removed on one side. N = number of animals in group, n = total number of behavioral trials analyzed. Left and right refer to turns made in response to wind puffs delivered from those sides. Note that increase in %ITs is greater as more GIs are deleted from one side.

tion in two ways. First we produced a turning bias with a GI lesion on one side and then reversed the behavior by subsequently lesioning GIs on the opposite side. Because we have not yet found it possible to reopen the abdomen and safely operate on the nerve cord a second time, we attempted to reverse an initial GI-3 lesion by removing the contralateral cercus. Cercal removal is known to reduce the wind responsiveness of vGIs on the side of the removal (Westin et al., 1977, and see earlier). We found that the initial turning bias of GI-3 deletion was reversed by subsequent cercectomy (Fig. 6). The second way we made bilateral lesions was more specific. We killed GI-3 with pronase bilaterally in one procedure and found that these animals displayed no turning bias but did have a lower probability of responding to wind, as predicted (G. T. Stubblefield and C. M. Comer, unpublished data).

In our modeling studies, simulated unilateral deletion of single GIs led to percentages of ipsiversive turning much higher than we have seen empirically. The reason for the large effect is simple; with only four pairs of cells controlling the response, the contribution of any one cell to the overall information set is significant. There are several possible explanations for the quantitative mismatch between simulated and empirical single-cell lesion results, and the most likely ones are discussed here. (1) There may be more wind interneurons (some in addition to the vGIs) involved in escape initiation. On the basis of their modeling studies, Camhi and Levy (1989) have suggested that perhaps all seven pairs of GIs are involved. However, as discussed in the previous section, if all seven pairs of GIs are presumed to be involved in mediating escape, normal behavior cannot be simulated. Real insects are flexible (e.g., Murphey, 1986) and given their abilities, specifically the abilities of cockroaches (e.g., Comer and Camhi, 1984) for plastic behavioral adjustments and recalibration or rewiring of neural circuits, they may quickly compensate for the loss of even a single GI. This remains possible and needs to be investigated. (2) The relatively small empirical deficits were the result of high wind stimuli (alternate

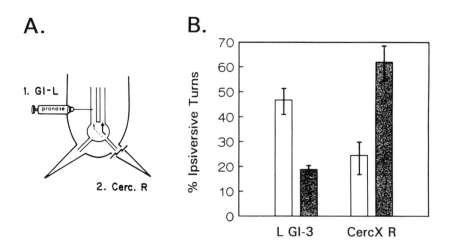

FIGURE 6. Effects of unilateral GI lesion can be reversed if wind-evoked activity is subsequently reduced on the intact side of the nerve cord. (A) Lesion protocol. Left GI-3 was first deleted by intracellular injection of proteolytic enzymes (pronase). After assessing the behavioral deficit, the right cercus was removed. (B) Open bars indicate percent ipsiversive turns (ITs) to winds from the animal's left side. Shaded bars show response to winds from the right side. After lesioning GI-3, there was a significant increase in the percent ITs in response to winds from the lesioned side. After the second lesion, the response to left winds returned to nearly normal levels, while ipsiversive turns to right winds were significantly elevated.

mechanosensory pathways were activated; see later). (3) The static temporal pattern employed in our simulations is an inadequate description of GI neural activity. This point has led us to analyze the temporal patterning within the GIs. There is clearly a fair amount of it (J. P. Dowd, unpublished), and this knowledge will allow finer analysis of the process of integrating wind-sensory information.

IV. Other Interneuronal Pathways Can Initiate Evasive Turning

Given the central role in escape that is generally assumed for the giant interneurons, it would be useful to remove *all* of them from an animal and then ask if any responses are still possible. Destruction of GIs with pronase is appropriately selective but cannot be used to lesion all the GIs, or even all of the vGIs, simultaneously. (Our current best is three GIs cleanly injected during one operation, e.g., Table I.) We therefore used two less selective techniques, but both have provided clear-cut conclusions.

In one group of animals we simply transected both connectives of the nerve cord in the abdomen (Comer *et al.*, 1988), and in others we used an electro-

Chapter V Multisensory Processing for Movement

cautery probe to burn the abdominal portion of the cord (Comer *et al.*, 1988; Stierle *et al.*, 1988; and I. E. Stierle, M. Getman, and C. M. Comer, in preparation). The first procedure, of course, blocks all axons ascending to the thorax; the second does not sever the cord but causes degeneration of all GIs and other unidentified axons. An example of a complete GI deletion is seen in Fig.7 (compare 7A with 7B). In both experiments, all animals were still capable of generating escape responses to wind following lesion. These responses to wind, although apparently normal in form, occurred at longer average latencies (Fig. 7C). It can be safely concluded that *more than one sensory pathway can initiate escape behavior. The additional pathway(s) revealed in these experiments is necessarily noncercal*, because it is evident in animals completely transected below the thorax. It also *must originate from thoracic or cephalic receptors*, and

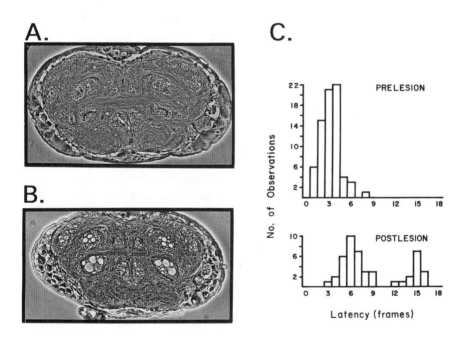

FIGURE 7. Escape responses persist after complete block of giant interneuron pathway. (A) Block produced by deletion of all GI axons from the ventral nerve cord. Cross section through abdominal ganglion showing degeneration of all GIs as a result of an electrocautery burn. (B) Cross section from control cockroach at same level as in 7A. Note the presence of the seven pairs of giant interneurons. (C) Frequency distributions of behavioral latencies of wind-triggered escapes. Top histogram shows responses of normal animals; bottom histogram gives responses of animals with GI pathway blocked by complete transection of the abdominal nerve cord. Part C from Comer *et al.*, (1988, Fig. 3).

the pathway(s) is necessarily non-GI, because in both experiments the GIs axons degenerated rostral to the site of lesion (Fig. 7A).

Analysis of the behavior of animals without any GIs immediately revealed several things. First, although such animals produce wind-triggered escapes, they do not respond to wind reliably. The frequency of responses dropped from almost 100% to about 30% after transection (Comer *et al.*, 1988, and Table II). The decline in responsiveness was less drastic after electrocautery (Stierle *et al.*, 1988, and I. E. Stierle, M. Getman, and C. M. Comer, in preparation). This lowered responsiveness to wind did not represent a deterioration of motor aspects of escape because these animals responded briskly with evasive turning and running when the pronotum was touched (Table II). Presumably the noncercal sensory receptors are not as wind sensitive as the cerci themselves. Second, subsequent antennal removal decreased responses to wind (Table II). We concluded that receptors associated with the antennae could explain responsiveness to wind, at least in part, in animals with the GI pathway blocked. The low reliability and long latency of escape responses *to wind* in these animals suggested that we might not be dealing with a true wind pathway at all, but rather wind activation of some other type of mechanosensory pathway. The touch-evoked evasive responses just described seemed suggestive in this regard because they appeared to be essentially the same motor output (pivot away and run) as that triggered by wind.

Further studies have now revealed that while both cercal and antennal receptors can trigger escape, antennal receptors can initiate escape responses only to very intense wind puffs. When a group of animals had their cerci removed (with antennae intact), they responded only about half the time to standard wind puffs of high peak velocity. They were essentially unresponsive to puffs of lower peak velocity (Fig. 8, upper right). However, with the antennae removed and the cerci intact, they were no less responsive to either high or low winds than normal (Fig. 8, bottom right). In sum, the antennae do not display any properties expected of a true wind

TABLE II. Escape Turns Produced in Response to Wind and Touch Following a Complete Block of the GI Pathway (Cord Cut) and Then Subsequent Removal of Both Antennae (Ant-X)

	Escape Turns (%)	
Lesion	Wind Puff	Touch Pronotum
Prelesion	97.3	—
Cord cut	30.2	83.3
Ant-X	14.1	90.9

Note that escapes are reduced but not eliminated by cutting abdominal cord, and remaining responses to wind are substantially reduced by antennal removal. Animals readily produced escape turns to touching pronotum throughout. Pooled data from seven animals.

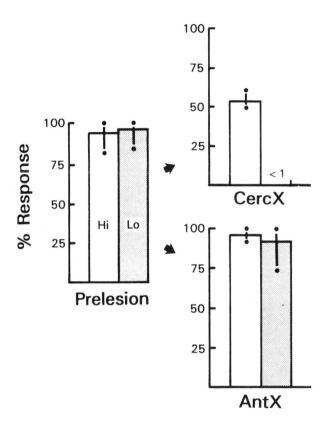

FIGURE 8. Sensitivity to wind is affected greatly by cercal, but not by antennal, removal. Histograms show responsiveness (% responses over all trials) to standard high-intensity wind (1.8 m/s peak velocity, open bars) or lower-intensity wind (1.2 m/s, shaded bars). Prelesion histogram (left) represents pooled data from 10 animals. Five animals had both cerci removed (CercX, top right) and five had both antennae removed (AntX, bottom right). Height of each bar represents the average % response across all animals; bars give range of percents in individual animals.

sensory system with respect to escape. Furthermore, evasive turning in response to winds of low peak velocities (real predators generate even gentler winds than we used) is totally dependent on the cerci.

From physiological studies, the antennae can be seen to possess properties consistent with a role as a touch-sensitive system (Burdohan and Comer, 1990). Large-amplitude units can be recorded from the neck connectives following stimulation of a cockroach with standard high-intensity wind puffs. These units are rapidly conducting, descending interneurons. By recording simultaneously from the neck connectives and abdominal connectives, it is possible to compare the wind responsiveness of these descending interneurons and the ascending GIs. High-intensity winds elicit impulses in both, whereas low-intensity winds activate the GIs but not the descending interneurons (Fig. 9A). Although interneurons of the antennal system thus are not particularly sensitive to wind, they respond reliably to touch and at extremely short latencies (i.e., less than 10 ms, Fig. 9B). Application of a touch stimulus to one antenna activates one or more large-amplitude units in each cervical connective (Fig. 9B).

The descending mechanosensory interneurons (DMIs) have now been recorded from and characterized in some detail (Burdohan and Comer, 1990; J. A. Burdohan, in preparation). There are several individually identifiable neurons with cell bodies in one of the head ganglia that are touch sensitive and have large-caliber axons descending into thoracic ganglia. We have recorded from at least six different DMIs; two have been studied in great detail because they have the largest axonal diameters in the cervical connectives. An example of one of these is shown in Fig. 10, where this descending interneuron (DMIa-1) is compared to an ascend-

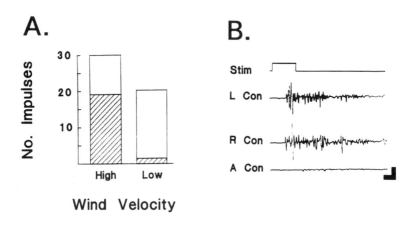

FIGURE 9 (A) Responsiveness of descending (antennal) interneurons compared to ascending (cercal) interneurons. Pooled data from three animals. Extracellular recordings were made simultaneously at cervical and abdominal levels, and the nerve cord was transected between the abdomen and thorax. All units greater than 25% of maximum amplitude were counted (within 70 ms of stimulus onset). Open bars, abdominal units (GIs); crosshatched bars, cervical units. Height of bars represents average number of units per trial ($n = 31$ trials at high peak velocity, $n = 49$ at low velocity). (B) Unit activity recorded in response to tap on left antenna. Top trace represents monitor of touch stimulus (voltage to piezoelectric crystal). Bottom three traces are simultaneous extracellular records. L Con, Left cervical connective; R Con, right cervical connective; A Con, abdominal connectives. Note that antennal touch does not activate GIs. Calibrations: vertical, 200 µV (except stimulus trace); horizontal, 5 ms. After Burdohan and Comer (1990, Fig. 2).

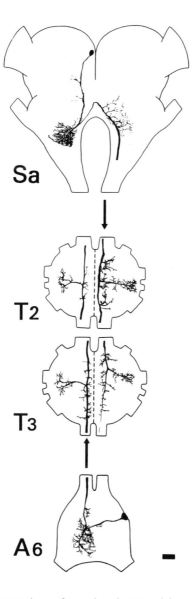

FIGURE 10. Structural comparison of cercal and antennal interneurons. Camera lucida reconstructions of two cells injected with cobalt hexamine and viewed in whole mount from the dorsal surface. Descending cell (DMIa-1) has axon on right side of thoracic ganglia for comparison with an ascending giant interneuron (left GI-1). Sa, Supraesophageal ganglion (brain); T2, mesothoracic ganglion; T3, metathoracic ganglion; A6, terminal abdominal ganglion. Other intervening ganglia not shown (arrows). Scale bar, 100 μm. Adapted from Stubblefield and Comer (1989, Fig. 5) and Burdohan and Comer (1990, Fig. 3).

ing giant interneuron (GI-1). Interestingly, the axonal diameter of this cell, and the other large-caliber DMIs that we have recorded from in the neck connectives, is in the same range as the caliber of the GIs in the abdomen, and thus they are, in a sense, descending "giants". Also, all the DMIs we have analyzed so far show extensive branching in each of the thoracic ganglia, in a way that roughly parallels that of the wind-sensory GIs (Fig. 10).

The specific sensory receptors that activate the DMIs have not yet been entirely worked out. What we know is that tapping an antennal flagellum is a potent stimulus and that this is probably due to stimulation of receptors on the flagellum (presumably exteroceptors), as well as receptors in the basal segments that respond to deflection of the whole flagellum with respect to the base (presumably proprioceptors) (J. A. Burdohan, unpublished observations). Finally, not only do these cells respond to touch at short latencies, but from measuring their axonal conduction velocities we calculate that the two DMIs with the largest-caliber axons should be able to activate the thoracic motor apparatus at impressively short latencies—perhaps 15–20 ms from touch application (Burdohan and Comer, 1990). These bilaterally paired DMIs have characteristics that make them likely candidates to explain evasive behavior in response to touch stimuli.

All of the results in this section have led us to the conclusion *not* that there are two wind pathways for triggering escape, but rather that there are at least two sensory modalities and two distinct interneuronal systems that can trigger and direct the cockroach escape response. A similar conclusion has been reached by Ritzmann and colleagues, based on a different line of experiments (Ritzmann *et al.*, 1991). Finally, our data indicate that behavioral and physiological analyses of the wind-sensory escape system must consider the nature of stimuli very carefully. Winds used in most previous behavioral studies, for example, were in the range at which noncercal receptors would be activated. Future work, especially that involving subtle manipulations such as single-cell lesions, will need to use stimuli appropriately specific to the wind-sensory system.

V. Context for Escape Initiation by Cercal and Antennal Pathways

The idea that escape is multisensory can be accepted only if multiple sensory cues can be demonstrated in the normal behavioral interactions of cockroaches with predators. It has already been shown that some predators, e.g., marine toads, generate winds when they strike at cockroaches, and that other obvious sensory cues (e.g., visual or tactile) are not required for that particular predator (Camhi *et al.*, 1978). What sort of predators might generate touch stimuli in addition to, or in place of, wind cues?

In considering this question, we turned to relatively small-bodied predators that should have figured prominently in the evolution of predator evasion systems of insects as they colonized terrestrial environments—arachnids and other insects. We have now used high-speed videography to examine the interactions of cockroaches with several species of spiders and with mantids (Comer *et al.,* 1989).

An attack by a wolf spider is reconstructed in Fig. 11A. This attack is typical of what we observed. Because the antennae are so long and moved about actively, there is a high probability that they will be contacted by a predator attempting capture. Note first that before contact the spider was in motion toward the cockroach for at least seven video frames (117 ms) and there was no movement by the cockroach during this time. The spider made contact with the right antenna on frame 8, and the cockroach began turning to the left between frames 8 and 9, *after* contact. This is unlike interactions with toads, where cockroaches usually responded within a few frames, often before a toad got its tongue out of its mouth, i.e., *before* contact (Camhi *et al.,* 1978; C. M. Comer and K. A. Murphy, unpublished). Over all trials we reconstructed, the time between physical contact with a striking spider and first movement by cockroaches was always short (mean = 1.1 frames, or about 18 ms *following* contact). Finally, the directionality of the response in Fig. 11A was also typical. Animals usually turned away from the side on which contact was made with an antenna during the strike.

Our experiments have not ruled out a role for sensory cues other than wind or touch in evasive behavior, but we have been able to demonstrate directly the importance of touch for certain predators. Perhaps the best example comes from studies in which we compared the effectiveness of escape in groups of cockroaches with either the antennae or cerci removed. Covering cercal wind receptors causes cockroaches to be caught more often by toads (Camhi *et al.,* 1978). We found a small but not statistically significant decrease in the rate of successful escape from spiders following cercal removal. However, antennal removal led to a significant decline in the percentage of successful escapes from spiders (Fig. 11B). Thus cockroaches with other major sensory receptors intact, such as cerci and eyes, could not effectively detect and escape from spiders after loss of antennal receptors.

VI. Integration of Touch-Sensory Information Compared with Integration of Wind-Sensory Information

The evasive behavior of cockroaches triggered by touch can be analyzed quantitatively with controlled stimuli. Brisk escape turns and runs can be elicited by punc-

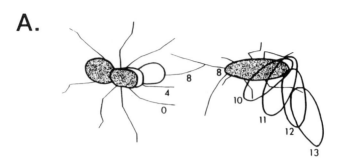

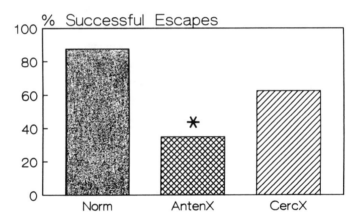

FIGURE 11. (A) Dynamics of spider–cockroach interactions implicate a short latency touch system. Shaded outlines indicate animals' initial positions, spider on left. Numbers indicate positions at successive video frames. Lunge of spider results in contact on frame 8. Cockroach began moving between frames 8 and 9. Duration of each frame, 8 ms. (B) Importance of antennae to the success of encounters with certain predators can be demonstrated. Histogram shows percentage of cockroaches' successful escapes from spiders in the normal condition (shaded bar), with antennae removed (crosshatched bar), and with cerci removed (striped bar). The antenX (but not cercX) condition is significantly different from the normal case.

tate stimulation of the antennae, the legs, the pronotum, etc. (Keegan *et al.,* 1991). In the following we will describe only our results for the antennae.

In Fig. 12A the orientation of turns elicited by tapping one antennal flagellum is summarized as a circular histogram. Note that the turns are overwhelmingly contraversive, and they are oriented with an average vector appropriate for escape. Consistent with all of the previously mentioned physiological and behavioral data, these responses were also of extremely short latency (the mean latency of the turns in Fig. 12A was 33 ms).

Since both ascending and descending systems produce short-latency contraversive turning responses, we would like to know if information descending within the DMI pathway is integrated in a manner similar to information ascending within

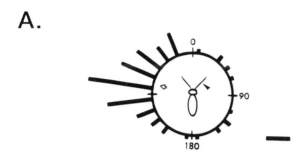

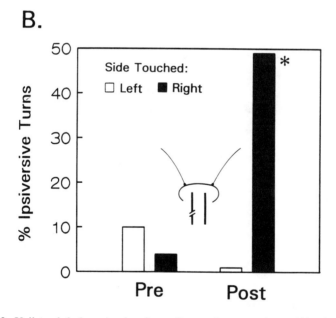

FIGURE 12. Unilateral lesions in the descending pathway produce shifts in turning responses to touch stimuli. (A) Circular histogram shows orientation of turns elicited by tapping one antenna. Open arrow shows average angle of turning. Scale bar at bottom right, 10 responses. (B) Percentage of ipsiversive turns in response to tapping one antenna after severing the left cervical connective. Open bars show turns when the left antenna had been touched. Filled bars show turns when the right antenna had been touched. There was a significant increase in ipsiversive turns when the right antenna was touched.

the GI system. We have begun this analysis by using lesions to disrupt the flow of information from antennal receptors to thoracic motor centers much in the same way that we lesioned the ascending pathway (Section III.B). We have found that completely severing one cervical connective does bias the direction of turns evoked by touching the antennae (Keegan et al., 1991). Just as with severing one abdominal connective, this causes ipsiversive turning for sensory inputs from one side only. However, the inappropriate turning responses show up not on the side of the lesion but rather on the side opposite the lesion (Fig. 12B). In addition, although the increased percentage of ipsiversive turns is quite marked and statistically significant, it does not represent a nearly complete bias in turning as seen for wind-evoked escape following unilateral abdominal lesions [Compare elevated %ITs in Figs. 3C (95%) and 12B (48%)]. This suggests that while the level of touch-evoked activity descending to the thorax is related to the direction in which an animal turns, the descending (DMI) system is not as lateralized as the ascending (GI) system.

We do not yet know how many DMIs normally participate in triggering and orienting escape turns to touch stimuli. The number may be more restricted than in the case of the wind system because a wind stimulus activates hairs on both cerci and all of the GIs, but touch can be restricted to a part of the body and therefore to a subset of DMIs with specific receptive fields. We are actively studying the physiology and anatomy of the DMIs so that we can describe the details of information flow within identified sensory and interneurons of this system.

VII. Conclusion

A first-order analysis of the cockroach escape system has almost been completed. The specification problem is understood, at least generally, for the wind-sensory control of escape. The solution to the problem appears to be similar within the descending touch-sensory system, but analysis of that system has just begun. The next phase in understanding cockroach escape will proceed to a level of analysis at which the flow of sensorimotor information through specific synapses will be described. This corresponds to the solution of what Georgopoulous (1990) has called the "implementation problem." Significant progress on the problem has already been made by Ritzmann's group (see Chapter VI, in this volume). Some exciting questions can now be addressed at the interface of sensorimotor processing in this system. For example, do the GIs and DMIs converge synaptically on the same premotor cells in the thorax? Do signals in the ascending and descending interneuronal pathways interact?

Chapter V Multisensory Processing for Movement

At the same time as these details are fleshed out, it will be necessary (and is now finally possible) to consider the operation of escape circuitry in more realistic contexts. Cockroaches undoubtedly evolved coping with situations in which stimuli signaling predatory strikes were complex. Our testing of neural coding and behavioral capacities will need to consider, for example, that stimuli are not restricted to the horizontal plane (as we currently present them), that they are often embedded in noise, and that they are rarely unimodal.

Using biologically realistic contexts, we may find that our definitions of the escape circuit will have to be expanded even more than they have been in the last 10 years. In a sense, Kenneth Roeder preceded us in this regard, as in so many other aspects of neuroethology. Roeder tested his ideas about the circuitry guiding behavior in the cockroach by building a neurally inspired mechanized cockroach, *Blatta electromagnetica* [described in an unpublished manuscript; see Hodgson (1990) for a description]. He did not succeed in producing a robot with appropriately insect-like behavior. He found, though, that if he went beyond thinking of circuitry in terms of reflex pathways (by "adding on" endogenously active units) he could generate more realistic behavior in his robot. Such expanded thinking will surely facilitate the interaction of neuroethology with robotics in the future.

Acknowledgments

The research summarized here has been supported by grants from the National Science Foundation. We are grateful to A. Don Murphy, Isabel Stierle, and Shuping Ye for providing comments on this manuscript.

References

BURDOHAN, J. A., and COMER, C. M. (1990). *Brain Res.* **535,** 347–352.
CAMHI, J. M., and LEVY, A. (1989). *J. Comp. Physiol.* **165,** 83–97.
CAMHI, J. M., and TOM, W., (1978). *J. Comp. Physiol.* **128,** 193–201.
CAMHI, J. M., TOM, W., and VOLMAN, S. (1978). *J. Comp. Physiol.* **128,** 203–212.
COMER, C. M. (1985). *Brain Res.,* **335,** 342–346.
COMER, C. M., and CAMHI, J. M. (1984). *J. Comp. Physiol.* **155,** 31–38.
COMER, C. M., and DOWD, J. P. (1987). *J. Comp. Physiol.* **160,** 571–583.
COMER, C. M., and STUBBLEFIELD, G. T. (1988). *Neurosci. Abstr.* **14,** 310.
COMER, C. M., DOWD, J. P., and STUBBLEFIELD, G. T. (1988). *Brain Res.,* **445,** 370–375.
COMER, C. M., GETMAN, M. E., MUNGY, M. C., and PLISHKA, J. (1989). *Neurosci. Abstr.* **15,** 349.
DOWD, J. P., and COMER, C. M. (1988). *Biol. Cybernet.* **60,** 37–48.

FRAENKEL, G. S., and GUNN, D. L. (1961). *The Orientation of Animals.* Dover Publications, New York.
GEORGOPOULOS, A. P. (1990). *Cold Spring Harbor Symp. Quant. Biol.* **55,** 849–859.
GEORGOPOULOS, A. P., SCHWARTZ, A. B., and KETTNER, R. E. (1986). *Science* **233,** 1416–1419.
HEETDERKS, W. J., and BATRUNI, R. (1982). *Biol. Cybernet.* **43,** 1–11.
HODGSON, E. S. (1990). *Am. Zool.* **30,** 403–505.
HUBER, F., KLEINDIENST, H. U., WEBER, T., and THORSON, J. (1984). *J. Comp. Physiol.* **155,** 725–738.
KEEGAN, A. P., MARA, E., and COMER, C. M. (1991). *Neurosci. Abstr.* **17,** 1245.
KNUDSEN, E. I., DU LAC, S., and ESTERLY, S. D. (1987). *Annu. Rev. Neurosci.* **10,** 41–65.
LEE, C., ROHRER, W. H., and SPARKS, D. L. (1988). *Nature* **332,** 357–360.
MURPHEY, R. K. (1986). *J. Neurobiol.* **17,** 585–591.
RITZMANN, R. E., and POLLACK, A. J. (1988). *J. Neurobiol.* **19,** 589–611.
RITZMANN, R. E., and POLLACK, A. J. (1990). *J. Neurobiol.* **21:**1219–1235.
RITZMANN, R. E., POLLACK, A. J., HUDSON, S. E., and HYVONEN, A. (1991). *Brain Res.* **563,** 175–183.
ROEDER, K. (1948). *J. Exp. Zool.* **108,** 243–261.
ROEDER, K. (1959). *Smithson. Misc. Coll.* **137,** 287–306.
SMITH, S. R., DOWD, J. P., WHEELER, B. C., and COMER, C. M. (1988). *Proc. Ann. Int. Conf. IEEE Eng. Med. Biol. Soc.* **10,** 1171–1172.
SMITH, S. R., WHEELER, B. C., DOWD, J. P., and COMER, C. M. (1991). *Proc. Ann. Int. Conf. IEEE Eng. Med. Biol. Soc.* **13,** 457–458.
STIERLE, I. E., GETMAN, M. E., and COMER, C. M. (1988). *Neurosci. Abstr.* **14,** 689.
STUBBLEFIELD, G. T., and COMER, C. M. (1989). *J. Morphol.* **200,** 199–213.
WESTIN, J., LANGBERG, J. J., and CAMHI, J. M. (1977). *J. Comp. Physiol.* **121,** 307–324.
WESTIN, J., RITZMANN, R. E., and GODDARD, D. J. (1988). *J. Neurobiol.* **19,** 573–588.

Chapter VI

The Neural Organization
of Cockroach Escape
and Its Role in Context-
Dependent Orientation

ROY E. RITZMANN
Departments of Biology and Neuroscience
Case Western Reserve University
Cleveland, Ohio

I. Introduction

For an animal to survive, it must be able to orient its movements toward goals and away from threats. This is also the case for many automated vehicles or robots. A large part of the control we exert over an automobile involves directing its movements toward our ultimate destination or away from other vehicles that might pose a threat. A system that allows an automobile to sense the impending impact of another vehicle and execute an avoidance action would be extremely beneficial. However, an avoidance system that only factors in the location of the immediate threat might cause more harm than good. A vehicle in close proximity could evoke a turning movement leading to a collision with a wall or another vehicle on the other side. Moreover, if the automobile is on a steep curve, or is in the process of passing another vehicle, an "escape turn" might be inappropriate. Thus, the decision to escape and the orientation of the ensuing movement must be made in the context of the vehicle's ever-changing situation.

The same problems must be solved continuously by most animals. Natural orientation systems have the advantage of millions of years of evolution. As in the automobile, appropriate animal behavior requires a great deal of information. An animal must consider its surrounding environment, the position of its limbs, and its physiological state before making a movement. In the case of escape systems, appropriate responses are critical to the survival of the animal, and they must be executed in a very short period of time.

The American cockroach *Periplaneta americana* escapes from predators by sensing the wind front created as the predator lunges (Camhi, *et. al.*, 1978). It initially turns away from the predator and then executes a more random run (Camhi and Tom, 1978). The escape system was once thought to be a simple reflex that maximized speed by utilizing relatively few neurons. However, as we have studied the behavior and underlying neural circuitry, we have come to appreciate the complexity of the system.

The cockroach must evaluate a variety of information in order to execute a proper escape movement. First, it must distinguish between real threats and harmless environmental stimuli such as breezes or stimuli from potential mates. Given that a real danger exists, the animal must orient its movements away from that stimulus. However, these movements must act in the context of the animal's present situation. Is it near a wall or in the middle of an open space? Is it in a particularly vulnerable situation, such as a well-lighted environment? What are the initial positions of its limbs? The escape decision must include these and many more bits of information. Moreover, the cockroach must act very rapidly. Any delay could be fatal.

This chapter describes the state of our understanding of the cockroach escape system. I will outline the behavior itself and then summarize the basic neural connections that are used to identify a threat and evoke the appropriate turn. I will then describe additional forms of input and modification which ensure that the response is appropriate for all internal and external factors. Except where background information is necessary, this chapter will concentrate on data accumulated within the last 10 years. Readers are referred to previous reviews for complete descriptions of earlier works (Camhi, 1980; Ritzmann, 1984). The components I will describe constitute a sophisticated predator avoidance system. The basic principles of this system could potentially be implemented in orientation systems of any autonomous agent, including crash avoidance systems for vehicles.

II. Behavior of the Escape Turn

What must a cockroach accomplish with its escape system? It must be able to detect faithfully the presence and location of threats. It must then use this information to direct escape movements that carry it out of harm's way. A lunging predator creates a wind front, which the cockroach detects with sensory apparatus located on its cerci, two antenna-like appendages on the rear of the animal's abdomen (Camhi and Tom, 1978; Camhi *et al.*, 1978). The ventral surface of each

Chapter VI   Neural Organization of Cockroach Escape

cercus is covered with very sensitive mechanoreceptive hairs. Winds as small as 12 mm/s evoke an escape response (Camhi and Nolen, 1981). The cockroach turns away from the wind source and then runs to safety. The directional nature of the turn ensures that the animal will at least start its escape by moving away from the lunging predator. The latency between wind stimulation and the onset of turning movements is approximately 58 ms for a standing cockroach (Roeder, 1967). This drops to 14 ms for an animal that is walking slowly at the time of stimulation (Camhi and Nolen, 1981). The actual turning movement takes between 20 and 30 ms (Nye and Ritzmann, 1992).

The principal parameter in determining turn direction is the location of the wind source relative to the cerci. Camhi and Tom (1978) performed a series of experiments to demonstrate that wind is the critical stimulus rather than other cues such as the large visual or auditory signals provided by the machine that generates the wind stimulus. For example, in one experiment, the wind machine was inactivated by turning off the fans that create air movement. A smaller, secondary wind source, which provided a much smaller visual image and produced no detectable sound signal, was placed in the arena opposite the primary wind machine. When both devices were activated simultaneously, the cockroach oriented to the wind coming from the smaller tube rather than to the larger and noisier wind machine. In another experiment, the cerci were displaced to the side so that the cockroach could either orient away from the wind source as perceived by the cerci or relative to the rest of its body. The animals consistently escaped relative to the cerci. It is important to note that, while these experiments clearly demonstrate that the directional properties of the wind puff constitute the primary determinant for escape direction, other parameters may still influence that decision.

In addition to wind direction, the cockroach also monitors acceleration. Escape responses are initiated consistently only if the wind puff accelerates at 0.6 m/s^2 or greater (Plummer and Camhi, 1981). In this way the animal can distinguish between nonthreatening wind stimuli in its environment and wind fronts generated by attacking predators.

Although the escape turn is directional in nature, it is not extremely precise. A wind from the right evokes a turn to the left, and a wind from the front evokes a turn to the rear. Greater precision is not required, nor is it observed. Indeed, there is considerable scatter around the turn angles that represent precise responses. This may actually be beneficial. In an escape system, precision may result in a predictable movement that could actually decrease the animal's chances for survival. Variability may result from the limits of the control system or some degree of randomness may be purposefully built into it. Alternatively, the apparent randomness may result from factoring many parameters, in addition to wind direction, into the orientation decision.

III. Leg Movements

Since the actual turning movements are completed in 20–30 ms, high-speed movie cameras or video systems are required to visualize the leg movements in any detail. With data from these systems, it is now clear that the escape turn is a unique behavior distinct from the conventional tripod gait associated with running (Camhi and Levy, 1988). Moreover, three types of turns are generated, depending on the front–back orientation of the wind source (Nye and Ritzmann, 1992).

The escape turn can be observed either in a free-ranging animal or in one that has been tethered to a rod and is walking on a slippery glass plate. Observations on free-ranging animals demonstrate how leg movements actually turn the animal away from a wind stimulus. However, a tethered preparation allows more precise quantitative analysis of each leg movement. A comparison of the data in these two conditions indicates that the active leg movements are essentially the same (Camhi and Levy, 1988).

The principal joints that are used in all turns are the coxa–femur (CF) joint and femur–tibia (FT) joint of each leg (Fig. 1A and B). In general, the CF joint develops forces in the anterior–posterior direction driving the animal forward or backward. The FT joints, especially in the prothoracic and mesothoracic legs, cause lateral movements that turn the animal's body to the right or to the left. In the simplest escape turns (type I), the prothoracic (T_1) and mesothoracic (T_2) legs extend posteriorly and laterally toward the wind source (Camhi and Levy, 1988; Nye and Ritzmann, 1992). The backward force of each leg occurs as the CF joint extends (Fig. $1C_1$). If the tarsi remain stationary, as is the case in freely moving animals, the CF movements push the animal's body forward (Fig. $1C_2$). In the T_2 legs (and probably also in the T_1 legs) the FT joint that is ipsilateral to the wind source invariably extends while the FT joint of the contralateral leg flexes. As a result, both legs push laterally toward the wind source. The resulting forces pivot the animal away from the stimulus. At the same time, both of the T_3 legs push backward as a result of coordinated extensions of CF and FT joints. This movement simply provides forward thrust. The combined movement of all of the legs drives the animal forward and away from the wind source.

The anterior–posterior orientation of the wind source influences the direction of movement of some joints (Nye and Ritzmann, 1992). In the T_2 legs, the FT joint movements remain essentially the same. However, with more anterior positions, the CF joints of the contralateral T_2 and T_3 legs tend to flex rather than extend (Fig. $1D_1$). Nevertheless, the tarsi remain in contact with the substrate. As a result, the leg that is contralateral to the wind source pulls the animal backward, while the ipsilateral leg pushes that side forward. The effect is similar to that seen in a rowboat when a rower pulls on one oar while pushing on the other. In this way, a larger turn angle is accomplished than occurs with the type I turn (Fig. $1D_2$). This is referred to as a type II turn.

With the wind placed at even more anterior positions, yet another turning strategy is revealed, which is referred to as a type III turn (Fig. $1E_1$). In a type III turn, *both* T_3 legs move forward as a result of CF flexion. Under free-ranging conditions, this actually pulls the animal backward (Fig. $1E_2$). However, the FT movements continue to turn the cockroach away from the wind source. With the animal's body drawn back and over its legs, it executes a second movement similar to a type I turn. The effect resembles the movements made as one attempts to back up a car and turn it around in a driveway. In free-ranging animals, type III turns can rotate the animal in excess of 90°.

Although the incidence of type II and type III turns increases as the wind source is positioned more anteriorly around the animal, the turning behavior remains a probabilistic event. Even with the wind placed directly in front of the animal, type I turns are still seen (Nye and Ritzmann, 1992). As with the variable nature of the turning direction, this could represent inherent limitations of the control circuit or it could be due to additional factors besides the wind source. Moreover, although I have described the turns as falling into three different categories, it is possible that they actually represent a continuum of movements as the CF joints move less and less in one direction and eventually switch to the opposite direction.

IV. Basic Circuit for the Escape Turn

The basic neural circuit is responsible for sensing wind stimulation, conducting that information to the motor control centers in the thoracic ganglia, and ultimately generating the appropriate motor responses. The neural components of this circuit and their connections have been documented by extensive observations, primarily employing dual intracellular recording and dye filling techniques (Fig. 2). There are two main sites of processing in this circuit. One is in the terminal abdominal ganglion, where sensory information is integrated by a population of giant interneurons. The other is in the thoracic ganglia, where directional wind information and other sensory data are integrated by thoracic interneurons and then used to direct motor activity.

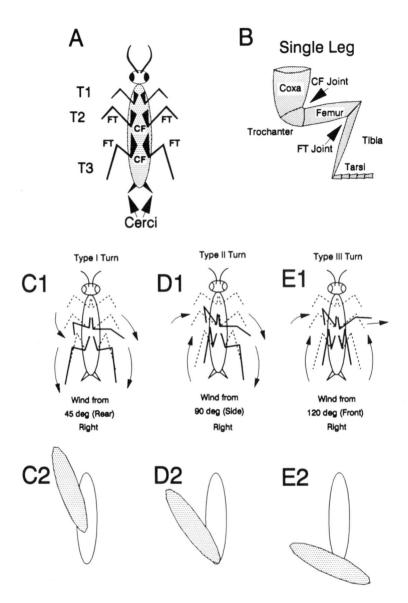

FIGURE 1. A description of the leg movements involved in escape turns. (A) The principal joints used in the escape turn are shown on a diagram of the cockroach. Only T_2 and T_1 legs have been analyzed to date. CF, coxa–femur joints; FT, femur–tibia joints. (B) A more detailed representation of the cockroach leg to show the location of CF and FT joints. C_1–E_1 summarize leg movements recorded in a tethered preparation where the legs slip on a lightly oiled glass surface. C_2–E_2 show the resulting body movements that would occur in a

Chapter VI Neural Organization of Cockroach Escape

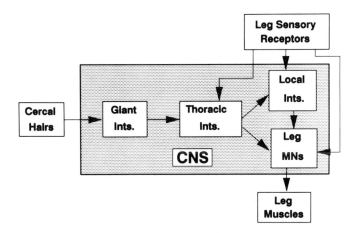

FIGURE 2. The basic escape circuit. This diagram describes the pathway between populations of neurons that make up the escape circuit. Cercal afferents project to giant interneurons within the central nervous system. The ventral giant interneurons project to type A thoracic interneurons. These produce motor effects either by direct connections with motor neurons or via local interneurons. Leg sensory receptors can influence the circuit via connections with type A thoracic interneurons, local interneurons, or in some cases direct connections to motor neurons.

A. Processing in the Terminal Ganglion

Mechanosensitive hairs located on the ventral surface of the cerci detect the wind puffs. The receptor hairs sit in asymmetrical sockets that confer a preferred plane of deflection upon each hair (Nicklaus, 1965). The sockets are arranged in columns such that every hair in a particular column has the same preferred direction of stimulation. Bipolar sensory neurons innervate each hair, and their axons project to the terminal ganglion through the cercal nerves. Intracellular recordings from the axons of these sensory neurons reveal receptive fields that are biased to one broadly tuned direction (Westin, 1979). Considering the entire population of sensory neurons in the cercal nerve, wind stimuli can be detected from any direction around the animal.

FIGURE 1. (cont.) freely moving animal. (C) With wind from the right rear, the cockroach usually responds with a type I turn. All CF joints extend, while the T_2 FT joints provide turning direction. As a result the animal moves forward and to the left. (D) With wind from the side, the cockroach makes type II turns. In this movement the contralateral CF joints flex, moving the leg forward. This will pull the animal forward on the contralateral side while pushing the animal back on the ipsilateral side. As a result, the cockroach pivots in place through a greater angle. (E) With wind from the front the animal flexes both CF joints, pulling the animal backward while it turns to the left. A subsequent type I turn completes a very large angle turn.

Within the terminal ganglion, cercal afferents make contact with the giant interneurons via cholinergic synapses (Sattelle, 1985). The pattern of synaptic connections with the ventral giant interneurons (vGIs) has been documented (Daley and Camhi, 1988). Again, this pattern reflects the columns of hairs on the cerci. Afferents within each column make essentially the same type of connection with each vGI. The strength of connections varies from column to column and the pattern of connectivity varies with each of the vGIs. Using the information on connectivity and strength of connection, it is possible to reconstruct the wind fields of each vGI (see Daley and Camhi, 1988; Beer and Chiel, this volume).

An individual GI can be identified either by its dendritic morphology within the terminal ganglion (Daley *et. al.*, 1981) or by the position of its axon relative to other GIs as they pass through the abdominal ganglia (Westin *et. al.*, 1977). The GIs can be divided into morphologically and physiologically meaningful subgroups. The axons of the four pairs of ventral giant interneurons (vGIs) reside in the ventral intermediate tract (VIT) of the abdominal nerve cord. Those of the three pairs of dorsal giant interneurons (dGIs) are found in the dorsal intermediate tract (DIT).

The number of action potentials recorded from a GI as wind is presented from different directions forms a reproducible wind field for each GI (Westin *et al.*, 1977). Much of the information on GI wind fields has been reviewed previously (Ritzmann, 1984). For our discussion, it is sufficient to point out the various reproducible wind biases of each of the vGIs (Fig. 3). GIs 2 and 4 are essentially omnidirectional. However, Camhi and Levy (1989) have presented evidence that, under certain stimulus conditions, a left–right bias can be detected in GI2 responses. GI1 wind fields are consistently biased toward winds originating from the side on which its axon is located. GI3 wind fields are biased toward wind from the front of the animal.

It is the activity within the vGIs that ultimately directs the escape turn. This can be demonstrated by lesioning individual vGIs and testing freely moving animals for turns that move the animal toward the wind source (Comer, 1985; see also Comer, this volume). Thus, in order to direct the turn, there must be at least some element within the circuit that integrates the information contained within the eight vGIs and uses this to direct activity in the appropriate leg motor neurons. The most logical site for such integration would be in the thoracic ganglia.

B. Processing Within the Thoracic Ganglia

The organization within each of the thoracic ganglia includes at least two levels of interneurons. Although stimulation of vGIs can evoke motor activity (Ritzmann and Camhi, 1978), this response is very weak, occurs at long latencies, and requires high-frequency trains of activity in the vGIs. Several laboratories have spent con-

Chapter VI Neural Organization of Cockroach Escape

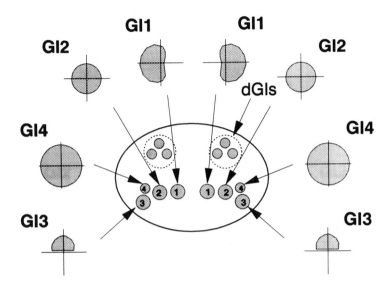

FIGURE 3. A summary of the wind fields of vGIs. A diagram of a cross section of an abdominal ganglion shows the typical location of vGI (1–4) and dGI axons. The wind fields for each vGI summarize data on number of action potentials evoked by winds from various different directions as plotted on polar coordinates (based on Westin et al., 1977). See text for details.

siderable time searching for direct connections between vGIs and leg motor neurons with no success. However, vGIs do make constant, short-latency connections with thoracic interneurons, which in turn activate leg motor neurons either directly or via additional thoracic interneurons (Ritzmann and Pollack, 1986).

1. Input to Type A Thoracic Interneurons. The thoracic interneurons that receive input from vGIs have been classified as type A thoracic interneurons (TI_As) (Westin et. al., 1988). So far, 13 TI_As have been identified. Since they have been found on either side of both the mesothoracic (T_2) and the metathoracic (T_3) ganglia, these individuals represent a population of at least 52 cells. Given that they probably exist in the prothoracic ganglion (T_1) as well, a conservative estimate would put the total population at between 76 and 100 interneurons. All but one of the interneuron types are intersegmental. The axon of each TI_A is located contralateral to its soma. The somata are located in one of two areas of the ganglion. In one subpopulation, which is referred to as the dorsal posterior group (DPG), the somata are found bilaterally in the dorsal, posterior region of the ganglion. In the other subpopulation the somata are located on the ventral surface on either side of the midline. These interneurons are referred to as ventral median cells (VMCs)

Perhaps the most interesting morphological feature of the TI_As is the presence of prominent ventral branches on one or both sides of the midline (Fig. 4A). We have referred to these as ventral median (VM) branches. Their location near the ventral intermediate tract, which contains the vGI axons, suggests a possible point of contact (Fig. 4B). This is supported by morphological studies that included double labeling of TI_As and vGI axons with differentially colored dyes (Casagrand and Ritzmann, 1991). These preparations revealed distinct areas of overlap between branches of vGI axons and VM branches (Fig. 4C). In addition, studies were performed in which one abdominal connective was lesioned in the posterior nerve cord of young nymphs. This procedure affects many axons, but, relative to the thoracic ganglia, the principal deficit is confined to the GI axons. Since the TI_As do not receive inputs from dGIs, the only cells of interest that were affected were the vGIs. After the animals grew to adulthood, individual TI_As were filled with dye and studied for changes in morphology. TI_As that normally have symmetrical VM branches were found to be decidedly asymmetrical (Fig. 5). The VM branch on the side that

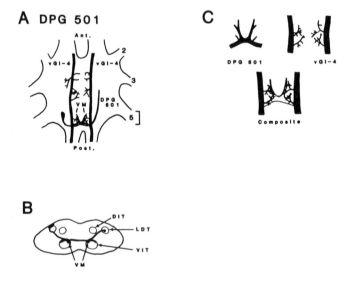

FIGURE 4. Cameral lucida drawings of TI_As filled with ethidium bromide and vGIs filled with Lucifer Yellow to show the overlap of vGI branches and VM branches of TI_As. (A) DPG 501 is shown with both vGI4s. VM branches are indicated. (B) A series of cross sections through the thoracic ganglion shows the dorsal–ventral path of the TI_A. This figure is a composite of several sections. Note the VM branches are located in the VIT, which contains axons of all vGIs. (C) Drawings focusing on the VM branch area of the TI_A, the vGI branches in this region, and a composite of the two as they appeared in the dual fill. Reprinted from "Localization of Ventral Giant Interneuron Connections to the Ventral Median Branch of Thoracic Interneurons in the Cockroach," by J. L. Casagrand and R. E. Ritzman, in the *J. Neurobiology.* © 1991 John Wiley & Sons, Inc.

Chapter VI Neural Organization of Cockroach Escape

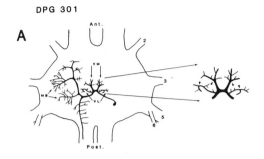

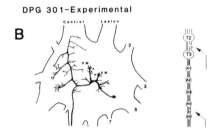

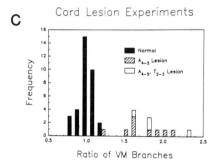

FIGURE 5. The effect of unilateral connective ablation on VM branches. (A) A camera lucida drawing of a typical bilateral TI_A, in this case DPG 301. The inset shows a higher-power picture of the region containing the VM branches. Note the symmetrical nature of the two VM branches. (B) An animal that had undergone lesion of connectives as a nymph has a distinctly asymmetrical VM branch architecture as an adult. Inset shows the operation that was performed. In this case the left side was lesioned both in A_4–A_5 and in the T_2–T_3 connectives. (C) Taken as a group, the ratio of left to right VM branches was around one (i.e., they were equal). In animals that experienced ablation of connectives the ratio of the control side to the ablated side was significantly greater than one, indicating that the VM branch on the operated side was significantly shorter than that on the control side. Reprinted from "Localization of Ventral Giant Interneuron Connections to the Ventral Median Branch of Thoracic Interneurons in the Cockroach," by J. L. Casagrand and R. E. Ritzman, in the *J. Neurobiology.* © 1991 John Wiley & Sons, Inc.

was lacking vGI axons was consistently shorter than the VM branch on the opposite side, further supporting the hypothesis that the vGI–TI$_A$ synapses occur preferen-

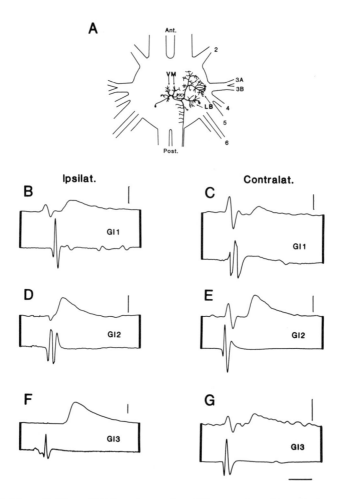

FIGURE 6. TI$_A$s with bilateral VM branches receive vGI inputs on both sides. All vGIs excite DPG 301 (formerly labeled Reverse J cell). (A) Whole mount of a DPG 301 cell showing bilateral VM branches. (B–G) Responses to intracellular stimulation of individual ipsilateral and contralateral vGI axons. In all records the top trace is an intracellular recording form a DPG 301. The bottom trace is an extracellular recording of the abdominal cord used to verify intracellular activation of the vGI axon. The GI that was stimulated is indicated above each abdominal cord record. The amplitude calibration at each record represents 4 mV for each trace except G, where it represents 2 mV. The time calibration represents 2 ms for all records. Reprinted from "Wind-Activated Thoracic Interneurons of the Cockroach: II. Patterns of Connection from Ventral Giant Interneurons," by R. E. Ritzman and A. J. Pollack, in the *J. Neurobiology.* © 1988 John Wiley & Sons, Inc.

Chapter VI Neural Organization of Cockroach Escape

tially on the VM branches (Casagrand and Ritzmann, 1991).

The VM branch morphology predicts the pattern of vGI connections to each TI_A. We have performed numerous experiments in which individual vGIs were stimulated intracellularly while recording from various identified TI_As. Out of these observations, we have arrived at a consistent pattern of vGI-to-TI_A connections (Ritzmann and Pollack, 1988). The pattern matches perfectly with VM branch morphology. TI_As with bilateral VM branches always receive inputs from

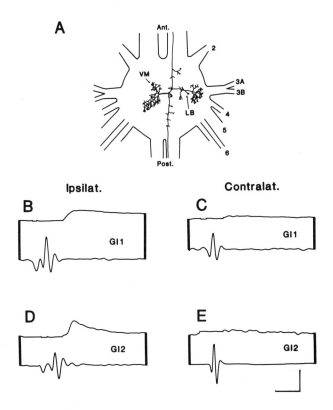

FIGURE 7. TI_As with unilateral VM branches only receive vGI inputs on the side of the VM branch. DPG 703 (formerly designated Cross cell) has a single VM branch on the side ipsilateral to its soma. It only receives vGI inputs on that side. (A) Morphology of DPG 703 in whole mount, showing location of VM branch and lateral motor branch (LB). (B–E) Intracellular records taken from DPGs 703 in response to intracellular stimulation of various vGI axons. Set up is as in previous figure. Note that only the GIs on the ipsilateral side (the same as the VM branch) evoked a response in this cell. Calibrations: 10 mV, 2 ms for all records. Reprinted from "Wind-Activated Thoracic Interneurons of the Cockroach: II. Patterns of Connection from Ventral Giant Interneurons," by R. E. Ritzman and A. J. Pollack, in the *J. Neurobiology.* © 1988 John Wiley & Sons, Inc.

vGIs on both sides of the ganglion (Fig. 6). TI_As with only one VM branch only receive vGI inputs on the side that contains the VM branch, whether that be ipsilateral to the soma or to the axon (Fig. 7). The wind response of each TI_A is also correlated to the VM branch morphology (Westin *et al.*, 1988). TI_As with bilateral VM branches have symmetrical responses to right and left wind. However, TI_As with only one VM branch have a stronger response to wind on the side that contains that branch. This is again regardless of whether the VM branch is on the soma side or on the axon side. Both of these observations follow from the assumption that the VM branches are the sites of synaptic contact with vGI axons.

It is important to note that all documented connections between vGIs and TI_As have been excitatory. Moreover, we have not detected any inhibitory connections between individual TI_As. We have found weak excitatory connections between left and right homologous pairs of bilateral TI_As. Since the wind fields of bilateral TI_As are not biased in a left–right orientation, weak excitatory interactions would serve only to potentiate or synchronize the signal and would not affect directionality. The total lack of mutual inhibition between left and right TI_As suggests that each TI_A acts as an independent entity, reaching a level of activity that is dependent on the pattern of vGI connections and the direction of wind activity. The output of the entire system then is a tally of activity in all of the TI_As. Thus, the TI_A network acts much like the congress of cells that has been proposed for primate motor control (Georgopoulos *et al.*, 1988).

2. Output From Type A Thoracic Interneurons. Are there morphological clues to the patterns of output from TI_As similar to the correlation between VM branch morphology and vGI input? The primary connections from the TI_As are made on the side opposite the soma (Ritzmann and Pollack, 1986). This is consistent with the assumption that the contralateral fiber, which extends to adjacent ganglia, is in fact the axon. All of the TI_As have one or more large branches that come off the primary axon and project to the dorsal lateral region of the ganglion, where most of the leg motor neurons are located. We have documented connections between TI_As and several motor neurons and local interneurons in this lateral region (Ritzmann and Pollack, 1990). In these data, we found yet another interesting correlation with morphology. In all of our records, TI_As in the DPG subgroup (dorsal somata) excite postsynaptic cells (Fig. 8), whereas TI_As in the VMC subgroup (ventral cell bodies) inhibit postsynaptic neurons (Fig. 9). There is precedence for this arrangement in insects. In the locust flight system, intersegmental interneurons with dorsal cell bodies are consistently excitatory, while those with ventral cell bodies located near the midline are inhibitory (Pearson and Robertson, 1987).

Chapter VI Neural Organization of Cockroach Escape

For both subgroups, the influence on motor activity follows several parallel pathways (Ritzmann and Pollack, 1990). Because most of the TI_As are intersegmental interneurons, they reach cells both in their ganglion of origin and in adjacent ganglia (Fig. 8). In addition, they evoke motor neuron activity either directly or via local interneurons. The local interneurons that are involved are morphologically similar to interneurons in locust that serve to group activity in several motor neurons into coor-

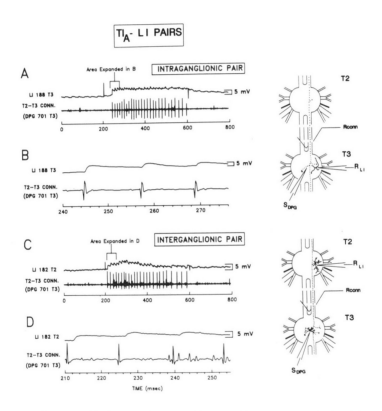

FIGURE 8 PSPs evoked in interganglionic and intraganglionic pairs are similar. (A and B) Recordings from local interneuron (LI) 188 (top traces) in T_3 in response to intracellular stimulation of DPG 701 in T_3 (see insets for recording setup). The area bracketed in A is expanded in B to show the consistent relationship between DPG 701 action potentials in the connective and PSPs in LI 188. (C and D) Results of a similar experiment in which an interganglionic pair was monitored. In this case DPG 701 in T_3 was stimulated in conjunction with a recording from LI 182 in T_2. Note that the responses in the two pairs are similar. The drawing of DPG 701 in the inset for the intraganglionic pair (A and B) shows only the primary branches to avoid obscuring the drawing of the motor neuron. Reprinted from "Parallel Motor Pathways from Thoracic Interneurons of the Ventral Giant Interneuron System of the Cockroach, *periplanta americana*," by R. E. Ritzman and A. J. Pollack, in the *J. Neurobiology.* © 1990 John Wiley & Sons, Inc.

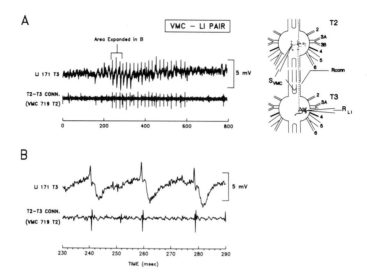

FIGURE 9. Action potentials of VMC interneurons evoke inhibitory PSPs (IPSPs). This experiment is similar to those in Fig. 8. However, here a member of the VMC subpopulation was stimulated. In this case, a VMC 719 interneuron in T_2 was stimulated and a local interneuron, LI 171, was recorded. Action potentials in VMC, monitored in the extracellular record from the T_2–T_3 connective (bottom trace), are associated with IPSPs in LI 171 (top trace). The segment indicated in A is expanded in B to show the consistent relationship. The short positive wave preceding each IPSP is probably an artifact resulting from capacitative crosstalk from the presynaptic record. Reprinted from "Parallel Motor Pathways from Thoracic Interneurons of the Ventral Giant Interneuron System of the Cockroach, *periplanta americana*," by R. E. Ritzman and A. J. Pollack, in the *J. Neurobiology*. © 1990 John Wiley & Sons, Inc.

dinated movements (Burrows, 1980; Siegler and Burrows, 1979). The strength of connections for each of these parallel pathways in the cockroach is similar. That is, intersegmental and intrasegmental pathways appear to be of equal strength, as are direct motor connections and polysynaptic pathways including local interneurons.

3. **Hypotheses on Orientation in the TI_A Circuit.** Given the patterns of connectivity that have been elucidated for both input and output pathways of the TI_As, we can begin to develop hypotheses on how the TI_As process wind information and generate the appropriate turning movements. The organization of motor connections could fall into one of two different patterns. The most straightforward pattern is a dedicated system. Each TI_A could be dedicated to one or more roles in generating the turn. This could be as major as commanding a complete turn to the right, or it could be a small subset of turning movements such as extending the femur–tibia joints. In the opposite extreme the pattern of output could be distrib-

uted among the entire population of TI_As. Under this form of organization, several TI_As might make motor connections that are not totally consistent with the turning movement. However, taken as a total population, the responses of all TI_As would average out to evoke an appropriate turn.

Given the pattern of connections from vGIs to TI_As and the information we have recently acquired on joint angle changes associated with the turning movements, we can predict the connections that would be made from various TI_As in a dedicated system. The TI_As can be segregated into smaller groups based on their pattern of vGI inputs and the side of the animal on which their motor effects occur. For example, if we consider a response to wind from the right rear, we will have to account for excitatory connections from four subgroups of TI_As (Fig. 10). Two of the subgroups include TI_As with VM branches on both sides of the midline. These include interneurons with axons on the right side and those with axons on the left side. Because cells with bilateral VM branches are excited equally well by wind from the right and wind from the left, we would expect excitation of these cells to cause joint movements that occur regardless of left or right wind direction. Such movements include extension of coxa–femur (CF) joints in both T_2 and T_3. These movements provide power to the turn but do not enter into directionality unless the wind comes from the front of the animal.

Left–Right directionality would arise from TI_As with unilateral VM branches. These interneurons receive inputs only from vGIs on the side on which the VM branch is located. The resulting directional wind bias suggests that these cells would be associated with movements that influence the direction of turning, *e.g.*, the femur

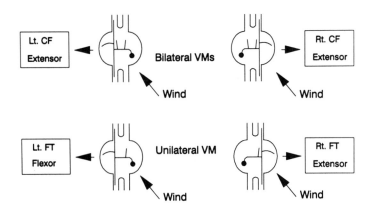

FIGURE 10. In a dedicated system, these excitatory cells must be accounted for: Cells with bilateral VM branches respond to wind from either direction and excite motor neurons that are active regardless of turn direction. Cells with unilateral VM branches act on motor neurons that provide directional movements.

–tibia (FT) joints of the legs in T_2. Two subgroups of unilateral TI_As must be considered. These are TI_As that have a VM branch ipsilateral to their axon and those that have a VM branch contralateral to their axon. Since direct motor outputs are restricted to the side on which the axon is located, we can restrict our discussion of motor output to that side. Given the example with which we are working (wind from the right rear), we need only consider TI_As with VM branches on the right side. The two relevant subgroups are those that respond to wind from the right and evoke motor output on the right and those that respond to wind from the right and evoke motor output on the left (Fig. 10). We know from our video analysis that wind from the right causes the FT joint of the right T_2 leg to extend. If the system has dedicated connections, we would, therefore, predict that TI_As with VM branches on the right side and ipsilateral to their axon would excite extensor motor neurons of the FT joint. Our video analysis would also predict that TI_As with VM branches on the right side but contralateral to the axon would excite flexor motor neurons of the left FT joints.

In addition to excitatory connections, we must account for inhibitory connections from TI_As. This is facilitated by the correlation between soma location and sign of connection. As a result, we would predict that TI_As with ventral somata would inhibit motor neurons that are antagonistic to those excited by similar TI_As with dorsal somata.

So far we have not detected any connections that would argue against this type of system. However, we have only begun to map the details of motor connections. What would we conclude if several TI_A-to-motor connections do not fit with known behavioral movements? Just such a situation has been documented in the leech bending reflex (Lockery and Kristan, 1990; see also Lockery and Sejnowski, Chapter XI, this volume). Lockery concluded that the control circuitry for leech bending is distributed rather than dedicated. Some of the connections between interneurons and motor neurons were not expected given the direction of movements that have been documented in the reflex. Nevertheless, using computer modeling techniques, Lockery demonstrated that, with proper connection weights, the output of the distributed network can account for the reflex activity.

The population of thoracic neurons that integrates vGI information is quite large. Therefore, a complete understanding of the nature of its control function will be greatly facilitated by the use of computer modeling techniques. With this in mind, we have entered into a collaborative effort to incorporate our circuitry data into a physiologically accurate computer model of the escape system (see Beer and Chiel, Chapter XII, this volume). The results of this exercise indicate that a control system based on the connections that have been documented by intracellular circuit analysis is capable of generating the leg movements that are observed in the escape turn. As more neurobiological data are collected and incorporated into the model, we plan to use it to formulate and test further hypotheses regarding the control circuitry.

V. Context-Dependent Behavior

In considering the complexity of the TI_A control circuit, one is moved to ask why the cockroach needs 100 interneurons to control the relatively simple behavior of turning into one of four different directions. Perhaps the answer is that the problem is actually much more demanding than that. Does the cockroach always turn away from the wind regardless of the situation? Or does it sculpt this basic escape turn to match a variety of environmental conditions? Our behavioral observations indicate that the animal factors a wealth of information into the orientation decision that underlies the escape turn.

A. *Additional Sensory Information*

Perhaps the most obvious cues that are added to the escape equation are additional sensory modalities that indicate environmental conditions. For example, if the cockroach is standing near a wall and receives a wind puff from the opposite side, a typical escape turn would cause it to run into the wall. If the animal is making antennal contact with the wall at the time of stimulation, this does not happen (Ritzmann *et. al.,* 1991). Rather it simply runs forward. A similar alteration has been observed in teleost fish (Eaton and Emberley, 1991). When the fish is near a wall, it alters the direction of its escape turn. If the cockroach is not making antennal contact at the time of stimulation, it will turn toward the wall. However, upon striking the wall with one antenna, the cockroach will, within 5 ms, reverse the direction of its turn. In either case, information from the antennae modifies the leg movements that generate the initial escape activity.

Recent work from Comer's laboratory provides a clue as to how this is possible (see Comer and Dowd, Chapter V, this volume). Tactile stimulation is capable of evoking escape turns similar to those that result from wind directed at the cerci. If the pathway that generates these turns includes the thoracic interneurons that integrate vGI activity, it could also be utilized to modify wind-evoked escape movements.

The TI_As are indeed excited by antennal stimulation (Fig. 11). This activity follows tactile stimuli at relatively high frequencies, indicating that the connection is fairly direct. Following this pathway, directional cues from the antennae could be incorporated into the TI_A population. As a result, the control circuit would already be biased to generate a turn away from the wall. A wind signal from the left carried by the vGIs would then be interpreted by the TI_As *in the context* of a predisposition to turn toward the left. The summed effects of these two cues would generate an escape movement in which the animal ran straight ahead.

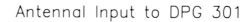
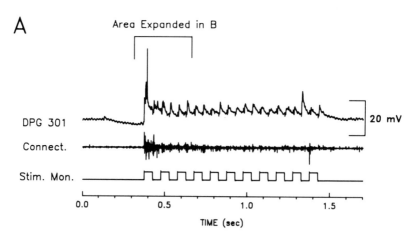
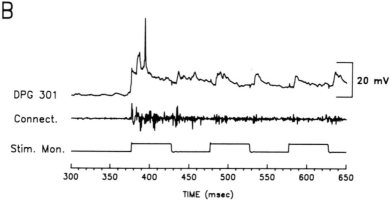

FIGURE 11. TI$_A$s respond to rapid movements of antennae. A pin attached to a piezoelectric crystal was used to deflect one of the antennae at a frequency of 10 Hz while recording from DPG 301. The top trace is the intracellular record from DPG 301. The middle trace is an extracellular record from the T$_2$–T$_3$ connective. The bottom trace is a monitor of the stimulus pulse that was used to deflect the piezoelectric crystal. The area indicated in A is expanded in B. Note that PSPs occur faithfully at both the beginning and end of each stimulus. The initial response is much stronger than the rest. It leads to an action potential in DPG 301. Action potentials in the thoracic connective that occur in synchrony with the stimulus represent potential presynaptic cells.

In the absence of antennal contact, the cockroach begins to make a turn to the right. Upon making contact with the wall, the antenna would then generate a tactile-mediated turn to the left, effectively reversing the turn. This is simply a sequence of wind-evoked and tactile-evoked responses. Since the TI_As are already driven to threshold by the wind stimulus, the tactile-evoked reversal would have an extremely short latency.

Additional sensory cues are not limited to antennal input (Ritzmann *et al.*, 1991). We have demonstrated that TI_As are sensitive to tactile activity over the entire cuticle, as well as auditory inputs and changes in ambient light. Thus, in addition to responding to wind from an approaching predator, the cockroach can factor in auditory cues such as those that might arise from a predator breaking branches. Anyone who has had the misfortune of inhabiting a cockroach-infested dwelling is aware that they have a greater tendency to escape when in the light than in the dark. Turning on a room light is generally met with the scurrying of roaches in every direction. Our electrophysiological data suggest that increased light evokes an excitatory response in at least some TI_As that could either generate a full-blown escape response or at least lower the threshold to wind-evoked escape. We are confident that further observations will uncover even more sensory modalities that may be factored into this decision.

B. Proprioceptive Activity

Proprioceptive activity represents a particularly important form of sensory information (see Zill, Chapter III, this volume). The leg movements of the initial turn occur so rapidly as to make guidance by negative feedback unlikely. Nevertheless, leg position at the time of stimulation can be extremely important. Camhi and Levy (1988) noted that the initial anterior–posterior position of the mesothoracic legs is correlated with the extent of lateral movement that it executes during the turn. They suggested that this may occur as a result of movement at the FT joint. Movement at the T_2 FT joint on the side ipsilateral to the wind source is indeed correlated with initial joint angle (Nye and Ritzmann, 1992). If the ipsilateral FT joint is flexed at the time of wind stimulation, it will extend farther than if it was already partially extended at the time of stimulation. The degree of initial flexion is directly correlated with the amount of extension that is generated during the escape response. Since the full movement occurs over the same duration of time regardless of initial joint position, a larger excursion represents an alteration in the rate of movement. This could be accounted for by tonic increases in motor activity or gain changes in local interneurons prior to stimulation.

At least two points in the basic escape circuit are amenable to proprioceptive input. The TI_As themselves receive what appear to be monosynaptic connections from various proprioceptive structures, including the femoral chordotonal organ (Murrain and Ritzmann, 1988). Many of the connections vary from TI_A to TI_A.

Such heterogeneous connections could alter the TI_A-to-motor pathways, thereby altering the drive to motor neurons.

In addition to these elements, the local interneuron circuitry that is interposed between the TI_As and motor neurons provides an excellent point for integrating proprioceptive information. Indeed, we found that mesothoracic FT joints are much more sensitive to proprioceptive input than are metathoracic FT joints. This suggests that the local interneuron circuitry that is dedicated to legs of a single segment may be more critical to the final turning movements than proprioceptive connections to the intersegmental TI_As.

We have not studied the proprioceptive connections to local interneurons in any detail. However, the morphological similarity between nonspiking local interneurons in the cockroach escape circuit and those that control leg movements in locust is striking (Siegler and Burrows, 1979). Given the wealth of information on processing mechanoreceptive information in locust local interneurons (Burrows, 1989; and Laurent, Chapter IV, this volume), it is reasonable to expect that similar information is also integrated in the local circuits of cockroach thoracic ganglia.

C. Physiological State

In addition to sensory cues, the escape system is susceptible to alterations in the animal's own physiological state. For example, as part of the cockroach's mating ritual, the male taps the female on the female's pronotum. From another source, such activity would normally evoke an escape response. However, in receptive animals, no escape occurs. Indeed, a cockroach rarely escapes when another cockroach runs into it, even if the first cockroach is moving at a very high speed. How does the cockroach distinguish between small predators such as spiders and other cockroaches? One possibility is that olfactory cues, such as pheromones, alter the physiological state of the system. Such information could increase or decrease the threshold for the escape response. To test this, we either removed an animal's antennae or rendered the antennae inoperable by coating them with fingernail polish (Watson and Ritzmann, 1992). As a result, the incidence of cockroach-induced escapes increased dramatically from 36% (intact animals) to 85% for either antenectomized or antenna-painted animals. This supports the notion that chemical cues detected by the antennae can alter the escape threshold.

It is also possible to cause a cockroach to enter a state of complete quiescence by providing a continuous tactile sensation to both antennae. If a plastic shield is placed in front of the cockroach so that it touches both antennae, the animal will enter a state in which no movements are discernible. All motion in legs, antennae, and mouthparts is eliminated. In this quiescent state, wind puffs are ineffective in evoking escape movements. Recordings made from the abdominal nerve cord under these conditions clearly indicate that the vGIs are activated in a normal way. Thus, the state must be induced either in the thoracic ganglia or in more anterior structures.

What might cause these alterations in the escape system? In crayfish, the threshold for the lateral giant–mediated escape response can be dramatically increased by pressure on the thoracic cuticle (Krasne and Wine, 1975). This may be caused by release of neuromodulators near the connections between sensory interneurons and the lateral giants. Bath application of serotonin can decrease synaptic responses in the lateral giants, whereas octopamine enhances them (Glanzman and Krasne, 1983).

A similar observation has been made in the cockroach. Bath application of either dopamine or serotonin can alter motor output. Motor activity, generated by wind puffs or by stimulation of the abdominal nerve cord, is augmented in the presence of dopamine and reduced in the presence of serotonin (Goldstein and Camhi, 1991). At least one site of action for neuromodulatory substances has been documented in the escape system. The strength of vGI-to-TI_A synapses is altered in the presence of neuromodulators (Casagrand and Ritzman, 1992). Bath application of octopamine or dopamine increases the amplitude of vGI-evoked excitatory postsynaptic potentials (EPSPs) in TI_As. Application of serotonin decreases the amplitude of the same EPSPs.

If neuromodulatory effects can be linked to physiological state, they could provide a mechanism for altering the escape system. A wealth of information is available on substances that are released by cockroaches during altered states (Bell, 1982). Pheromones responsible for courtship have been isolated. Indeed, the active element of sex pheromone, periplanone B, has been synthesized. Internally, stress-induced factors have been isolated from the hemolymph and from nervous tissue (Beament, 1958). These substances are released in large quantities when the animal is placed in extremely stressful situations, such as forced walking for long periods of time. Injection of one such substance, factor S, results in paralysis of the recipient insect. It is possible that the quiescent state induced by antennal contact may be a more gentle form of stress. In a situation where the cockroach perceives itself cornered and with little hope of escape, it may be more beneficial to simply shut down active movements for a period of time rather than draw attention to itself with an ineffective escape response.

VI. Conclusion

The escape system of the cockroach was once thought to be a simple reflex. The data that have now been accumulated by several laboratories paint a much more complex picture. The cockroach does not simply jump and run in response to a threatening wind puff. Rather it senses the direction from which the threat came and then orients its escape path away from that source. The principal cues for this decision are the directional properties of the wind puff. However, the cockroach interprets the wind information provided by the vGIs in the context of a wealth of additional information, including sensory cues monitoring its immediate surroundings, the position of its limbs, and its present physiological state. Thus the cockroach possesses all of the requirements for a successful crash avoidance device that were listed in the introduction.

An understanding of how the cockroach performs these calculations and executes a successful escape behavior in a remarkably short time may shed light on how similar control circuits could be constructed and implemented in a vehicle or in a robot. Nobody would recommend construction of a car that moves like a cockroach. However, by studying a variety of invertebrate systems and their neural control circuits, one can identify solutions that might inspire new engineering principles. For example, rate-dependent sensors could detect a dangerous situation in the same way as acceleration-sensitive hairs on the cerci differentiate between wind puffs originating from lunging predators and those coming from nonthreatening sources. Directional properties of the TI_A circuit could inspire systems that would detect the location of a source of danger or identify targets to be pursued. Context-dependent biasing within such a control circuit could factor in properties such as barriers and other surrounding vehicles as well as internal properties such as speed and the state of various engine properties. This, then, is an example of how information gained from neurobiological experiments could be utilized in the world of technology.

Acknowledgments

The author would like to thank all of the students and colleagues who have contributed to the work done in his laboratory. Of particular note is Alan J. Pollack, who has been a friend and collaborator for many years. Most if not all of the work done in the author's lab is either directly or indirectly linked to the efforts of Mr. Pollack. This manuscript was reviewed by Drs. Randall Beer, James T. Watson, and Joanne Westin in addition to Alan Pollack. Much of the work discussed was supported by NIH grant NS-17411 to R.E.R. and ONR grant N00014-90-J-1545 to Randall Beer.

References

BEAMENT, J. W. L. (1958). *J. Insect. Physiol.* **2,** 199–214.
BELL, W. J. (1982). In Bell, W. J., and Adiyodi, K. D. (eds.), *The American Cockroach* (p. 371). Chapman & Hall, New York.

Chapter VI Neural Organization of Cockroach Escape

BURROWS, M. (1980). *J. Physiol.* **298**, 213–233.
BURROWS, M. (1989). *J. Exp. Biol.* **146**, 209–227.
CAMHI, J. M. (1980). *Sci. Am.* **243**(6), 158–172.
CAMHI, J. M., and LEVY, A. (1988). *J. Comp. Physiol.* A**163**, 317–328.
CAMHI, J. M., and LEVY, A.(1989). *J. Comp. Physiol.* A **165**, 83–97.
CAMHI, J. M., and NOLEN, T. G. (1981). *J. Comp. Physiol.* **142**, 339–346.
CAMHI, J. M., and TOM, W. (1978). *J. Comp. Physiol.* **128**, 193–201.
CAMHI, J. M., TOM, W., and VOLMAN, S. (1978). *J. Comp. Physiol.* **128**, 203–212.
CASAGRAND, J. L. (1992). *J. Neurobiol.* (in press).
CASAGRAND, J. L., and RITZMANN, R. E. (1991). *J. Neurobiol.* **22**, 643–658.
COMER, C. M., (1985). *Brain Res.* **335**, 342–346.
DALEY, D. L., and CAMHI, J. M. (1988). *J. Neurophysiol.* **60**, 1350–1368.
DALEY, D. L., VARDI, N., APPIGNANI, B., and CAMHI, J. M.(1981). *J. Comp. Neurol.* **196**, 41–52.
EATON, R.C., and EMBERLEY, D. S. (1991). *J. exp. Biol.* **161**, 469–487.
GEORGOPOULOS, A. P., KETTNER, R. E., and SCHWARTZ, A. B. (1988). *J. Neurosci.* **8**, 2928–2937.
GLANZMAN, D. L., and KRASNE, F. B. (1983). *J. Neurosci.* **3**, 2263–2269.
GOLDSTEIN, R. S., and CAMHI, J. M. (1991). *J. Comp. Phsyiol.* A **168**, 103–112.
KRASNE, F. B., and WINE, J. J. (1975). *J. Exp. Biol.* **63**, 433–450.
LOCKERY, S. R., and KRISTAN, W. B. (1990). *J. Neurosci.* **10**, 1816–1829.
MURRAIN, M. P., and RITZMANN, R. E. (1988). *J. Neurobiol.* **19**, 552–570.
NICKLAUS, R., (1965). *Z. Vergl. Physiol.* **50**, 331–362.
NYE, S. W., and RITZMANN, R. E. (1992).*J. Comp. Physiol.* A (in press).
PEARSON, K. G., and ROBERTSON, R. M. (1987). *Cell Tissue Res.* **250**, 105–114.
PLUMMER, M. R., and CAMHI, J. M. (1981). *J. Comp. Physiol.* **142**, 347–357.
RITZMANN, R. E. (1984). In Eaton, R. C. (ed.), *Neural Mechanisms of Startle Behavior* (pp. 93–31). Plenum, New York.
RITZMANN, R. E., and CAMHI, J. M. (1978). J. Comp. Physiol. **125**, 305–316.
RITZMANN, R. E., and POLLACK, A. J. (1986). *J. Comp. Physiol.* A **159**, 639–654.
RITZMANN, R. E., and POLLACK, A. J. (1988). *J. Neurobiol.* **19**, 589–611.
RITZMANN, R. E., and POLLACK, A. J. (1990). *J. Neurobiol.* **21**, 1219–1235.
RITZMANN, R. E., POLLACK, A. J., HUDSON, S. E. and HYVONEN, A. (1991). *Brain Res.* **563**, 175–183.
ROEDER, K. D. (1967). *Nerve Cells and Insect Berhavior.* Harvard University Press, Cambridge, Ma.
SATTELLE, D. B. (1985). In *Comprehensive Insect Physiology, Biochemistry and Pharmacology* (vol. 11, pp. 395–434). Pergamon, New York.
SIEGLER, M. V. S., and Burrows, M. (1979). *J. Comp. Neurol.* **183**, 121–148.
WATSON, J. T., RITZMANN, R. E. (1992). *International Cong. Neuroetho. Abst.*
WESTIN, J., (1979). *J. Comp. Physiol.* **133**, 97–102.
WESTIN, J., LANGBERG, J. J., and CAMHI, J. M. (1977). *J. Comp. Physiol.* **121**, 307–324.
WESTIN, J., RITZMANN, R. E., and GODDARD, D. J. (1988). *J. Neurobiol.* **19**, 573–588.

Chapter VII

Acoustic Startle: An Adaptive Behavioral Act in Flying Insects

RONALD R. HOY
Section of Neurobiology and Behavior
Cornell University
Ithaca, New York

I. Introduction

A. *Startle as an Adaptive Behavioral Act*

Most animals have evolved some sort of startle behavior. Such behaviors evolved in the context of the never-ending contest between predator and prey. Given the widespread occurrence of startle or escape behavior throughout the animal kingdom, there can be little doubt about its adaptive or survival value. In the context of this book, I am not aware that designers of locomotory robots include startle responses in their designs. I would argue that in a world where Murphy's law has virtually Newtonian universality, it might not be a bad idea for roboticists to take a close look at how animals cope with threats to their well-being, by sensing "warning signals." Our understanding of startle behavior comes primarily from mammalian studies. However, I want to show for the analysis of startle, as for other behavioral acts, that the insect has appealing features as a model system.

This volume is mainly on terrestrial locomotion in insects. Although we may rightly admire the mechanistic bases of walking, it must be borne in mind that the singular adaptation that really permitted the radiation of insects into supremacy in species diversity and number is their ability to fly. Accordingly, I will discuss startle behavior in flying insects. Escape from predators is as much a problem for flying insects as for terrestrial, walking insects. By day, insects on the wing are preyed upon by birds; indeed, some species of birds (e.g., flycatchers) make their living as insectivores. Taking to the wing at night reduces an insect's risk from predation by birds, but nocturnally flying insects are confronted by another highly efficient predator—insectivorous bats, which detect and locate their prey by echolocation.

Bats are among the most diverse of mammals in terms of numbers of species. They prey on a wide variety of flying insects, ranging in size from small gnats and mosquitoes to large moths and even praying mantises. Their ultrasonic biosonar signals, which commonly consist of a frequency sweep from high to low frequencies, lie far above the range of human hearing and in some species exceed 200 kHz, but more commonly encompass the range from 20 to 100 kHz (Fenton and Fullard, 1981; Griffin, 1958). Such high frequencies permit the detection of even small insects.

As might be expected, predation pressure from bats, which are found in the fossil record back as far as 50 million years (Novacek, 1985), has resulted in the evolution of countermeasures in the insects. Many insects have evolved auditory organs that are sensitive to ultrasound as well as flight steering responses to ultrasound that underlie startle or evasive maneuvers that permit them to escape from approaching bats.

In this review, I will describe the acoustic startle response (ASR) or escape responses that have evolved in several orders of insects. I will discuss the neural bases of these ASRs, where such information is available. I also refer the interested reader to other reviews (Hoy *et al.*, 1989; May, 1991).

II. The Acoustic Startle Response

A. *Definition*

The ASR is a phylogenetically primitive behavioral act commonly found in most animals that have a sense of hearing. It undoubtedly evolved as an antipredator behavior. The acoustic stimulus that elicits an ASR is generally intense (loud) and unexpected. It must also be abrupt (have a rapid onset to peak intensity), and consequently its power spectrum will have peaks at high frequencies. In terrestrial animals, the startle response itself is highly stereotyped and predictable for any given species and can range from subtle contractions of facial muscles, for example, to obvious postural reactions, such as "freezing," to jumping off the ground into the air. Flying animals, when startled, may abruptly alter their flight course and steer away from the sound, or cease flying altogether and fall or dive toward the ground. In all cases, the essense of the acoustic startle reaction is its very short latency, which varies depending on species, from under 10 ms to a few hundred milliseconds. The ASR may be elicited by merely a single, suprathreshold stimulus presentation.

B. Phylogenetic Occurrence

Although it is likely that any animal that can hear has an ASR, the response has been studied primarily in mammals (Landis and Hunt, 1939; Hoffman and Ison, 1980; Davis and File, 1984). Because it is a highly stereotyped behavior and can be elicited in the controlled environment of a laboratory, the ASR has long been a phenomenon of interest to physiological psychologists. Their favorite experimental subjects, human beings and rats, provide most of the data on the ASR. Among terrestrial animals, hearing is a prominent characteristic only among vertebrates and insects. The ASR has been widely studied in mammals but not much in other vertebrates. Among invertebrates the study of the ASR has been investigated primarily in insects, especially in the past decade (Nolen and Hoy, 1986a,b; May *et al.*, 1988; Yager *et al.*, 1990). It could be argued that that classic behavior of neuroethology, the escape response of the cockroach, is also an ASR. However, the adequate stimulus for cockroach startle is actually a wind puff delivered to the cerci, not sound. Nonetheless, as a *startle response,* the cockroach escape behavior is one of the best analyzed of any, which is clearly demonstrated by the work of Ritzmann and his colleagues and reviewed in this book.

C. The Mammalian ASR

Studies of the mammalian ASR have set the standards of research in this field, and I will list a few results to provide a basis for comparison with the insect studies. The interested reader should consult the reviews, cited above, of Davis, Hoffman, and Ison, from whom I draw the following observations. (1) The amplitude of the ASR is related to the intensity of the stimulus: the response is graded and not all-or-none. The graded nature of the response permits measurement of startle plasticity. (2) The stimulus must be abrupt: in rats, the stimulus must reach a peak of 90 dB (SPL) within 12 ms of onset to elicit startle (Fleshler, 1965). (3) A startling stimulus must be brief: stimulus durations longer than 8 ms were no more effective in eliciting startle than briefer ones. (4) Rats are startled more by high-frequency sounds than by low-frequency ones. (5) Startle may be modality specific. In rats, auditory but not visual stimuli (light flashes) are effective in eliciting startle, and the same is apparently true for humans (Meier-Ewert *et al.*, 1974); but in pigeons, visual stimuli but not auditory stimuli elicit startle. (6) The latency of the ASR in mammals is very short. In rats, using electromyography to record the earliest muscle reactions of an ASR, latencies of 5–7 ms have been recorded (Ison *et al.,* 1973). In humans, latencies on the order of 15 ms can be recorded in the jaw muscles of a startled subject (Meier-Ewert *et al.,* 1974). (7) The ASR of mammals shows several kinds of plasticity, including habituation, sensitization, and prepulse inhibition (Davis and File, 1984).

Finally, the full-blown ASR in mammals is no simple localized reflex reaction but is a global one involving nearly every muscle in the body (Davis and File, 1984). Startle in humans can be described as a stylized "hunching over" reaction and in rats as an exaggerated "crouch." In either case, the startle posture has been characterized as flexor dominated. However, Davis points out that analysis of myographic recordings reveals considerable extensor activity as well. In any event, it is important to keep in mind that the ASR is a *global* response and that references to the acoustic startle "reflex" imply an oversimplification of the motor act, which is both highly coordinated and yet very stereotyped within a given species. This predictability and stereotypy may give the impression of a simple motor act. The ASR is not. The only thing simple about the ASR may be the eliciting stimulus.

Davis and his colleagues have also made great strides in defining the major acousticomotor neural systems that underlie the mammalian ASR (Davis and File, 1984). They have identified five key stages of neural processing: the posteroventral cochlear nucleus, the dorsal and ventral nuclei of the lateral lemniscus, the nucleus reticularis pontis caudalis, the reticulospinal tract, and lower motor neurons of the spinal cord. Moreover, they have implicated a number of different neurotransmitters that appear to modulate acoustic startle, and these operate at various of the key central nuclei. This is a remarkable achievement and one that provides us with a working neural circuit for mammalian startle. With this material as background, I will now turn to the ASR in insects.

III. Acoustic Startle in Insects

A. *General Features*

By comparison with the mammalian ASR, the insect ASR has not been as systematically investigated. The emphasis in most insect studies has been more teleological because the ultrasonic ASR of flying insects is presumed to be an adaptation to a specific predator: the echolocating bat. However, where direct comparisons can be made, the ASRs of mammals and insects exhibit many features in common. In insects, the ASR amplitude is graded and related to stimulus intensity (May *et al.*, 1988). Startling stimuli can be simple but must be abrupt, brief, and of high frequency (Nolen and Hoy, 1986a,b). The ASR is of short latency and is global in that it involves virtually every appendage in the insect's body (Moiseff *et al.*, 1978). And, as in mammals, the ASR of insects shows at least some degree of plasticity—it habituates (May and Hoy, 1991); other kinds of behavioral plasticity have not yet been tested. It will be clear that ultrasound elicits an ASR from a wide variety of nocturnally active, flying insects, covering six different insectan orders to date. The motor expression of the ASR varies from group to group. Variability on the sensory side is also common. Within some orders (Lepidoptera and Orthoptera), ultrasound-sensitive hearing organs have evolved independently several times. Nonetheless, in all these insects, the startle response to bat-like ultrasound signals is taxon specific, stereotyped, of short latency and would be adequate to serve as an evasive maneuver.

The neural analysis of the ASR in insects has just begun and is described below. It is in the neural analysis of the ASR that the investigation of the auditory insects is appealing. The ASR of insects is really no more simple than it is in mammals, and both ASRs have more features than differences in common. So why study the ASR in insects? There is the presumption that the neural circuit underlying the insectan ASR can be defined in terms of identified neurons, including auditory receptors, interneurons (including modulatory interneurons), and motor neurons. This knowledge at the cellular level would surely yield insight into the design and modification of neural circuits for startle behavior, whether they reside in mammals or insects. However, an equally valid reason for investigating the ASR of insects is that it has much to tell us about the evolution of behavior, in this case a coevolutionary predator–prey story: the tens of millions of years old interaction between bats and insects.

B. *The Many Forms of the Ultrasound ASR in Insects*

It seems likely that the ASR is an evasive response by nocturnal, flying insects to predatory bats that use ultrasonic biosonar to echolocate and detect prey. Ideally, one would want to observe bat–insect interactions directly under natural conditions, in the field, in every species of insect shown to have an ASR in the laboratory. In practice, this is difficult because natural encounters occur in the dark and high in the air. Nonetheless, beginning with the classic work of Roeder and his associates with moths (Roeder, 1967), bat–insect encounters have been documented in which a flying insect exhibited evasive maneuvers in response to an approaching bat emitting biosonar signals. Besides moths, green lacewings

(Miller, 1975, 1983) and praying mantises (Yager *et al.,* 1990) have been observed in the field to evade bats. Other insects, including those just mentioned, have been studied in the laboratory and found to exhibit startle or evasive steering in response to ultrasound. In the laboratory, the insect is tethered and placed in front of the moving air stream of a wind tunnel, which elicits sustained flapping flight. During tethered flight, insects respond to sound stimulation with a variety of responses, including startle (Popov et al., 1975; Moiseff *et al.,* 1978; Nolen and Hoy, 1986a,b).

Acoustic stimulation with ultrasound in the range of 20–50 kHz and above threshold levels (30–60 dB SPL, depending on species) elicits two categories of responses: steered and unsteered. Steered responses are typically oriented away from the source of ultrasound; these are directional. Unsteered responses are not directional and can be as simple as a halt in wing beating and dropping or the performance of an unpredictable maneuver (with respect to the direction of the stimulus) such as diving or looping under full power of flight. Some insects are capable of making both kinds of maneuvers, depending on the intensity of ultrasound stimulation. Thus, moths (Roeder, 1967), crickets (Nolen and Hoy, 1986a,b), and locusts (Robert, 1989) steer directionally away from ultrasound unless the levels of stimulation are very intense (typically over 90 dB SPL), which may elicit unpredictable evasive maneuvers. Other insects, such as the green lacewing (Miller, 1975, 1983) and the katydid, *Neoconocephalus ensiger* (Libersat and Hoy, 1991), respond to ultrasound not by steering but by ceasing to flap their wings and simply dropping. The praying mantis, which does not have a directional sense of hearing (Yager and Hoy, 1989), responds by unpredictable loops and dives (Yager *et al.,* 1990).

However variable, these locomotory responses all occur with short latency, in the range of 30 to 150 ms from the onset of ultrasound stimulation. This latency is characteristic of startle escape responses to "dangerous" stimuli in many animals and not just for the ASR (Eaton, 1984).

In addition, this short latency is not a reflexive response to the mere presence of acoustic stimulation: the ASR is frequency specific. Tethered, flying crickets respond not only to ultrasound but also to stimuli of lower frequencies, especially in the range 3–6 kHz, corresponding to the peaks in the power spectrum of the reproductive calling songs of these insects. Crickets steer *toward* these lower frequencies, but they do so with much longer latencies, typically in the range of several hundred milliseconds to seconds (Moiseff *et al.,* 1978). Moreover, whereas a *single* pulse of suprathreshold ultrasound elicits an ASR, a single or even a dozen pulses of 5 kHz are not sufficient to elicit positive phonotaxis. Finally, the ultrasonic ASR takes precedence over positive phonotaxis to 5 kHz. A single pulse of ultrasound delivered to a cricket in the midst of steering toward a 5-kHz signal elicits a negatively phonotactic ASR (Nolen and Hoy, 1986a,b). It is clear that in crickets ultrasound is "special" and has priority access to the steering pathways. Thus, the ultrasound ASR in crickets has a short response latency, can be elicited by a single sound pulse, and takes priority over other ongoing flight steering. This is what one would predict of an antipredatory evasive response.

C. Acoustic Startle in Crickets

A detailed study of the aerodynamic changes that underlie the ASR in crickets was made by May (May, 1991; May et al., 1988; May and Hoy, 1990a,b). These measurements were made on tethered, flying crickets, *Teleogryllus oceanicus*. The insect was suspended in a wind stream to induce flapping flight. It was then stimulated acoustically with a pulse of ultrasound. Measurements were made of the contribution of the movements of the insect's wings, legs, and abdominal "rudder" to the ASR by assessing the changes induced by the ultrasound in the cricket's steering reactions. The aerodynamically relevant movements were measured by changes in pitch, roll, and yaw. In brief, a single pulse of suprathreshold ultrasound (e.g., 30 kHz, 50 ms duration and 5 ms rise/fall time, 60 dB SPL) causes the flying cricket to increase briefly its wing beat frequency (fly faster) and pitch downward. Roll and yaw were directed *away* from the ultrasound speaker—negative phonotaxis. All these reactions were made with the short response latencies typical of startle. Moreover, the ASR is graded. The response latency was related to the intensity of the ultrasound stimulus: the more intense the stimulus, the shorter the latency. In addition, the amplitude of the response increased with stimulus intensity.

As might be expected, wing movements accomplish much of the steering in the ASR. However, they are not the whole story. Even if the fore and hind wings were amputated and the abdomen glued to a stiff bar to prevent its movement, responses to suprathreshold ultrasound could still be measured by changes in the *yaw plane* in our crickets "flying" on their tethers. When stimulated by a pulse of ultrasound, the insect could twist its entire thorax, and this was sufficient to generate a small but measurable yaw movement. Thoracic twisting could be prevented by gluing it to a rigid rod, and this procedure completely eliminated yaw movements in response to ultrasound.

Even more surprising was May's discovery that an important part of the ASR included a fleeting swing of the hind leg into the path of the beating hind wing (the wing opposite to the sound source). Moreover, this motor act had a significant aerodynamic consequence: the hind leg swing *amplified* the yaw response to ultrasound in that the latency of yaw movement was reduced and its amplitude increased (May and Hoy, 1990b).

Like the mammalian ASR, the ASR in crickets is also a global, multisegmental response. It involves movements in virtually every appendage in the insect's body, from the antennae at the tip of the head, and includes movements of the head, thorax, legs, wings, abdomen, and, in females, the ovipositor at the tip of the abdomen.

IV. Neural Analysis of Acoustic Startle

A. Auditory Organs and Receptor Cells

The simplest auditory organ that mediates an ASR is surely the moth ear (Roeder, 1967). Noctuid moth tympanal organs, located in the thorax, have only two auditory receptor cells, designated A-1 and A-2, which differ primarily in sensitivity, the latter being 10 dB SPL less sensitive than the former. Depending on species, both cells are sensitive to a broad band of ultrasound from 20 to over 80 kHz, which encompasses the dominant energy peaks in the biosonar pulses of bats. Typical thresholds of the A-1 cell are on the order of 30–40 dB SPL. There seems little doubt that hearing organs evolved in these lepidopteran insects as an adaptation to bat predation, although there is evidence that some moths use ultrasonic signals in intraspecific communication, during courtship (Sanderford and Conner, 1990).

Field crickets and katydids have tympanal organs in the tibial segment of their forelegs. Hearing in these insects mediates not only predator avoidance but also intraspecific communication for reproduction. Correspondingly, their tympanal organs contain more auditory receptors, up to 70 cells (Schwabe, 1906; Michelsen and Larsen, 1985). In the field cricket *T. oceanicus,* the tympanal organ may be sensitive to a range of frequencies from 4 to nearly 100 kHz, (Moiseff *et al.,* 1978). The proportion of receptor cells allotted to specific frequencies is not known, but Oldfield and co-workers (1982, 1986) demonstrated that the auditory receptor cells in the tympanal organ are arranged by linear tonotopy: low-frequency receptors at the proximal end of the organ and high-frequency receptors at the distal end. In katydids, tonotopic organization is even preserved in the prothoracic ganglion, where the auditory afferent axons terminate in discrete zones, depending on frequency sensitivity (Romer *et al.,* 1988). Considerable work has gone into showing that the auditory organs of crickets and katydids are sensitive to the direction of sound owing to their structure–function organization as a pressure gradient detector (Michelsen and Larsen, 1985).

The ear of the praying mantis is a curiosity. It is located in a deep cleft, right between the hind legs on the insect's midline axis (Yager and Hoy, 1986). The auditory organ of *Mantis religiosa* contains about two dozen receptor cells. Recording

Chapter VII   Acoustic Startle: An Adaptive Behavioral Act   147

from the whole auditory nerve reveals a sensitivity in the range from 20 to 60 kHz. The lowest threshold, for about 25 kHz, is about 60 dB SPL and is considerably higher than that found in moths and even crickets. An auditory organ with a midline location might not be expected to exhibit directional sensitivity, because there would be no separation of the hearing organs upon which directional sound sources could impinge differentially. Direct measurements of auditory neurons demonstrated that the mantis ear is not directionally sensitive (Yager and Hoy, 1989).

Vying with the praying mantis for unexpected places to find an ear is the green lacewing, *Chrysopa carnea.* Miller (1975, 1983) showed that each ear is located in a fluid-filled vein, near the base of each wing. About two dozen receptor cells are sensitive to a broad band of ultrasound, and the sensitivity of the organ is in the range 25–50 kHz. Little is known about the functional organization of receptors.

B.   *The Central Auditory Pathways of Audition and the ASR*

The neural analysis of ultrasound acoustic startle has been primarily investigated in the field cricket, *T. oceanicus,* because it is robust enough to maintain flight behavior even after the surgery that is necessary to permit intracellular recording from its nervous system. Furthermore, the ultrasonic ASR can be readily elicited from such preparations. Only in field crickets has there been a systematic investigation of ultrasound-sensitive neurons in central nervous system, including thoracic ganglia and the brain. Ultrasound-sensitive interneurons in moths (Boyan and Fullard, 1988), praying mantises (Yager and Hoy, 1989), and katydids (Libersat and Hoy, 1991) have been identified, but the neural basis of the ASR has been most thoroughly investigated in the field cricket, *T. oceanicus,* and I will discuss data from this species.

1. **Int-1, a Key Interneuron.**   One of the first auditory interneurons to be identified in crickets, int-1, was found to be highly sensitive to ultrasound (Moiseff and Hoy, 1983) and was therefore suspected to be involved in the ultrasound steering response. Nolen (Nolen and Hoy, 1984) did the critical experiments demonstrating that int-1 was actually involved in the ASR. Nolen was able to impale interneuron int-1 with a micropipet in a minimally dissected cricket that was still able to perform flight-like behavior (the wings were removed, but the tethered cricket performed behavior typical of flight). Such preparations responded to ultrasound by making steering movements, which were monitored by myogram electrodes embedded in abdominal muscles involved in steering. This permitted Nolen to stimulate or suppress int-1's excitability to ultrasound and observe the effect of stimulation (or suppression) of this *single cell* on the ASR (Fig. 1). We found that int-1 exerts

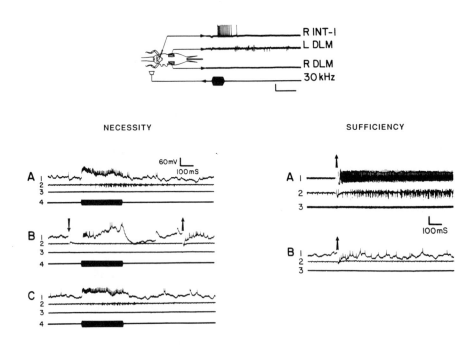

FIGURE 1. Auditory neuron, int-1, is both necessary and sufficient to initiate the ultrasound acoustic startle response (ASR). In (a), simultaneous recordings were made from the right int-1 (intracellular record) and both left (L) and right (R) abdominal dorsal longitudinal steering muscles (electromyographic recordings). The response to a single sound pulse of 30 kHz, 50 ms played from a loudspeaker on the cricket's right. The sound pulse first activates the ipsilateral, right int-1; about 50 ms later, an EMG burst is evoked in the contralateral (L) DLM, which is correlated with a steering movement to the left. The ipsilateral (R) DLM trace is silent. After the experiment, lucifer yellow was injected into int-1, revealing its morphology, seen at the top, right. In the camera lucida drawing, S = int-1's cell body; Ax = axon, which projects to brain; M and L refer to medial and lateral dendritic fields, respectively. (b) shows that activity of int-1 is necessary to initiate the ASR. Bi is the same experiment as shown in (a): 30 kHz sound pulse evokes a burst of spikes first in int-1 and then in the contralateral steering muscles. Bii: at the downward pointing arrow, int-1's excitability was depressed by hyperpolarizing current (−15 nA) injected just prior to the delivery of the ultrasound pulse. Depressing int-1's excitability "cancelled" the response in the abdominal steering muscles. Biii: control pulse delivered just after the hyperpolarization above, to show that without hyperpolarization ultrasound restores int-1's ability to evoke activity in steering muscles. (c) Neural activity induced by electrical stimulation, but not by sound, is sufficient to evoke steering. Ci: the right int-1 was electrically stimulated to discharge, by

extraordinary control over the expression of the ASR. A cricket, performing flight-like behavior, can be induced to elicit an ASR by merely exciting int-1 by injecting electrical current through the microelectrode, *but in the absence of ultrasound acoustic stimulation.* Moreover, the latency of the myogram reaction is on the order of 50–75 ms, the same as it is to an ultrasound stimulus. This demonstrated that activity in int-1 was *sufficient* to initiate the ASR. Conversely, when int-1's response to ultrasound was suppressed by injection of hyperpolarizing current, the ASR was also suppressed, even in the presence of an otherwise adequate ultrasound stimulus. This demonstrates that activity in int-1 is *necessary* to initiate the ASR.

It may be surprising that so much decision-making "power" is invested in a single second-order interneuron. However, it is important to note that int-1 only initiates the ASR; it makes no direct connections with motor neurons in the thorax or abdomen, for example. Its axon terminates in the brain, where it excites an unknown number of higher-order interneurons that descend to the segmental motor centers that drive the actual appendages of steering (Fig. 2). Perhaps the most important outcome of these experiments was the finding that electrical stimulation of int-1 was sufficient to intitiate startle or steering only when the cricket's flight motor was active. If the cricket was not performing flight behavior, no amount of excitation of int-1 would elicit steering movements. We interpreted this as demonstrating the *context dependence* of int-1: its ability to initiate an ASR depends on a prior ongoing behavioral context (flight behavior) when int-1 is activated by either ultrasound or electrical stimulation. This can be modeled as a logical AND-gate, in which the outcome (steering) occurs only if activity of int-1 (due to acoustic or electric stimulation) is temporally combined with activity of the flight motor.

2. **Brain Neurons.** The next obvious step in a physiological analysis of the ASR was to record from neurons that are postsynaptic to int-1 in the brain (Brodfuehrer and Hoy, 1990). In contrast to the simplicity seen at the level of second-order interneurons, the situation in the brain was humbling, and the research has only just begun. We have thus far described 21 different ultrasound-sensitive interneurons. They can be divided into three categories based on their morphology and physiology. Class I: consists of ultrasound-sensitive neurons with descending axons, which could presumably activate locomotory steering centers. Class II and class III interneurons are ultrasound-sensitive neurons whose axons project entirely

FIGURE 1. (cont.) anode break excitation, at a high spike rate (over 400 spikes/sec), which elicited a discharge in the contralateral steering muscles of the abdomen. In Cii anode break elicits a low spike rate in int-1, which is not sufficient to initiate activity in abdominal steering muscles. In all traces, the top trace (1) is int-1 activity, trace 2 is the contralateral EMG recording, trace 3 is the ipsilateral EMG recording, and trace 4 (where shown) is the ultrasound pulse. Reprinted with permission from the American Association for the Advancement of Science, from Thomas G. Nolen and Ronald R. Hoy in *Science,* **226,** 992-994 (1984). © 1984 by AAAS.

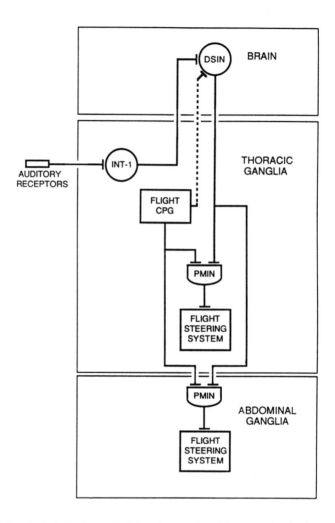

FIGURE 2. Hypothetical circuitry underlying ultrasound avoidance steering in flying crickets. Ultrasound-sensitive auditory receptors in the cricket's hearing organ stimulate the ipsilateral int-1. Int-1 makes an ipsilateral projection in the brain, where it is seen to activate a contralateral steering "command" system that descends to the steering motor centers in the thorax and abdomen. The flight CPG is seen to play a crucial role in flight steering, because it must be coactive with ultrasound stimulation before flight steering is initiated. The flight CPG and activity from int-1 are seen to form an AND gate. The site(s) where the ANDing operation are unknown and are drawn here as a series of distributed convergence sites. Reprinted by permission of John Wiley & Sons, Inc., from Higher Brain Functions: Recent Explorations of the Brain's Emergent Properties, edited by Steven P. Wise. © 1987 John Wiley & Sons, Inc.

within the brain; these are local-circuit neurons. Class II neurons differ from class III neurons in that the former are *excited* by ultrasound stimulation, whereas the latter are *suppressed* by ultrasound (suppression is measured against a level of spontaneous activity characteristic of these local circuit cells). Although ultrasound sensitivity suggests a role in the ASR, their activity has not yet been shown to be causally linked to it. Classification is only the first step in identifying the role of these cells in the ASR, and work continues on their characterization.

3. **Segmental Motor Centers.** Work has also begun on characterizing the ASR at the level of the motor system (Miles *et al.*, 1992). Electromyographic (EMG) data show that activity in a metathoracic leg abductor muscle, M126, is correlated with the metathoracic leg swing (described above) in the ASR. The response latencies decreased and the duration of the EMG spike burst increased with intensity of ultrasound stimulation. During normal tethered flight (no acoustic stimulation) M126 is silent; however, ultrasound elicits a burst of spikes, temporally correlated with the ASR expressed as a leg swing. EMG spikes are elicited with a latency as short as 20 ms, which can be compared with the shortest latency observed for the leg swing itself, 30 ms. Muscle 126 appears to be innervated by two motor neurons (C. I. Miles, unpublished). It will be essential to identify the premotor interneurons that drive M126 motor neurons directly and relate them to upstream steering neurons. We hope that by investigating the ASR from both ends, the acoustic input and the motor output, and working centrally, that we will be able to describe the neural circuitry that underlies this multisegmental but highly coordinated acousticomotor reaction.

V. Habituation of Acoustic Startle

A. *Is Habituation Adaptive?*

A general feature of the mammalian ASR is that it is graded and it habituates (Davis and File, 1984). One might wonder about the adaptive value of habituation in the ASR, which we have emphasized, after all, is an evasive or protective act in response to threat by potential predators. In the first place, one could answer that if the startle/escape reaction is not effective in response to the first few presentations of the startling stimulus, the time window for effective escape will have closed: the predator will have won. In the second place, a general feature of all closed stimulus–response chains seems to be habituation; i.e., response wanes in amplitude to repeated presentations of the *same stimulus;* habituation appears to be ubiquitous and general.

Now come the qualifications. Habituation is a much-studied form of behavioral plasticity because it lends itself to investigation under the scrupulously controlled conditions of the laboratory. The physical position of the experimental subject, the environmental context of the experiment, and, most important, every measurable parameter of the stimulus can be controlled and monitored by the investigator. Indeed, such conditions must be met and stated up front in the "materials and methods" section of any publication on the phenomenon. This may be all to the good for systematic studies of the act of habituation, but such conditions are unlikely to be replicated in nature. Rarely will a startling stimulus from a predator be "presented" unchanged from one emission to the next. Predators would likely be moving toward the prey, for example, and so both the location of the sound source and the its intensity may vary from one emission to the next. Repetitions of the signal are also likely to vary in the frequency and time domains. All of these variations in repetitions of the signal will work against habituation by the receiver of the signal. In fact, small variations in a repeated series of an identical acoustic signal, such as changing its location of emission, its intensity, or its frequency content or its temporal properties, may be sufficient to *dishabituate* the response. This may all be obvious to the reader, but one must always keep in mind such differences when thinking about the habituation as described from laboratory studies and when applied to the unpredictable conditions in nature. Caveats aside, however, I will now describe the habituation of the ASR from laboratory studies in tethered, flying crickets. I do so because a hallmark of the mammalian ASR is habituation.

B. Habituation of the ASR in Crickets

Drawing on the well worked out experimental paradigms from studies in mammals, our study of habituation in crickets (May and Hoy, 1991) utilized a structured, nine-point criterion for defining habituation developed by Thompson and Spencer (1966). We selected the swing of the hind leg in response to suprathreshold ultrasound (see above) as the behavioral act with which to investigate habituation because it is a robust behavior and easy to observe and measure. The tethered cricket was suspended in a wind stream, which induced flapping flight. It was stimulated acoustically with pulses of ultrasound (typically 20 kHz, 50 ms duration with 5 ms onset and offset) that could be varied in both intensity and rate at which pulses could be delivered (interpulse interval, or IPI rate). The standard stimulus intensity was set at 10 dB SPL above an animal's response threshold to a single pulse and maintained at that level throughout the set of nine experiments using the Thompson and Spencer criteria for habituation. The latency and amplitude of the leg swing were measured as a dc voltage shift in the output of a photocell that registered the shadow cast by the hind leg as it swung out from the body

Chapter VII Acoustic Startle: An Adaptive Behavioral Act

in response to sound. The tethered, flying cricket was illuminated from above with a fiber-optic lamp. Measurements from 50 different animals provided our data set.

We found that habituation occured in the ASR and that eight of the nine Thompson and Spencer criteria could be demonstrated (May and Hoy, 1991). Specifically, (1) the amplitude of the ASR decreases in response to repetitive stimuli (Fig. 3) and follows a negative exponential function; (2) the response amplitude is spontaneously restored after a period of rest from stimulation; (3) a series of habituation sessions induces more rapid and stronger habituation; (4) increasing the stimulus repetition rate (decreasing the IPI) produces more rapid and stronger habituation; (5) stimuli of weaker intensity (relative to threshold) produce more rapid and stronger habituation than more intense stimuli; (6) habituation shows stimulus generalization (the response is similar for frequencies of 20 and 40 kHz—both are in the ultrasound range); (7) a novel stimulus produces dishabituation (Fig. 3); and (8) the dishabituating stimulus itself shows habituation after repeated trials. Thompson and Spencer's ninth point, that habituation can wane through the zero point and even go negative, does not apply to crickets. For this to occur the hind leg would have to cross over the body to the other side, and it is morphologically constrained from doing so.

Our findings on dishabituation bear further comment. As might be expected, interpolating a sound pulse of a different "behavioral category" into a series of habituating ultrasound pulses breaks the habituation and restores the response amplitude to normal. Thus, a 5-kHz sound pulse causes dishabituation. Viewed teleologically, 5 kHz has a different behavioral "valence" in that 5 kHz is the dominant frequency of the mating call of *T. oceanicus,* whereas 20 kHz is found in the biosonar calls of bats. However, the dishabituating stimulus need not be categorically different in frequency. Dishabituation can also be produced by the same ultrasonic 20-kHz pulse used to induce the habituating response by simply switching the location of the stimulus for a single pulse, say to the speaker on the insect's left side (dishabituating side) from the speaker on the right side (habituating side). We have not systematically worked out the just-noticeable differences in the stimulus that induce dishabituation, but we expect that any stimulus that can be perceived as different in behavioral tests could serve as a dishabituating (novel) stimulus. In fact, we are now using habituation as a means of testing the acuity of discrimination in crickets.

C. Summary of the ASR in Insects

(1) An ultrasound ASR can be demonstrated in at least five different orders of nocturnally active insects; we interpret this as reflecting the selection pressure to hear the biosonar signals of insectivorous bats. (2) Like the ASR in mammals, the ASR in insects is not an all-or-none response but is graded in character: increasing the stim-

ulus intensity heightens the response. (3) Where it has been investigated, the neural basis of the ASR need not involve interneurons with "giant axons," as found in the escape systems of many invertebrates (Bullock, 1984) and beautifully exemplified in the work of Ritzmann (this volume). Rather, like the startle/escape system of teleost fishes (Eaton, 1984) startle in crickets is mediated by "nongiant" interneurons (Nolen and Hoy, 1984). This is an important point, for there has been a tendency to view startle/escape behaviors in invertebrates as being linked to giant neuron neural systems, thus setting them apart from what is perceived to be the more complex systems

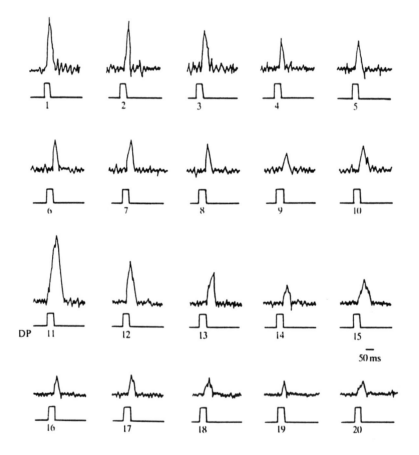

FIGURE 3. Habituation of the ultrasound ASR in flying crickets, measured by habituation of the amplitude of the metathoracic leg swing. Each numbered segment shows the leg swing to a series of 20 ultrasound stimuli (20 kHz), with an interpulse interval of 750 ms; the intensity of the ultrasound was 10 dB over the threshold. Waning of the leg swing amplitude over repetitive stimuli is shown. Between pulses 10 and 11, a dishabituating pulse (DP) was

Chapter VII Acoustic Startle: An Adaptive Behavioral Act

of mammals and other vertebrates. (4) Like the ASR in mammals, its insectan counterpart shows habituation and dishabituation of a classical form.

These several points of similarity in the ASRs of insects and vertebrates should encourage further investigation of the underlying neural systems in insects. It should be possible not only to construct neural circuits out of identified sensory neurons, motor neurons, and interneurons but also to identify the role of neurotransmitters and neuromodulators that act to modify these circuits, such as may occur in habituation. Certainly, there is excellent evidence for the role of biogenic amines in modifying mammalian startle (Davis and File, 1984), and biogenic amines are well known to play modulatory roles in the neural systems of invertebrates (Harris-Warrick and Marder, 1991).

VI. Startling Insights for Roboticists?

Animals that have a sense of hearing—and this includes most vertebrates and many insects—use their ears not only for communication with their own kind but also to detect threats in their environment, such as predators. As observed earlier, startle responses can be mediated by other sensory modalities as well. The widespread occurrence of startle reactions attests to their adaptiveness. However, are there any clues here for the designer of robotic vehicles? Even those built to walk and navigate terrain as would a cockroach? After all, a robot has no predators, and

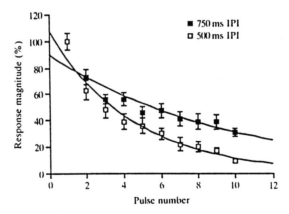

FIGURE 3. (continued) presented; this pulse was identical to the habituating stimulus except that it was presented from the left speaker, whereas all other pulses shown were presented from the speaker on the cricket's right side. A curve of the response amplitude as a function of pulse number is shown on this page for two IPIs, 500 ms and 750 ms. As expected from Thompson and Spencer (1966), habituation is stronger and faster with decreasing IPI. From May and Hoy (1991).

a threatening situation might only be an uneven terrain that might cause it to tip over. Whether or not it would be adaptive for robots to have a startle response depends on the job required of the robot. How much unpredictability must the machine cope with? Surely, locomoting through an unpredictable terrain could be accomplished by using continually active sensors that feed environmental information back to motor elements. These would be analogous to tonically active spindle or joint receptors in the mammalian limb. More severe instabilities might be dealt with through a second tier of sensors, activated only if the continually active array could not restore stability. These might be analogous to phasically active receptors, perhaps with higher sensory activation thresholds. Continuing this line of argument, one could conceive of another tier of sensors that might be monitoring other, contextual aspects of the sensory environment but with their activation thresholds set at levels that would indicate the presence of a situation that threatens the integrity of the system. These might require higher power consumption to operate and would be initiated only by a "threatening" stimulus. This would be analogous to a startle system. Activating a startle system might result in a shutting down of normal operations or even in taking some evasive action. The form of the startle response would depend on what the machine's job is, what the designer had in mind for it. Whether or not such considerations have any relevance for the enterprising roboticist, it is worth keeping in mind that startle behaviors have been around probably as long as multicellular organization, or at least when animals could first be divided into predators and prey.

Acknowledgments

I thank the organizers of this conference, Roy, Randy, and Tom, for inviting me and putting on such a superb meeting. My real debt is to my doctoral students and postdocs, whose labors are reflected in these pages and who should all be coauthors—they are the real heroes. My research has been funded by a Javits Neuroscience/Pepper Award from the NINDS, NS 11630; a Hatch Grant from New York State, 191403; and a grant from the Eppley Foundation for work on the mantis.

References

BOYAN, G. S., and FULLARD, J. H. (1988). *J. Comp. Physiol.* **164,** 251.
BRODFUEHRER, P. D., and HOY, R. R. (1990). *J. Comp. Physiol.* **166,** 651.
BULLOCK, T. H. (1984). in Eaton, R. C. (ed.), *Neural Mechanisms of Startle Behavior* (p. 1). Plenum, New York.

Chapter VII   Acoustic Startle: An Adaptive Behavioral Act

DAVIS, M., AND FILE, S. E. (1984). in Peeke, H. V. S., and Petrinovich, L. (eds.) *Habituation, Sensitization, and Behavior* (p. 287). Academic Press, Orlando.
EATON, R. C. (ED.) (1984). *Neural Mechanisms of Startle Behavior.* Plenum, New York.
FENTON, M. B., and FULLARD, J. H. (1981). *Am. Sci.* **69,** 266–275.
FLESHLER, M. (1965). *J. Comp. Physiol. Psychol.* **60,** 200.
GRIFFIN, D. R. (1958). *Listening in the Dark.* Yale University Press, New Haven.
HARRIS-WARRICK, R. M., and MARDER, E. (1991). *Annu. Rev. Neurosci.* **14,** 39.
HOFFMAN, H. S., and ISON, J. R. (1980). *Psychol. Rev.* **87,** 175.
HOY, R. R. (1989). *Annu. Rev. Neurosci.,* **12,** 355–375.
HOY, R. R., NOLEN, T. G., and BRODFUEHRER, P. D. (1989). *J. Exp. Biol.* **146,** 287–306.
ISON, J. R., MCADAM, D. W., and HAMMOND, G. R. (1973). *Physiol. Behav.* **10,** 1035.
ISON, J. R., and RETITER, L. A. (1980). *Physiol. Psychol.* **8,** 345.
LANDIS, C., and HUNT, W. A. (1939). *The Startle Pattern.* Farrar and Rinehart, New York.
LIBERSAT, F., and HOY, R. R. (1991). *J. Comp. Physiol.,* **169,** 507.
MAY, M. L. (1991). *Am. Sci.* **79(4),** 316–328.
MAY, M. L., BRODFUEHRER, P. D., and HOY, R. R. (1988). *J. Comp. Physiol.* **164,** 243–249.
MAY, M. L., and HOY, R. R. (1990a). *J. Exp. Biol.* **151,** 485–488.
MAY, M. L., and HOY, R. R. (1990b). *J. Exp. Biol.* **149,** 177–189.
MAY, M. L., and HOY, R. R. (1991). *J. Exp. Biol.* **159,** 489–499.
MEIER-EWERT, K., GLEITSMANN, K., and REITER, F. (1974). *Electroencephalogr. Clin. Neurophysiol.* **36,** 629.
MICHELSEN, A., and LARSEN, O. N. (1985). in Kerkut, G., and Gilbert, L. (eds.), *Comprehensive Insect Physiology, Biochemistry, and Pharmacology.* Pergamon, New York.
MILES, C. I., MAY, M. L., HOLBROOK, E. H., and HOY, R. R. (1992). *J. Exp. Biol.,* in press.
MILLER, L. A. (1975). *J. Insect Physiol.* **21,** 205–219.
MILLER, L. A. in Huber, F., and Markl, H. (eds.), *Neuroethology and Behavioral Physiology* (p. 251). Springer-Verlag, Berlin.
MILLER, L. A., and OLESEN, J. (1979). *J. Comp. Physiol.* **131,** 113–120.
MOISEFF, A., and HOY, R. R. (1983). *J. Comp. Physiol.* **152,** 155–167.
MOISEFF, A., POLLACK, G. S., and HOY, R. R. (1978). *Proc. Natl. Acad. Sci. USA* **75,** 4052–4056.
NOLEN, T. G., and HOY, R. R. (1984). *Science,* **226,** 992–994.
NOLEN, T. G., and HOY, R. R. (1986a). *J. Comp. Physiol.* **159,** 423–439.
NOLEN, T. G., and HOY, R. R. (1986b). *J. Comp. Physiol.* **159,** 441–456.
NOVACEK, M. (1985). *Nature* **315,** 140.
OLDFIELD, B. P. (1982). *J. Comp. Physiol.* **147,** 461–469.
OLDFIELD, B. P., KLEINDIENST, H. U., and HUBER, F. (1986). *J. Comp. Physiol.* **159,** 457–464.
POPOV, A. V., and SHUVALOV, V. F. (1977). *J. Comp. Physiol.* **119,** 111–126.
POPOV, A. V., SHUVALOV, V. F., and MARKOVICH, A. M. (1975). *J. Evol. Biochem. Physiol. USSR* **11,** 398–404.
PRAGER, J. (1976). *J. Comp. Physiol.* **110,** 33–50.
ROBERT, D. (1989). *J. Exp. Biol.* **147,** 279.
ROEDER, K. D. (1967). *Nerve Cells and Insect Behavior.* Harvard University Press, Cambridge, MA.
ROMER, H., MARQUART, V., and HARDT, M. (1988). *J. Comp. Neurol.* **275,** 201–215.

SANDERFORD, M. V., and CONNER, W. E. (1990). *Naturwissenschaften.* **77,** 345.
SCHWABE, J. (1906). *Zoologica* **50,** 1–154.
THOMPSON, R. F., and SPENCER, W. A. (1966). *Psychol. Rev.* **173,** 16–43.
YAGER, D. D., and HOY, R. R. (1986). *Science* **231,** 727–729.
YAGER, D. D., and HOY, R. R. (1989). *J. Comp. Physiol. A* **165,** 471–493.
YAGER, D. D., and MAY, M. L. (1990). *J. Exp. Biol.* **152,** 41–58.
YAGER, D. D., MAY, M. L., and FENTON, M. B. (1990). *J. Exp. Biol.* **152,** 17–39.

Chapter VIII

Organization of Goal-Oriented Locomotion: Pheromone-Modulated Flight Behavior of Moths

EDMUND A. ARBAS

Arizona Research Laboratory Division of Neurobiology
and Department of Physiology
University of Arizona
Tucson, Arizona

MARK A. WILLIS AND RYOHEI KANZAKI[1]

Arizona Research Laboratory Division of Neurobiology
University of Arizona
Tucson, Arizona

[1] Present address: Institute of Biological Sciences, University of Tsukuba, 1-1-1 Tennodai, Tsukuba-shi, Ibaraki 305, Japan.

I. Introduction

Successful locomotion to a target requires that an organism or a robot (1) produce appropriate forces to overcome gravity and propel the body, (2) stabilize the orientation of the body during locomotion, e.g., through bilateral and longitudinal coordination of appendages, and (3) orient the resulting movements in a direction appropriate to navigate to the goal. The latter task becomes considerably more difficult in an unpredictable environment, or when a discrete representation of the target's location in space is not available. Sensory cues of several modalities may have to be integrated to control the orientation of locomotion, and the control system must be adapted to overcome the indeterminate nature of a changing environment.

Many animals find unseen targets such as mates, homesites, or food by tracking odors. These chemical cues may form gradients or trails or, when borne in air or water currents, may form turbulent plumes (Schöne, 1984; Murlis and Jones, 1981; Atema, 1987; Westerberg, 1990; Murlis *et al.,* 1992). Locomotion guided by chemical gradients or trails may involve simple behavior like tropotaxis, i.e., turning toward the side providing higher-intensity stimulus to single or pairs of sensors (Schöne, 1984, p. 59), or klinokinesis, i.e., changing the rate of turning with increasing intensity of the stimulus (Schöne, 1984, p. 60). Controllers producing such behavior have been embedded in a simulated cockroach, *Periplaneta computatrix* (Beer, 1990, pp. 126–139), and have been evolved using a genetic algorithm in dynamical neural networks controlling a chemotactic agent (Beer and Gallagher, 1992). Orientation in a turbulent plume is more complex for agents that are suspended in the moving current and requires multimodal control systems and responses both to chemical signals and to the air or water current (anemo- or rheotaxis). Our goal in this chapter is to survey current efforts aimed at understanding the mechanisms, both behavioral and neural, underlying goal-oriented locomotion in turbulent odor plumes and to stimulate further efforts to understand mechanisms of adaptive control in biological and engineered systems.

II. Olfactorily Guided Tracking in Fluid Media

A. Olfactorily Guided Locomotion

Several species of fish (Mathewson and Hodgson, 1972; Emanuel and Dodson, 1979; Johnsen, 1981; Døving et al., 1985; Hasler and Scholz, 1983) and birds (Hutchison and Wenzel, 1980; Stager, 1967; Grubb, 1974) and a large number of flying insect species (Cardé 1986) locomote toward, and successfully locate, distant unseen sources of odor. In many cases, the paths traveled through air or water on the way to odor sources are very similar to each other and are complex zigzags (Hutchison and Wenzel, 1980; Johnsen and Hasler, 1980) (Fig. 1).

Certain Procellariiform sea birds like petrels and shearwaters have been observed to approach sources of food odor (Grubb, 1972), nesting colonies (Grubb 1979), and individual nest sites (Grubb, 1974) at night, exclusively from downwind. These approaches sometimes describe a zigzagging path (Hutchison and Wenzel, 1980; Hutchison *et al.,* 1982) (Fig. 1A). Neuroanatomical studies revealed atypically large olfactory bulbs in the brains of these birds, compared to other birds (reviewed by Bang and Wenzel, 1985). The common turkey vulture also has an atypically large olfactory system and approaches hidden prey items exclusively from downwind while sometimes performing "quartering" turns back and forth across the wind (Stager, 1964).

Many fish produce similar odor-modulated tracks when responding to the odor of prey (Pawson, 1977; Johnsen and Teeter, 1985; Mathewson and Hodgson, 1972), sexually receptive females (Emanuel and Dodson, 1979), and home territory (Johnsen and Hasler, 1980; Døving *et al.,* 1985). Young codfish (Pawson, 1977)

(Fig. 1B) and certain sharks (Mathewson and Hodgson, 1972; Johnsen and Teeter, 1985) exhibit upcurrent zigzagging swimming in response to the odor of prey. Emanuel and Dodson (1979) showed that male rainbow trout swimming in a flow tank immediately oriented into the current and zigzagged upstream when a plume of female oviduct extract was introduced into the tank. Stimulating the trout with a moving striped pattern beneath the floor of a still-water optomotor tank (simulating a water current) in the presence and absence of odor demonstrated that odor triggers a positive rheotaxis that is controlled by visual feedback. Coho salmon returning to their home streams also show similar optomotor rheotaxis, as well as upstream zigzagging, along boundaries between odor-bearing and clean water (Johnsen and Hasler, 1980) (Fig. 1D).

Although identified in many animals, this type of odor-modulated orientation behavior finds its greatest expression among the insects (e.g., Cardé, 1986; Baker, 1985; Payne *et al.*, 1986). Males of many species of moth are attracted to receptive females for mating by complex blends of airborne sexual pheromones (Baker 1985) (Fig. 1C). Female moths locate the larval host plant for oviposition by flying upwind in plumes of odor emanating from the plant (Haynes and Baker, 1989; Willis and Arbas, 1991a) (Fig. 1C). Parasitic wasps locate their caterpillar hosts by flying upwind in plumes of odors released by the caterpillar's feeding activity on its host plant (Zanen *et al.*, 1989; Drost *et al.*, 1986). Successful location of a point source of odor (prey, territory, mating partner) in all of the species mentioned appears to result from the combination of rapid responses to external cues (quality and quantity of the olfactory signal, water or air current direction and velocity) and integration of these with self-generated behavior patterns triggered by odors (but see below for alternative views). The mechanisms involved in source location in any of these species are not completely understood, yet a positive anemotaxis (or rheotaxis) triggered by encounters with the odor signal appears to be common to them all.

B. The Information Content of Odor Plumes

For many years, molecular diffusion was assumed to be the dominant process generating the structure of odor plumes. Plumes were thought to comprise a longitudinal concentration gradient of odor, growing steeper nearer to the source. Several models formulated to predict the "active space" of an odor plume, i.e., the region where the odor is present in quantities and mixtures that elicit a behavioral response, were based on a time-averaged distribution of the odor (Sutton, 1953; Bossert and Wilson, 1963). However, it was recognized even early on that odor plumes are not homogeneous cone-shaped envelopes of odor, but rather are fenestrated and irregular structures comprising filaments of concentrated odor inter-

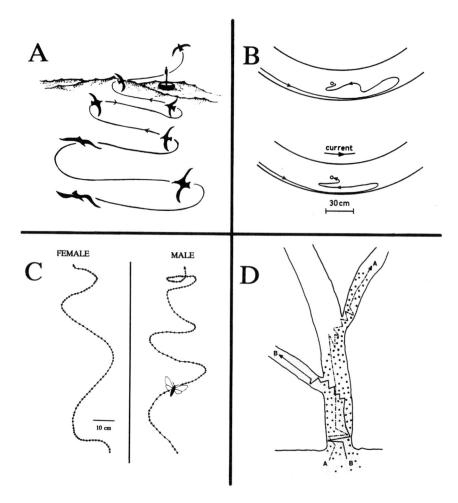

FIGURE 1. Odor-modulated locomotion in fluid media. (A) Diagrammatic representation of the flight pattern of procellariiform birds approaching a source of food-related odor. Reprinted with permission from Hutchison and Wenzel (1980). (B) Tracks of cod swimming with the current in an annular tank change to swimming against the current when presented with aqueous extract of squid flesh issuing from a pipe (open circle). Reprinted with permission from Pawson (1977). (C) Typical flight tracks reconstructed from video recordings of female and male moths, *M. sexta*, flying upwind to odor sources. The male was flying toward a pheromone source loaded with a hexane extract of a female pheromone gland, and the female was flying toward a single freshly cut tobacco leaf. The filled circles on each flight track mark the positions of the moths every 1/30 of a second. The last point of each measured flight track was 10 cm downwind from the odor source. After Willis and Arbas (1991a). (D) Schematic diagram of paths by which salmon imprinted upon their separate home-stream odors (represented by large dots in A and small dots in B) traveled up a dendritic river system. Locomotory behavior included initial bank-to-bank lateral movements

Chapter VIII Organization of Goal-Oriented Locomotion

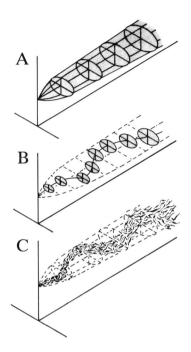

FIGURE 2. The structure of odor plumes. (A) Time-averaged Gaussian plume model, showing the principal axes and the source positioned at some height. (B) A meandering plume model with concentration of each disc distributed normally about the meandering centerline. (C) A more accurate representation of a real plume. Reproduced with permission from the *Annual Review of Entomology*, Vol. 37. © 1992 by Annual Reviews Inc.

spersed with clean air (Wright, 1958). Tests of several models of time-averaged plumes indicated that they underestimate, sometimes by several orders of magnitude, the distance at which the earliest behavioral responses can be evoked by actual plumes (Elkinton *et al.*, 1984). Our best current understanding (Fig. 2) is that the turbulence of the medium, not diffusion, generates and carries the plume and determines its internal structure (Murlis *et al.*, 1992).

Findings on the temporal and spatial characteristics of odor plumes in air (Murlis and Jones, 1981; Murlis *et al.*, 1990) and water (Moore and Atema, 1991) suggest that plume structure may carry information about the proximity of the

FIGURE 1. (cont.) near the mouth of the stream, positive rheotactic swimming where odor was distributed bank-to-bank, and zigzagging swimming where scented and unscented water plumes existed. In the absence of their imprinting odor, fish in both channels showed negative rheotaxis and swam downstream. Reprinted with permission from Johnsen (1981).

odor source. Stationary detectors placed in a plume at different distances from the source would record (1) higher peak and mean concentrations in any given encounter with odor filaments nearer to the source and lower concentrations farther downwind; (2) more frequent and briefer encounters with odor near to the source and less frequent, longer duration encounters farther downwind; and (3) a steeper rise in the concentration of odor from onset to peak concentration in a given encounter near the source, and a less rapid rise in concentration farther downwind (Murlis *et al.,* 1990; Moore and Atema, 1991). For moving detectors, such as those carried by responding animals, the frequency and length of encounters with odor filaments will depend on the velocity of the current carrying the plume and the velocity of the animal.

Moths alter their flight steering and velocity as they approach the source of an odor plume, appearing to "home in" on the target (Marsh *et al.,* 1978; Sanders, 1985; Willis and Arbas, 1991a). This change in behavior can include decreasing the ground speed, narrowing the overall width of the zigzagging flight track, and counterturning more frequently. Similar behavior is also exhibited by moths challenged to fly to different odor sources emitting progressively higher concentrations of pheromones, where other aspects of the plume are kept nearly constant (Cardé and Hagaman, 1979; Kuenen and Baker, 1982). This suggests that the alteration of flight parameters on approaching the odor source may be in response to changes in odor concentration.

Most insect pheromones and host plant odors responsible for attraction are multicomponent blends, and the concentrations and ratios of the components may be additional important determinants of the insect's behavior. Two main hypotheses have been proposed to explain the effect of blends on the orientation to odors in wind: (1) the complete blend of active components elicits all behaviors at a lower threshold (i.e. greater distance) than incomplete blends, or those with inappropriate ratios of components (Baker and Cardé, 1979; Linn *et al.,* 1986; 1987; Linn and Roelofs, 1989); (2) some subset of components (or all of the components at unspecified ratios) mediates earlier behaviors in source location (e.g., optomotor anemotaxis or zigzagging flight), while the complete blend at the correct ratios triggers later behaviors (e.g., landing or copulatory responses) (Baker *et al.,* 1976; Cardé *et al.,* 1975; Cardé and Charlton, 1985; Bradshaw *et al.,* 1983; Nakamura and Kawasaki, 1977).

Further studies involving controlled manipulation of blends and plume structure are required to demonstrate which aspects of odor quality and of spatial and temporal information in odor plumes animals actually use to guide their locomotory performance.

III. Pheromone-Modulated Flight of Moths

A. Behavioral Arousal and Appetitive Search

Flight and mating behavior of moths like the sphinx moth, *Manduca sexta,* is under circadian control, possibly through fluctuating levels of blood-borne amines (Lehman, 1990). *M. sexta* become active near dusk. Both males and females fly about, and virgin females soon alight on plants and emit a multicomponent pheromone blend (Tumlinson *et al.,* 1989) to attract males. Previously mated females do not call but rather seek out host plants for oviposition (Sasaki and Riddiford, 1984), also through olfactorily mediated flight. Circadian periodicity of mating activity is common in chemical communication systems in insects (Cardé and Baker, 1984) and in many cases seems to be responsible for species isolation (Roelofs and Cardé, 1974).

Although flight patterns of male and female moths prior to encountering pheromonal or host plant odors have generally not yet been studied, it has been suggested that they may take the form of active search strategies. One strategy for locating an odor plume carried by the wind might be to move directly across the wind until a plume is encountered. Several other models have been proposed for flight paths that would optimize the likelihood of encountering odor plumes in wind or water currents (Sabelis and Schippers, 1984; Dusenbery, 1989a, 1989b, 1990). The major underlying assumption of these models is, however, that animals treat odor plumes as time-averaged odor distributions (Fig. 2). Yet many observations and experiments indicate that animals do not react to time-averaged odor stimuli (Baker and Haynes, 1987; Baker and Vogt, 1988). In one of the few cases in which this issue has been studied, male gypsy moths, *Lymantria dispar,* were found to fly with no preferred orientation with respect to the wind direction prior to encountering an odor plume (Elkinton and Cardé, 1983).

If inactive *M. sexta* males are stimulated with the pheromone blend during the active period of their circadian cycle, they become aroused, fan their wings to raise thoracic temperature (a behavior termed preflight warmup), and then take flight and orient upwind in the pheromone plume. Males that encounter pheromones while already in flight alter their behavior by executing maneuvers that result in a characteristic flight track observed only during upwind flight toward an odor source.

B. Tracking the Signal to Its Source

Most of our understanding of odor-modulated flight behavior derives from studies of moths flying in odor plumes generated in laboratory wind tunnels (e.g,. Baker *et al.,* 1984; Cardé and Hagaman, 1979; Willis and Arbas, 1991a) (Fig. 3A). Field studies have generally confirmed, and often extended, observations made under laboratory

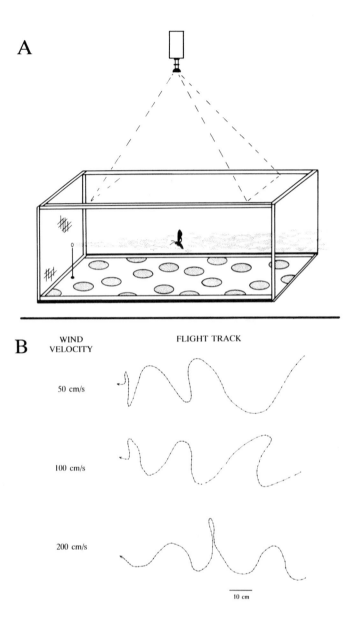

FIGURE 3. (A) Schematic diagram of a laboratory wind tunnel with details of the apparatus used during behavioral recordings. The wind flows through the tunnel from left to right, carrying the pheromones, which evaporate from a small filter paper source. The wind forms a filamentous plume (represented by stippling). Male moths are introduced into the plume at the downwind end. They become aroused, initiate flight, and fly upwind to the pheromone

Chapter VIII Organization of Goal-Oriented Locomotion 167

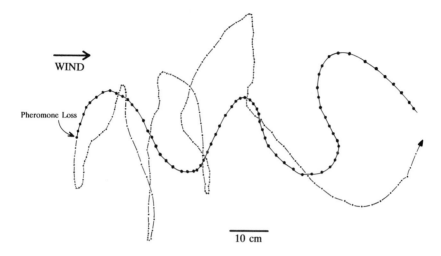

FIGURE 4. Stereotyped behavior changes predictably upon odor loss. Upwind flight tracks in odor show characteristic zigzagging turns illustrated here for a male flying upwind in a pheromone plume (large dots). When the pheromone source was quickly removed and the flying moth lost contact with the odor (arrow), the structure of the flight track changed to a characteristically different pattern (small dots). Upwind progress halted almost immediately, the time between turns increased slowly, and the track became wider. The moth faced upwind even as it was carried downwind. Wind speed = 100 cm/s. Dots indicate the moth's position every $\frac{1}{30}$ of a second.

conditions (Murlis *et al.*, 1982; David *et al.*, 1983; Willis *et al.*, 1991). Upon contacting a pheromone plume in flight, moths turn, orient, and progress upwind (in most cases, the direction to the odor source) (David *et al.*, 1982; Elkinton *et al.*, 1987). This upwind progress is controlled primarily through visual feedback from the apparent movement of the visual surround during locomotion and has been termed optomotor anemotaxis (Kennedy, 1939). Each moth's steering, air speed, and altitude also are thought to be primarily under optomotor control. As the moths progress toward the odor source, they also typically perform left–right counterturns that result in a zigzagging flight track (Fig. 3B). If contact with the plume is lost for any reason en route, upwind progress ceases, air and ground speeds sometimes change, and the excursions across the wind become wider (Fig. 4). This is termed casting flight and

FIGURE 3. (cont.) source in a characteristic zigzagging pattern. A spotted floor pattern provides visual cues to enable a successful performance of the oriented behavior in the laboratory. Flight tracks are video-recorded from above. (B) Flight tracks (as viewed from above) reconstructed from videotapes of males flying toward a source of female pheromones in three different wind velocities. Dots indicate the position of the moths every $\frac{1}{30}$ of a second.

TABLE I. Synopsis of orientation mechanisms hypothesized to operate during odor-modulated upwind flight.

Preiss and Kramer, 1986	Kennedy (and others), 1974–1986	Baker, 1990
Optomotor anemotaxis initiated by contact with the odor plume. (overall bias toward upwind movement).	**Optomotor anemotaxis** initiated and continuously modulated by contact with the odor plume (overall bias toward upwind movement).	**Optomotor anemotaxis** initiated by contact with the odor plume (overall bias toward upwind movement).
Set point for steering 0° (due upwind).	**Set point** for steering at some angle to the wind (to maintain an average "preferred" track angle).	**Set point** for steering 0° while in contact with odor, 90° upon odor loss (continuously variable).
Ground speed maintained at some small upwind value by feedback from longitudinal visual flow.	**Ground speed** maintained at an average "preferred" setting by feedback from longitudinal image flow.	**Ground speed** decreases upon contact with odor, increases upon odor loss.
Mechanosensory input not mentioned.	**Mechanosensory input** integrated with optomotor feedback to control flight maneuvers.	**Mechanosensory input** not mentioned.
Countertuning the result of noise (error) signals, caused either internally or environmentally (i.e., compensatory turns). Generated by a turning response proportional to transverse visual slip.	**Counterturning** generated internally with temporal regularity. Initiated (and modulated) by contact with odor plume.	**Counterturning** generated internally, "suppressed" during contact with odor, and expressed fully in response to odor loss.

is thought to be a search strategy to enhance relocation of a lost odor plume (David et al., 1983). If a moth reacquires contact with the plume as a result of changing these flight parameters, narrower zigzagging with upwind progress will resume. If not, the flight pattern may regress into downwind displacement while casting (Baker and Haynes, 1987; Willis and Baker, 1988, Willis and Arbas, 1991a) and eventually become disorganized and unoriented with respect to the wind.

C. Models of the Behavior

Results of a wide variety of experiments and observations have been incorporated into models designed to explain how upwind zigzagging flight is generated and what meaning it may have for goal-oriented locomotion. Our interpretations of the main assumptions and parameters included in the three most actively debated models are presented in Table I. As mentioned previously, the central unifying element in all the models is the existence of an odor-triggered optomotor anemotaxis that polarizes the net movement of moths in the upwind direction.

1. Optomotor Anemotaxis and Control of Ground Speed. Optomotor anemotaxis is a mechanism by which animals locomoting in fluid media determine, and orient their movements to, the direction of a current by monitoring the movements of visual flow fields. Directional responses to wind (anemotaxis) steered by using visual cues (optomotor) were first examined experimentally by Kennedy (1939), using the orientation of mosquitoes flying up a plume of human breath in a wind tunnel. Moths were also found to use visual cues to regulate upwind flight in pheromones in a series of wind tunnel experiments using a striped treadmill floor pattern to simulate increases and decreases in the apparent wind velocity (Kennedy and Marsh, 1974), and this behavioral response was subsequently termed optomotor anemotaxis (Kennedy, 1977). The results of these experiments led to the development of a two-part model stating that (1) upon contact with odor, moths turn into the wind and progress directly upwind as long as they are in contact with the odor, and (2) upon odor loss, upwind progress ceases and an internal program of self-steered counterturns is performed, resulting in zigzagging across the wind with an ever-widening trajectory (casting flight) (Kennedy, 1977). Since, in a filamentous odor plume, the insect experiences only intermittent contact with pheromones, there would be continuous switching back and forth between upwind flight and counterturning across the wind, resulting in a zigzagging flight track. When, in later experiments, male moths were presented with homogeneous clouds of pheromones, stimuli that should cause flight straight upwind according to this model, they were unable to maintain upwind progress but continued to counterturn (Kennedy *et al.,* 1980, 1981; Willis and Baker, 1984). These results show that moths *require* the odor signal to be intermittent for their optomotor control system to maintain upwind displacement (Baker *et al.,* 1985).

The two-part model was later elaborated on after additional studies (Marsh *et al.,* 1978; Kennedy *et al.,* 1980, 1981; Kennedy, 1986) to incorporate the idea of a variable set point in the optomotor feedback control system (Table 1). The model now states that, upon contact with pheromones, the set point for steering the course will be altered to some upwind angle greater than 0° (due upwind) and the optomotor steering that results will maintain some "preferred" track angle (Fig. 5) with respect to the wind. In this model, optomotor steering continuously compensates for externally generated changes in visual flow fields (i.e., those caused by fluctuations in wind speed and direction), and the set point is also modulated by the

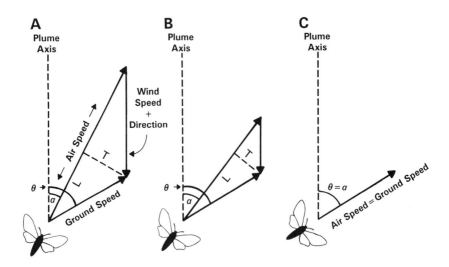

FIGURE 5. The moth's flight track can be decomposed into velocity and steering parameters by the triangle of velocities method. (A) The wind speed and direction in the wind tunnel are known and the ground speed and track angle (θ) are measured from videotape, enabling the calculation of air speed and course angle (α), and the transverse *(T)* and longitudinal *(L)* vectors of visual slip. The triangle in (B) demonstrates the effect on velocity and steering parameters of reducing the wind speed. Under these conditions, the moths maintain their average ground speed and track angle (Willis and Arbas, 1991a). The drawing in (C) illustrates the consequences of suddenly stopping air flow in the wind tunnel. Air speed = ground speed in the absence of wind, and the track angle with respect to the standing plume indicates the course steered by the moth. Wind-induced *T* slip (and anemotactic steering) vanishes if the moth's orientation is aligned with the track.

instantaneous pattern of encounters with pheromones or clean air in the turbulent odor plume, as well as by mechanoreceptive and proprioceptive information on air flow around the moth (Marsh et al., 1978).

The Kennedy et al., model requires a complex set of computations from the insect for maintenance of a preferred ground speed. To maintain both ground speed and track angle within some narrow range based on optomotor feedback, the moth must continuously regulate its airspeed and its course angle (heading) (Fig. 5). If either variable were altered independently, it would cause the other track parameter to fall out of the range of preferred values (Marsh et al., 1978). During odor-modulated flight, the visual system of moths, indeed of all insects, is thought to decompose ground pattern movements into longitudinal *(L)* and transverse *(T)* components of image flow that can be calculated using the triangle of velocities method (Ludlow, 1984; David, 1986; Preiss and Kramer, 1986a,b) (Fig. 5). It has been suggested (Ludlow, 1984; David, 1986) that track angles and ground speed are maintained at set

Chapter VIII Organization of Goal-Oriented Locomotion

values by the moth's maintaining a constant relationship between transverse and longitudinal image flow. The most precise relationship proposed is $(T^2+L^2)^{1/2}$, but moths may not be this precise, and $T+L$ may suffice (David, 1986). However, all experimental evidence to date indicates that moths do not maintain these proposed relationships throughout the entire duration of the flight track (Willis and Cardé, 1990; Willis and Arbas, 1991a). Rather, the longitudinal image flow appears to feed back with some complex relationship to control the ground speed, and the transverse component of image flow feeds back into the steering system.

An alternative view on the mechanisms underlying zigzagging flight was proposed by Preiss and Kramer (1986a,b) (Table I), based primarily on studies with tethered moths flying in a flight simulator. Male gypsy moths were stimulated by different visual patterns designed to simulate optomotor feedback that would be obtained during a variety of flight paths in the presence and absence of female pheromones. The authors concluded that the control system governing the steering of course angle is set to 0°, or due upwind, when the moth is in contact with pheromones. The actual course angles steered are scattered around due upwind in a Gaussian distribution because of uncertain determination of wind direction by the monitoring of visual slip during small deviations from 0°. The distribution of track angles over a real flight track was reasoned to be bimodal because any small deviation in course angle from due upwind results in significant drift due to the action of the wind on the moth's wings and body. Turns back toward the wind line were thought to be elicited by optomotor reflexes activated by the transverse component of visual slip. Preiss and Kramer (1986a) stated that oriented upwind flight is achieved by two independent control circuits influenced by the L and T vectors of image flow over the eyes. Steering is governed by the detection of the T component of slip, and ground speed is determined by a control system that permits a small positive value of L slip. Results obtained after selectively occluding portions of the compound eyes have shown that during tethered flight in pheromone, male gypsy moths control their ground speed with optomotor cues from the frontal sector of the compound eyes and their altitude with optomotor cues from lateral sectors (Preiss and Kramer, 1983; Preiss and Futschek, 1985). The control of pitch, roll, and yaw torques during flight is also localized to particular regions of the compound eyes (Preiss, 1991).

A third model, proposed by Baker (1990) is similar in several ways to the original model of Kennedy and Marsh (1974), but does not invoke steering at any set angle with respect to due upwind; rather, it proposes that the set point of the optomotor steering control system is determined moment to moment by contact with filaments of pheromones or clean air. Contact with an odor filament causes immediate steering toward 0°; loss of odor causes the set point for steering to change to values resulting in a track angle approximately across the wind (±90°). In this model, the ground speed is regulated by the instantaneous concentration of pheromones (Baker and Haynes, 1987). Upon flying out of the plume into clean air, the male may regain contact with pheromone by "oscillating to and fro (casting) with increasingly high airspeed (and groundspeed) and an ever decreasing reversal tempo to increase the air that is scanned.... A quick upwind surge upon recontacting pheromone will again result in direct progress toward the source before the next encounter with a large parcel of clean air occurs." (Baker, 1990, p. 21). Thus, ground speed becomes higher with a lower concentration of pheromones, but locomotion is steered more across the wind line. Upon contact with a higher concentration of pheromones, ground speed may be reduced, but locomotion is directed upwind, resulting in a surge toward the odor source.

2. **Counterturning.** A major unresolved issue concerns the presence, origins, and function of counterturning during odor-modulated flight, largely due to the great variation in flight performances exhibited within and across species. The simplest explanation of counterturning was proposed in the Preiss and Kramer (1986a) model, which assumes that turns are *reactions* to error signals caused by external perturbation (e.g., turbulence), or "internal noise functions" that cause orientation to deviate from 0° (Preiss and Kramer, 1986b, p. 78). That such internal asymmetries in motor performance exist and are corrected by exteroceptive feedback was also recognized in early studies of locust flight (Wilson, 1968). Deviations from 0° activate compensatory turns back toward due upwind when the threshold for visually detecting T slip is surpassed. These compensatory turns are proportional to the deviation from the "intended" heading; hence slight deviations result in minor adjustments, while larger deviations result in large turns back toward 0°. The prediction from this model, then, is that the temporal structure of the behavior will derive from the timing of externally induced error signals and the (unspecified) nature of the internal noise function.

Several lines of evidence argue against this interpretation. Naive moths exposed to pheromones in still air produce counterturns in flight, although overall progression is unpolarized (Baker, 1985; Baker *et al.*, 1984). If wind is stopped after moths have begun oriented flight up an odor plume, i.e., after they have been able to compute wind velocity and direction from visually monitored drift, zigzagging flight continues, often with the appropriate orientation to the standing plume (Baker *et al.*, 1984; Baker, 1985; David and Kennedy, 1987; Willis and Cardé, 1990; M. A. Willis, unpublished observations). It is important to note that under conditions of zero wind there is no wind-induced slip; thus the triangle of velocities collapses so that the course vector equals the track vector (Fig. 5C). Therefore, the measured zigzagging path can be interpreted only as the intended course of the moth. In addition, a finer-grained analysis of maneuvers that make up individual flight tracks and the motor discharges to wing muscles that underlie them during free flight (Willis and Arbas, 1991b) indicates that moths may make corrective maneuvers that stabilize the zigzagging path itself.

The Kennedy *et al.,* model proposes the activation, by contact with odor, of a motor program for repetitive turning. Hence, by this model, counterturning is *active* behavior, timed by the constraints of a pattern-generating mechanism and modulated by the instantaneous concentration of odor and timing of encounters with odor filaments. The integration of temporally regular counterturns and optomotor steering at an angle to the wind results in the typical zigzagging flight track. Upon loss of odor, the set point of the optomotor steering system is reset to result in the counterturning that characterizes casting flight.

The anemotactic turning model of Preiss and Kramer (1986a) was elaborated on by Kaissling and Kramer (1990). This model proposes that the anemotactic turning tendency activated by T slip may interact with an odor-independent internal turning tendency in a way that enables the insect to maintain a set course at an angle to the wind. Reversals of the sign of the internal turning tendency that occur spontaneously, or are elicited by changes in odor concentration, would initiate turns across the wind.

The Baker model invokes an internal turn generator that is activated upon initial contact with the odor plume and modulated by instantaneous contacts with pheromones or clean air. Baker (1990) offers several possible explanations of why different moths may exhibit different degrees of counterturning, or none at all, under similar stimulus conditions. (1) Not all species of moth necessarily integrate counterturning with upwind flight. However, counterturning across the wind line *is* expressed as *a response to odor loss* in all moths tested thus far. Thus, the tracks of these moths, while generally centered around a course of 0°, may show some modulations indicating irregular turns caused by activation of the internal counterturning generator upon straying out of the pheromone plume or upon encountering a large packet of clean air within the plume. The turns would arise from the incipient activation of casting flight cut short by the arrival of another odor filament. The key here is that the timing of turns will be determined by the microstructure of the pheromone plume, i.e., the timing of encounters with pheromones or clean air. (2) Moths that do integrate counterturning with upwind flight in pheromones may exhibit a form of behavioral low-pass filtering. The reaction time with which moths convert to casting flight upon loss of odor varies considerably across species: 0.15 s for male *G. molesta* (Baker and Haynes, 1987), 0.3–0.5 s for male *Antheraea polyphemus* (Baker and Vogt, 1988), 0.4 s for male *Amyelois transitella*

(Haynes and Baker, 1989), and about 0.5 s for *M. sexta* (Willis and Arbas, 1991a). Perhaps not coincidentally, these reaction latencies correspond remarkably well to the known interturn durations for these species. These data suggest the presence of a specific time period during which a filament of pheromones must be received for upwind progress to continue. If a filament is not encountered during this period, the optomotor steering and velocity responses change rapidly, resulting in flight across the wind. Baker (1990) suggests that these species-specific differences in behavioral response times to odor loss may underlie differences in the character of flight tracks recorded from different moths under similar conditions. Hence, species with a short behavioral time constant would exhibit a zigzagging flight track timed by the alternate encounters with filaments of pheromones and parcels of clean air, repeatedly switching from upwind surges to casting flight and back. Species with a long behavioral time constant would exhibit relatively straight flight tracks in an identical plume based on their inability to switch to casting flight quickly enough before the next encounter with a pheromone-laden filament. The implication is that the sensory systems of both short- and long-time-constant species may still be able to resolve the temporal structure of the plume, but "flicker fusion" may occur based on motor response time. By this mechanism, only encounters with parcels of clean air lasting longer than the species-specific reaction time would result in turns and modulation of the flight track. The temporal structure of the behavior would still be determined by external cues, but with a cut-off at a particular frequency. (3) Moths that do counterturn in the presence of pheromones would exhibit frequency modulation governed by intermittent contact and loss of pheromone filaments. Entry into a parcel of clean air would result in relaxation of the turning tempo into the low-frequency turns typical of casting flight, whereas contact with pheromones would accelerate turning frequency in proportion to the concentration of pheromone encountered, biasing the overall movement toward due upwind. In this mechanism (as in several of the models described earlier), the temporal structure derives from a blending of the internally determined pattern and external modulation. This mechanism also accounts for the observations that the rate of counterturning increases when moths are challenged to fly in plumes of increasing concentration of pheromones (Kuenen and Baker, 1982; Cardé and Hagaman, 1979) and that some moths also counterturn more frequently as they come near the odor source, where the concentration of pheromones is likely to be highest (Willis *et al.,* 1991; Willis and Arbas, 1991a).

What functions might counterturning serve for the insect? In reviewing pheromone-modulated zigzagging, Kennedy (1983) suggested that counterturning enhances the likelihood of encountering pheromone filaments while orienting up the plume. Another proposal is that counterturning is important because it carries

Chapter VIII Organization of Goal-Oriented Locomotion

the moth outside the time-averaged plume boundaries, enabling the moth time to clear its receptors, thus avoiding adaptation to the odor plume (Farkas and Shorey, 1974). This would, in effect, provide a self-imposed temporal break in the sampling of the odor signal similar to sniffing in mammals (e.g., Lynch and Granger, 1990) or flicking of olfactory antennules in crustaceans (Schmitt and Ache, 1979). An interesting possibility that remains to be tested is that counterturning and steering at an angle to the wind is an active mechanism used to "capture" wind-induced drift from which the insect is able to compute wind velocity and direction (Kennedy, 1983, 1986; Cardé, 1984).

3. Mechanosensory Information. There exists a well-developed literature on the importance of mechanoreceptive and proprioceptive input in control of flight in insects, notably locusts (e.g., see Wendler, 1983; Gewecke and Wendler, 1985; Rowell, 1988; Möhl, 1989). Many similar inputs may be of equal importance to the generation and control of flight in moths and other Lepidoptera (Niehaus and Gewecke, 1978; Gewecke and Niehaus, 1981; Niehaus, 1981; Arbas and Hildebrand, 1986; Orona and Agee, 1987; Dombrowski, 1991; E. A. Arbas and M. A. Willis, unpublished observations). Mechanosensory information has generally been dismissed as having no role in the control of orientation during flight because air flow is thought always to be directly front to back over the wings and body and over the main wind-sensitive structures on the head, the cephalic hair fields and antennae (e.g., Schöne, 1984, p.249). In addition, wind is defined as movement of the supporting air over the ground, not over the insect (Marsh *et al.,* 1978) and, under steady-state conditions, the insect will be carried in the wind "as if in a giant box of air" (Schöne, 1984, p. 241). Air flow along the moth's longitudinal body axis will stimulate a large number of mechanoreceptors, including the Johnston organ and Böhm bristles located in the basal segments of the antennae (Böhm, 1911; Vande Berg, 1971; Niehaus and Gewecke, 1978; Stengl *et al.,* 1990) and mechanosensory head hairs normally hidden from view among facial scales (Arbas and Hildebrand, 1986). By analogy with similar structures in locusts and other insects (Weis-Fogh, 1949; Gewecke, 1974, 1975), these may be important in *M. sexta* for regulating air speed. Like other Lepidoptera (Niehaus, 1981), *M. sexta* with antennae removed are unable to execute coordinated free flight (Willis and Butler, in preparation). Reattaching severed antennae to their proximal stumps restores the physical load on the aforementioned receptor structures and normal flight ability.

Various corrective steering reactions are elicited by changes in the angle of air flow over sensory structures on the head of locusts under open-loop conditions (Camhi, 1970a,b; Arbas, 1986). Some similar reactions, such as curling of the abdomen, have also been observed in *M. sexta* (E. A. Arbas, unpublished observations) under similar conditions. These reactions may be elicited by air current stimuli to antennal receptors and mechanosensory head hairs and also may be activated by optomotor pathways. Such reactions may be used to correct for deviations in the insect's attitude (i.e., in the pitch, roll or yaw planes) with respect to the direction of air flow over the body. Furthermore, both locusts (Baker *et al.*, 1981; but see Rainey, 1985) and moths (M. A. Willis, unpublished observations) have been observed in free flight with their body axes oriented at an angle to the direction of travel in very low wind (locusts) or still air (moths), indicating that the insects often tolerate sideslip during normal flight. When there is sideslip, air flow at an angle over the head and body will provide directional information to mechanosensors. Similarly, mechanosensory pathways may provide information about overall wind direction to the insect during turbulence, changes in wind velocity (gusts), or active maneuvering when air flow around the insect is complex, especially in larger insects like *M. sexta,* where significant inertial effects become important (e.g., Nachtigall, 1989).

IV. Neural Substrates

Processing of olfactory information by receptor neurons and olfactory brain neurons near the input end of the system has been reviewed in depth (Christensen and Hildebrand, 1987b; Homberg *et al.*, 1989; Kaissling and Kramer, 1990; Hildebrand *et al.*, 1992). We present only a brief overview here, focusing on the insect that we use in our own studies, the sphinx moth, *M. sexta.*

A. *Olfactory Signal Discrimination*

Pheromones and plant odors evoke behavioral responses in *M. sexta* by stimulating receptor neurons located in olfactory sensilla on each antenna (Sanes and Hildebrand, 1976a,b; Lee and Strausfeld, 1990). Cells responsive to pheromones reside in a particular class of male-specific trichoid sensilla on the antennal flagellum. Each pheromone-responsive sensillum contains two receptor cells (Sanes and Hildebrand, 1976a,b; Keil, 1989; Kaissling, *et al.*, 1989; Lee and Strausfeld, 1990): one that responds specifically to *(E,Z)-10,12-hexadecadienal* (E10,Z12-16:AL or bombykal), the major component of the pheromone blend (Starratt *et al.*, 1978; Tumlinson *et al.*, 1989), and another that is unresponsive to physiological concentrations of E10,Z12-16:AL but is stimulated by one of two other components of the pheromone blend, either *(E,E,Z)-10,12,14-hexadecatrienal* (E10,E12,Z14-16:AL) or *(E,E,E)-10,12,14-hexadecatrienal* (E10,E12,E14-16:AL) (Tumlinson *et al.*, 1989; Kaissling *et al.*, 1989). The specificity of receptor neuron responsiveness to

Chapter VIII Organization of Goal-Oriented Locomotion 177

pheromonal odors carries on into olfactory circuits of the brain, where certain classes of interneurons respond only to pheromonal odors and anatomically define apparent sex pheromone foci by their projection patterns to target neuropil. Thus, the pheromone-processing pathways within the central nervous system (CNS) of *M. sexta* appear to make up a labeled-line system (see below and Hildebrand *et al.*, 1992) beginning with these narrowly tuned neurons at the input end. Receptor neurons may signal the onset of an encounter with a filament of pheromone by a burst of action potentials that is conducted to the CNS and also may have other influences on patterning of behavior that are determined by response latencies and prolonged discharges (Kaissling and Kramer, 1990).

Among the other sensilla of the antennal flagellum are many with receptor cells sensitive to plant volatiles such as the leaf aldehyde *(E)-2-hexenal* (E2-6:AL) (Schweitzer *et al.*, 1976; Matsumoto and Hildebrand, 1981). Receptor cells of this type are found in the antennae of both sexes. Far less is known about receptors to plant volatiles on the antennae, and CNS studies indicate that interneurons near the input end of the system that receive inputs from these receptors are broadly tuned and responsive to a variety of compounds. This suggests that discrimination of plant odors may be based on a distributed, across-fiber pattern rather than narrowly tuned, specialized receptor cells (Christensen and Hildebrand, 1987b; Hildebrand *et al.*, 1992; B. Waldrop, T. A. Christensen, J. G. Hildebrand, unpublished observations).

In addition to olfactory receptor cells, the antenna also bears receptors of other modalities, including hygroreceptors, mechanoreceptors, and thermoreceptors, in its flagellum and basal segments (Böhm, 1911; Vande Berg, 1971; Lee and Strausfeld, 1990; Stengl *et al.*, 1990).

B. Olfactory Processing Pathways in the Brain

Axons of antennal olfactory receptor cells enter the brain via the antennal nerve and terminate in the ipsilateral antennal lobe (AL), the first-order olfactory center (Camazine and Hildebrand, 1979). Each AL in *M. sexta* comprises a central zone of coarse neuropil surrounded by an array of about 60 spheroidal "ordinary" glomeruli bordered by three groups of neuronal somata—lateral, medial, and anterior—totaling about 1200 cells (Hildebrand *et al.*, 1980; Matsumoto and Hildebrand, 1981; Christensen and Hildebrand, 1987b; Homberg *et al.*, 1988). Male ALs also contain a prominent, sexually dimorphic macroglomerular complex (MGC) (Matsumoto and Hildebrand, 1981; Hansson *et al.*, 1991). The axons of the pheromone receptor cells of the male antenna project exclusively to the male-specific MGC, while the axons of other olfactory receptor cells terminate in the "ordinary" glomeruli (Camazine and Hildebrand, 1979; Matsumoto and Hildebrand, 1981; Schneiderman et al., 1984).

Two structural classes of AL neurons integrate olfactory information and, in some cases, information from the other sensory modalities: neurons with their arbors confined to the AL (local interneurons) and neurons with axons projecting from the AL to other regions of the brain (projection neurons) (Matsumoto and Hildebrand, 1981; Christensen and Hildebrand, 1987a; Homberg et al., 1988, 1989; Kanzaki et al., 1989).

Some local interneurons of the antennal lobe respond to inputs from the antennal nerve with short latencies and therefore may receive primary olfactory input from receptors within the antennal lobes (Matsumoto and Hildebrand, 1981; Christensen and Hildebrand, 1987a; T. A. Christensen, unpublished results). Others receive inputs at longer latencies, apparently reflecting polysynaptic input pathways. Immunocytochemical staining shows some of the local neurons to be GABAergic (Hoskins et al., 1986), making them candidates for providing the inhibitory component of mixed excitatory–inhibitory synaptic input recorded following stimulation with pheromones in some projection neurons (Christensen and Hildebrand, 1987a).

Projection neurons of the AL in M. sexta can be distinguished structurally by uni- or multiglomerular arborizations within the AL and axons that project through one of five different tracts to the protocerebrum (PC) (Homberg et al., 1988, 1989; Kanzaki et al., 1989). The principal categories are projection neurons with dendritic arbors confined to the MGC and projection neurons with arbors in one or more ordinary glomeruli. Projection neurons that innervate the MGC may be classified further into structural and functional categories based on responsiveness to pheromone components and branching in toroid and cumulus-shaped regions of the MGC (Hansson et al., 1991).

After primary processing in the AL, olfactory information is relayed by the AL projection neurons to several regions of neuropil in the PC, including the calyces of the ipsilateral mushroom body and the lateral PC, for higher-order processing (Kanzaki et al., 1991a). The mushroom bodies are thought to play a significant role in olfactory learning and memory in insects (e.g., Erber et al., 1987) and may contribute to some of the computations made by moths during oriented zigzagging flight. In these areas, additional processing takes place by PC neurons that, in turn, transmit the processed information to other brain regions. Many olfactory PC neurons innervate a neuropil region lateral to the central body on each side of the PC, the lateral accessory lobe (LAL). Each LAL is linked by its constituent neurons to the ipsilateral lateral PC, where projection neurons from the AL terminate, as well as to other regions of the PC. The LALs are also linked to each other by bilateral

Chapter VIII Organization of Goal-Oriented Locomotion

neurons with arborizations in each LAL. The LAL appears to be an important region of convergence of olfactory neurons from other regions of the PC (Kanzaki, 1989; Kanzaki *et al.*, 1991a,b) (Fig. 6).

The LALs and also certain adjacent neuropil regions in the ventral PC are innervated by branches of neurons that respond to olfactory stimuli and have axons that descend in the ventral nerve cord (Kanzaki *et al.*, 1991b). The activity of these descending neurons is of particular interest because the information they carry represents the integrated multimodal output of brain circuits that may act on thoracic motor circuitry to affect behavior.

C. Coding by AL Neurons of Odor Quality, Quantity, and Temporal Structure

As described earlier, male moths respond behaviorally to pheromonal odors only when the appropriate components of the natural blend are present in the plume

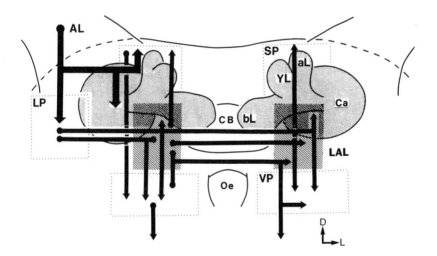

FIGURE 6. Schematic representation of higher-order olfactory pathways in the brain of male moths, *M. sexta*. Lines with arrowheads show brain regions linked by antennal lobe projection neurons (AL) and olfactory protocerebral neurons characterized by intracellular recording, dye injection, and reconstruction from serial sections (Kanzaki *et al.*, 1989, 1991a,b). Boxes outlined by dotted lines depict prominent target neuropil areas in the protocerebrum defined by the projection pattern of olfactory neurons (SP, superior protocerebrum; LP, lateral protocerebrum; VP, ventral protocerebrum). The diagonal lined boxes depict the lateral accessory lobes (LAL). Lightly shaded areas depict the bilateral mushroom bodies of the brain (bL, beta lobe; YL, Y lobe; aL, alpha lobe; Ca, calyx of the mushroom body). CB, central body; Oe, oesophageal foramen; D, dorsal; L, Lateral.

(Tumlinson et al., 1989), in the proper proportions (Linn et al., 1986), and when the odor stimulus is intermittent (Baker et al., 1985).

1. **Odor Quality.** Pheromone-responsive receptor neurons associated with antennal olfactory sensilla are specialized to detect specific components of the natural pheromone blend (Kaissling et al., 1989). The two principal components of the 12-component natural blend (E10,Z12-16:AL and E10,E12,Z14-16:AL), which are effective in eliciting oriented zigzagging flight in *M. sexta* (Tumlinson et al., 1989), were also found to be the most potent stimulators of responses recorded intracellularly in AL neurons (Christensen et al., 1989). AL neurons were identified that act as (1) "pheromone generalists" (i.e. they respond equally well when any of several individual components or the complete pheromone blend stimulates the antenna) or (2) "pheromone specialists" (i.e., they respond to some specific components and not others, or show excitatory responses to some components and inhibitory responses to others, and mixed excitatory/inhibitory responses when stimulated with blends) (Christensen and Hildebrand 1987a, 1988; Christensen et al., 1989). AL projection neurons of some other moth species appear to be more broadly tuned to major pheromone components shared by their own pheromone blend and that of related, sympatric species (Christensen et al., 1991). Thus, although processing of olfactory information may be carried out in different species by neurons that are structurally very similar, discrimination or identification of odor blends may occur in different ways in the different neural networks.

2. **Odor Quantity.** Dose–response relationships recorded from AL projection neurons in response to stimulation with pheromone blend or individual components indicate that concentration of the stimulant may be encoded in peak firing rates, as well as in the latencies and durations of their responses (Kanzaki et al., 1989; Christensen and Hildebrand, 1990). Where responses are excitatory, higher concentrations of odorant typically elicit phasic increases in firing to higher peak rates, with the rise to peak occurring sooner (Kanzaki et al., 1989). Some neurons show saturation at lower concentrations of odorant than other neurons (Christensen and Hildebrand, 1990). Where responses are inhibitory, higher odorant concentrations elicit prolonged reductions in firing rate (Kanzaki et al., 1989). Phasic responses of higher-order neurons typically did not follow stimulus intensity with such fidelity, where odorants at different concentrations were tested (Kanzaki et al., 1991a,b).

3. **Temporal Structure of the Plume.** Behavioral studies have shown that the odor stimulus received by moths must be intermittent for proper oriented locomotion to occur (Baker et al., 1985). Electrophysiological experiments using pulsatile

stimuli indicate that male moths can resolve intermittent pheromonal signals at high frequencies at the level of the peripheral receptors (Baker *et al.*, 1989; Kaissling, 1986; Rumbo and Kaissling, 1989) and in the CNS (Christensen and Hildebrand, 1988; Christensen *et al.*, 1989). Recordings from AL projection neurons showed that fusion frequencies for pulsed inputs vary significantly in different classes of AL neurons. One class of AL neurons, the pheromone-specialist types that receive two polarities of inputs (i.e., excitatory inputs when antennae are stimulated with dienal components and inhibitory inputs when antennae are stimulated with trienal components or vice versa) are able to follow pulsed inputs up to 10 Hz (Christensen and Hildebrand, 1988). Other AL projection neurons follow only lower frequencies of pulsatile stimulation and then, at some threshold, fuse their firing into a single burst.

D. *Higher-Order Processing of Pheromonal Information*

Responses of olfactory protocerebral neurons and of descending neurons to brief pheromone blend stimuli fell into two general classes: (1) brief excitation that recovered to background levels in <1 s after the stimulus and (2) long-lasting excitation (LLE) that outlasted the stimulus by > 1 s and, in many cases, as long as 30 s (Kanzaki *et al.*, 1991a,b) (Fig. 7). The LLE-type responses were elicited preferentially in PC neurons by pheromonal stimuli, including individual pheromone components, whereas in descending neurons only blends of pheromone components or an extract of the female lure gland elicited LLE. Barring a single exception (Kanzaki *et al.*, 1991a), all neurons exhibiting LLE recorded to date had arborizations within the LALs. Some descending neurons also showed conditional responses in which identical pheromone-blend stimuli had different effects depending on the state of firing of the descending neuron. Initial exposure to the pheromonal stimulus produced an increase in firing rate typical of LLE. Rather than increasing or showing saturation upon a subsequent identical stimulus, the rate of firing declined. The rate of firing then gradually increased to the higher level of the previously elicited LLE over the next several seconds, when another similar stimulus again caused a decline in firing rate (Kanzaki *et al.*, 1991b). It appears that the normally intermittent odor filaments of a natural plume would cause a significant state-dependent modulation or suppression of the firing of such neurons. Loss of odor would cause a restoration of LLE that would wane over some longer time course without additional pheromonal stimuli. Olfactory descending neurons were also encountered that exhibited LLE but not state-dependent responses (Kanzaki *et al.*, 1991b).

Visual stimulation elicited excitatory or inhibitory responses in all olfactory descending neurons tested in *M. sexta*. There were "on" or "off" responses in neu-

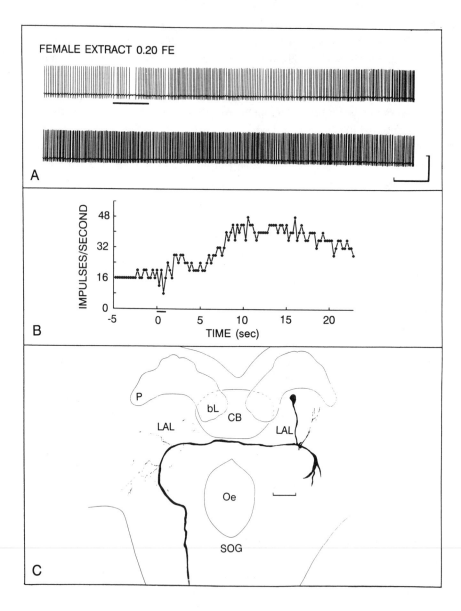

FIGURE 7. Physiology and morphology of a descending neuron that responded to pheromonal stimuli with long-lasting excitation (LLE). (A) Antennal stimulation with female extract elicited LLE lasting > 20 s. **FE,** female equivalent. Scale = 1 s, 40 mV. (B) Plot of the firing frequency of the neuron shown in **A** measured in 250-ms bins. (C) Reconstruction of the Lucifer Yellow–stained descending neuron (in frontal view) from

rons showing LLE and reduction of spontaneous firing with tonic increases in illumination (Kanzaki et al., 1991b). Neurons responding to olfactory stimuli with brief excitation also showed excitatory responses to increases or decreases in illumination. The fact that sustained high illumination tended to inhibit background firing of descending neurons that integrate pheromonal information might be relevant to the observation that *M. sexta* males are inactive and unresponsive to pheromone blend in the daytime. In silkmoths *Bombyx mori,* visual stimuli are effective in changing the firing state of certain olfactory interneurons (Olberg, 1983). Moreover, some of these neurons responded to both mechanosensory stimuli applied to the antennae and directional visual stimuli (Olberg, 1983). Pheromonal stimuli were shown to activate or amplify visual responses in several descending neurons of gypsy moths *L. dispar* (Olberg and Willis, 1990).

Long-lasting increases in firing elicited by pheromone components or visual stimuli have been described in a class of axons termed "flip-flopping" interneurons in male *B. mori* (Olberg, 1983). These neurons exhibited conditional responses, in that stimuli applied when a neuron was in a state of low-frequency firing elicited accelerated firing, while identical stimuli applied when the neuron was in a state of high-frequency firing caused decelerated firing (hence the term flip-flop). The conditional responses of *M. sexta* descending neurons differ from the flip-flopping of descending neurons in *B. mori* in several ways. In *B. mori,* long-lasting increases in firing could be activated in flip-flopping neurons by the primary component of the pheromone blend, *(E,Z)-10,12-hexadecadienol* (E10,Z12-16:OH or bombykol), which also activates the entire sequence of zigzagging walking by which the flightless males approach signalling females (Olberg, 1983). In *M. sexta,* individual pheromone components failed to elicit LLE, but blends of the major pheromone components or female extract were effective (Kanzaki et al., 1991b). In *B. mori,* both high and low states of firing elicited by pheromone components were stable and lasted as long as 4 min. In *M. sexta,* LLE elicited by pheromone blends can last several tens of seconds, but the conditional reduction of firing elicited by subsequent pheromone-blend stimuli spontaneously reverts to the state of high frequency firing after several seconds (Kanzaki et al., 1991b).

Another group of descending neurons has been identified in *B. mori* that showed LLE responses to stimulation of the antenna by E10,Z12-16:OH but did not give conditional responses similar to flip-flopping (Kanzaki and Shibuya, 1986). Because the dose–response relationship between E10,Z12-16:OH stimuli and the

FIGURE 7. (cont). which the recordings shown in **A** were obtained. The axon left the subesophageal ganglion (SOG) in the connective contralateral to the soma. CB, central body; P, peduncle; bL, beta lobe of the mushroom body; LAL, lateral accessory lobe; OE, oesophageal foramen. Scale = 100 μm. Reprinted from Kanzaki et al., (1991b).

prolonged firing of these descending neurons closely resembled a similar relationship between pheromonal stimuli and production of wing fluttering, Kanzaki and Shibuya (1986) suggested that these neurons may have a role in initiation and maintenance of the "mating dance" of *B. mori*.

Brief pulses of pheromone-blend stimuli cause long-lasting activation of flight motor activity in intact and dissected *M. sexta* (Willis and Arbas, 1991; Kanzaki and Arbas, 1990), but a direct relationship between LLE in descending neurons and the activation of motor activity has yet to be established. As mentioned earlier in relation to the AL PNs, caution is advised when comparing details of neural networks across species. Although many features of cells and circuits may be shared in common, details of the computations performed may differ.

E. Motor Outflow During Pheromone-Modulated Flight

Although a great deal is known about insect flight (e.g., for reviews see Gewecke and Wendler, 1985; Goldsworthy and Wheeler, 1989), the premotor and motor circuitry generating flight behavior in *M. sexta* and other Lepidoptera remains largely unexplored, with the exception of a few studies in which motor neurons of several flight muscles were characterized structurally (Cassaday and Camhi, 1976; Rind, 1983; Kondoh and Obara, 1982; Tsujimura, 1989). By contrast, a great deal is known about motor outflow during tethered (Kammer, 1971; Kammer and Nachtigall, 1973; Dombrowski, 1991) and free flight (Willis and Arbas, 1991b; and see below) and about the structure, development, and neuromuscular physiology of certain flight muscles (Rheuben and Kammer, 1987; Rheuben, 1985; Klaasen and Kammer, 1985).

To begin to establish a link between the activity of brain circuits and descending neurons with behavior, we have initiated studies of the activity produced by certain wing muscles during pheromone-modulated zigzagging flight in a wind tunnel (Willis and Arbas, 1991b), and during tethered flight (Kanzaki and Arbas, 1990).

1. Motor Outflow during Free Flight. Electromyographic (EMG) activity recorded with long, ultralight leads was synchronized with a video record of the zigzagging flight track of moths flying freely to a pheromone source in a laboratory wind tunnel (closed-loop conditions) (Willis and Arbas, 1991b) (Fig. 8). We have recorded from up to five muscles: the left dorsal longitudinal muscle (DLM), the principal wing depressor; the left and right third axillary muscles (AxM), which may be important in the remotion and promotion of the wings during maneuvering (Kammer, 1971; Rheuben and Kammer, 1987); and the left and right first basalar muscles (BaM), which generate lift and thrust through their actions as direct wing depressors. Parameters measured from the EMG recordings were burst

Chapter VIII Organization of Goal-Oriented Locomotion

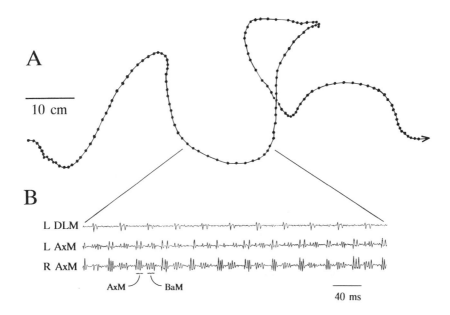

FIGURE 8. Motor outflow during upwind zigzagging flight in a wind tunnel. (A) Flight track reconstructed from a video record of a male moth flying in a plume of pheromone blend. Dots mark the position of the moth every $\frac{1}{30}$ of a second. (B) A segment of EMG recording obtained during this flight track from leads placed in the left dorsal longitudinal muscle (L DLM), the principal wing depressor, and the left (L) and right (R) third axillary muscles (3rd AxM). These latter two traces also picked up the activity of the nearby first basalar muscles on the corresponding sides via the volume conductor. Each cycle of activity in the DLM trace represents one wing beat cycle.

period (ms) and burst duration (ms) for all five muscles and the phase of AxM and BaM activity with respect to each cycle of DLM activity.

Modulation of DLM burst frequency (i.e., wing beat frequency) often, but not always, underlies the modulation of the ground speed of the flying moth (Willis and Arbas, 1991b). In concert with the modulation of DLM activity, the phase of AxM and BaM activity also shifts symmetrically on both sides of the body, undergoing phase advances when wing beat frequency declines and phase delays when wing beat frequency rises. Characteristic asymmetries between bilateral AxMs were identified during zigzagging turns in free flight. The most typical is the appearance of an additional spike at a later phase of the wing beat cycle than the primary burst of the AxM on one side of the insect but not the other (M. A. Willis and E. A. Arbas, in preparation). This asymmetry in firing pattern is usually associated with a slight yaw of the body axis to the side of the double burst, or an actual

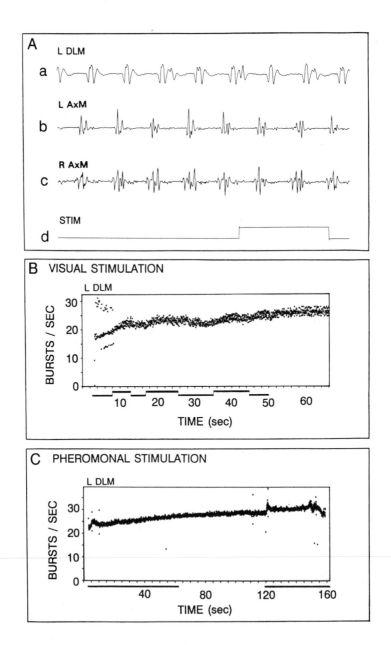

FIGURE 9. Olfactory and visual influences on motor outflow during tethered flight. (A) EMG recordings from the left DLM and both left (L) and right (R) third AxM during preflight warm up. Motor activity was activated by puffs of pheromone blend in the air stream that passed over the antennae of a male moth. (B) Plot of the wing beat frequency (each point

Chapter VIII   Organization of Goal-Oriented Locomotion

turn. We have also observed a similar pattern in the BaMs. These adjustments in firing pattern may occur during only one or a few wing beat cycles in a turn.

2. Motor Outflow during Tethered Flight. We have also recorded the activity of these muscles, in the same individuals, during tethered flight activated by visual stimuli or pheromone blends in a flight simulator (open-loop conditions) (Kanzaki and Arbas, 1990). The apparatus delivers a continuous current of filtered air to the tethered insect via a small wind tunnel equipped with ports into which pulsed odor stimuli can be injected. The mouth of the wind tunnel is surrounded by an array of > 125 green light-emitting diodes (similar to a design used to study visually controlled flight in locusts; Baader, 1991) controlled by a computer to simulate the visual flow fields for forward, backward, or turning movements, as well as roll. In some insects (particularly heavily dissected ones) we elicited flight by injecting or bath applying chlordimeform [*N-(2-methyl-4-chlorophenyl)*-N′,N′-*dimethylformamidine*, 10^{-9} to 10^{-7} mol/insect dissolved in acetone and diluted in saline].

Stimulation of tethered insects with either simulated visual flow fields or pheromone puffs in the absence of visual stimuli elicited preflight warm-up (Fig. 9), fictive takeoff, and flight. Activating the light-emitting diode array in a fashion that simulated turning elicited postures and adjustments in motor outflow that we interpret as attempts to turn. Specifically, the head was rolled and the abdomen curled in the direction of visual flow, the wing on the inside of the intended turn was remoted during wing beating, and the opposite wing was promoted. In association with changes in the remotion and promotion of the wings, the phases of firing of the AxMs on the two sides of the moth assumed an asymmetrical relationship with respect to the left DLM (Fig. 10). Reversing the sign of image flow led to postural adjustments signifying an attempted turn to the opposite side and a shift in the firing phases of the two AxMs reciprocal to that previously observed. Steering postures and phase-shifted firing patterns were maintained for many tens of wing beat cycles under these (open-loop) conditions. Activation of corrective steering by visual stimulation in our studies and others (Dombrowski and Wendler, 1989) is a demonstration of the anemotactic steering proposed by Preiss and Kramer (1986a) under open-loop conditions.

We have also stimulated preparations with pulsed pheromones in the absence of visual stimuli under open-loop conditions. In several preparations, brief

FIGURE 9. (cont.) = 1 cycle) revealed by motor bursts in the L DLM during warm-up elicited by moving stripes and wind stimulus without pheromone blend. Solid lines beneath plot indicate direction of visual pattern movement. Solid line: down = right to left; up = left to right. Wind was continuous. (C) Plot of wing beat frequency during warm-up activated by puffs of pheromone blend in the continuous air stream (*solid lines beneath plot*) without visual stimulation.

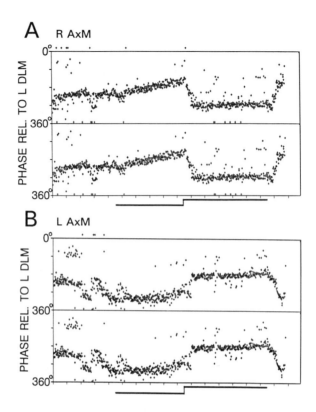

FIGURE 10. Double phase plots of R third AxM activity onset relative to L DLM activity (A) and of the L third AxM activity relative to L DLM activity (B) during spontaneous tethered flight (*each point = 1 wing beat cycle*), attempted left turning (*visual pattern moving right to left—solid line down*), and attempted right turning (*visual pattern moving left to right—solid line up*). Spontaneous tethered flight was activated by injection of 10^{-8} moles of chlordimeform (100-μl bolus in saline).

pheromonal stimulation followed by continuous clean air flow initiated motor activity that followed the sequence preflight warm-up, fictive takeoff, and flight. When pulsed pheromones were injected into the air flowing over moths in fictive flight, changes in motor outflow were observed that took the form of intermittent double bursts of AxMs per wing beat cycle and rhythmic phase shifts relative to DLM firing that were not phase locked to the pulsatile pheromone stimuli. We are continuing to study these adjustments in motor activity occurring after pheromonal stimulation in the absence of visual feedback to determine whether they can reasonably be interpreted as attempts by the moth to turn. If they are, they would be a demonstration of an internal turning tendency activated by pheromones.

V. Concluding remarks.

It is clear that we have only begun to understand many of the details of how moths achieve successful, goal-directed locomotion by orienting in pheromone plumes. Yet certain components of the behavior seem to be fairly clear. Multimodal sensory information about movement (visual, perhaps mechanosensory) and odor quantity, quality, and the temporal structure of the plume (olfactory) is integrated with patterned, self-generated motor outflow (wing beating for lift and thrust, counterturning) to result in coordinated upwind flight to the target. Similarly, we have only begun to elucidate the neural implementation of the behavior, but recent progress in recording motor outflow during free and tethered flight and the ability to record fictive motor outflow from reduced preparations (Kanzaki and Arbas, 1990) provide the opportunity to characterize neural mechanisms and to assess their function in a behaviorally relevant context.

Acknowledgments

We thank Drs. T. C. Baker, J. H. Belanger, R. F. Chapman, T. A. Christensen, and R. M. Olberg for helpful comments on the manuscript and the Office of Naval Research for bringing together the participants who have made this book possible. Our work was supported by NIH grants DC-00348 and NS-07309.

References

ARBAS, E. A., (1986). "Control of Hindlimb Posture by Wind-Sensitive Hairs and Antennae during Locust Flight," *J. Comp. Physiol. (A)* **159,** 849–857.

ARBAS, E. A., and HILDEBRAND, J. G. (1986). "Wind-Sensitive Head Hairs: A Mechanosensory Pathway Activating Flight in the Hawkmoth *Manduca sexta,*" *Soc. Neurosci. Abstr.* **12,** 857.

ATEMA, J., (1987). "Distribution of Chemical Stimuli," Atema, J., Fay, R. R., Popper, A. N., and Tavolga, W. N. (eds.), "Sensory *Biology of Aquatic Animals"* (pp. 29–56). Springer-Verlag, New York.

BAADER, A., (1991). "Simulation of Self-Motion in Tethered Flying Insects: An Optical Flow Field for Locusts," *J. Neurosci. Methods* **38,** 193–199.

BAKER, P. S., GEWECKE, M., and COOTER, R. J. (1981). "The Natural Flight of the Migratory Locust, *Locusta migratoria* L. III. Wing-Beat Frequency, Flight Speed and Attitude," *J. Comp. Physiol. (A)* **141,** 233–237.

BAKER, T. C. (1985). "Chemical Control of Behavior," Kerkut, G. A., and Gilbert, L. I (eds.) *Comprehensive Insect Physiology, Biochemistry and Pharmacology.* (pp. 621–672). Pergamon Press, Oxford.

BAKER, T. C. (1990). "Upwind Flight and Casting Flight: Complementary Phasic and Tonic Systems Used for Location of Sex Pheromone Sources by Male Moths," Doving, K. B. (ed.), *Proceedings of the 10th International Symposium on Olfaction and Taste* (pp. 18–25). Oslo.

BAKER, T. C., and CARDÉ, R. T. (1979). "Analysis of Pheromone-Mediated Behaviors in Male *Grapholitha molesta,* the Oriental Fruit Moth (Lepidoptera: Tortricidae)," *Environ. Entomol.* **8,** 956–968.

BAKER, T. C., and HAYNES, K. F. (1987). "Manoeuvers Used by Flying Male Oriental Fruit Moths to Relocate a Sex Pheromone Plume in an Experimentally Shifted Wind-Field," *Physiol. Entomol.* **12,** 263–279.

BAKER, T. C., and VOGT, R. G. (1988). "Measured Behavioural Latency in Response to Sex-Pheromone Loss in the Large Silk Moth *Antheraea polyphemus,*" *J. Exp. Biol.* **137,** 29–38.

BAKER, T. C., CARDÉ, R. T., and ROELOFS, W. L. (1976). "Behavioral Responses of Male *Argyrotaenia velutinana* (Lepidoptera: Tortricidae) to Components of Its Sex Pheromone," *J. Chem. Ecol.* **2,** 333–352.

BAKER, T. C., WILLIS, M. A., and PHELAN, P. L. (1984). "Optomotor Anemotaxis Polarizes Self-Steered Zigzagging in Flying Moths," *Physiol. Entomol.* **9,** 365–376.

BAKER, T. C., HANSSON, B. S., LÖFSTEDT, C., and LÖFQVIST, J. (1989). "Adaptation of Antennal Neurons in Moths Is Associated with Cessation of Pheromone-Mediated Upwind Flight," *Proc. Natl. Acad. Sci. USA* **85,** 9826–9830.

BAKER, T. C., WILLIS, M. A., HAYNES, K. F., and PHELAN, P. L. (1985). "A Pulsed Cloud of Sex Pheromone Elicits Upwind Flight in Male Moths," *Physiol. Entomol.* **10,** 257–265.

BANG, B. G., and WENZEL, B. M. (1985). "Nasal Cavity and Olfactory System. in King, A. S., and McLelland, J. (eds.), *Form and Function in Birds* Vol. 3 (pp. 195–225). Academic Press, London.

BEER, R. D. (1990). *Intelligence as Adaptive Behavior.* Academic Press, San Diego.

BEER, R. D., and GALLAGHER, J. C. (1992). "Evolving Dynamical Neural Networks for Adaptive Behavior," *Adaptive Behavior* **1,** 91–122.

BÖHM, L. K. (1911). "Die antennalen sinnesorgane der Lepidopteren," *Arb. Zool. Inst. Wien* **14,** 219–246.

BOSSERT, W. H., and WILSON, E. O. (1963). "The Analysis of Olfactory Communication among Animals," *J. Theor. Biol.* **5,** 443–469.

BRADSHAW, J. W. S. BAKER, R., and LISK, J. C. (1983). "Separate Orientation and Releasing Components in a Sex Pheromone,"*Nature* **304,** 265–267.

CAMAZINE, S., and HILDEBRAND, J. G. (1979). "Central Projections of Antennal Sensory Neurons in Mature and Developing *Manduca sexta,*" *Soc. Neurosci. Abstr.* **5,** 492.

CAMHI, J. M. (1970A). "Yaw-Correcting Postural Changes in Locusts," *J. Exp. Biol.* **52,** 519–531.

CAMHI, J. M. (1970B). "Sensory Control of Abdomen Posture in Flying Locusts," *J. Exp. Biol.* **52,** 533–537.

CARDÉ, R. T. (1984). "Chemo-orientation in Flying Insects," in Bell, W. J., and Cardé, R. T., (eds.), *Chemical Ecology of Insects* (pp. 111–126). Sinauer Associates.

CARDÉ, R. T., (1986). "Epilogue: Behavioral Mechanisms," in Payne, T. L., Birch, M. C. and Kennedy, C. E. J. (eds.), *Mechanisms in Insect Olfaction* (pp. 175–186). Clarendon Press, Oxford.

Chapter VIII Organization of Goal-Oriented Locomotion

CARDÉ, R. T., and BAKER, T. C. (1984). "Sexual Communication with Pheromones, in Bell, W. J., and Cardé, R. T. (eds.), *Chemical Ecology of Insects* (pp. 355–386). Sinauer Associates.

CARDÉ, R. T., and CHARLTON, R. (1985). "Olfactory Sexual Communication in Lepidoptera: Strategy, Sensitivity and Selectivity," *Symp. R. Entomol. Soc. Lond.* **12**, 241–264.

CARDÉ, R. T., and HAGAMAN, T. E. (1979). "Behavioral Responses of the Gypsy Moth in a Wind Tunnel to Airborne Enantiomers of Disparlure,"*Environ. Entomol.* **8**, 475–484.

CARDÉ, R. T., BAKER, T. C., and ROELOFS, W. L. "Ethological Function of Components of a Sex Attractant System for Oriental Fruit Moth Males, *Grapholitha molesta* (Lepidoptera: Tortricidae)," *J. Chem. Ecol.* **1**, 475–491.

CASSADAY, G. B., and CAMHI, J. M. (1976). "Metamorphosis of Flight Motor Neurons in the Moth, *Manduca sexta*," *J. Comp. Physiol. (A)* **112**, 143–158.

CHRISTENSEN, T. A., and HILDEBRAND, J. G. (1987a). "Male-Specific, Sex Pheromone-Selective Projection Neurons in the Antennal Lobes of the Moth *Manduca sexta*," *J. Comp. Physiol. (A)* **160**, 553–569.

CHRISTENSEN, T. A., and HILDEBRAND, J. G. (1987b). "Functions, Organization and Physiology of the Olfactory Pathways in the Lepidopteran Brain. in Gupta, A. P. (ed.), *Arthropod Brain: Its Evolution, Development, Structure and Functions.* (pp. 457–484). John Wiley, New York.

CHRISTENSEN, T. A., and HILDEBRAND, J. G. (1988)."Frequency Coding by Central Olfactory Neurons in the Sphinx Moth *Manduca sexta*," *Chem. Senses* **13**, 23–30.

CHRISTENSEN, T. A., and HILDEBRAND, J. G., (1990). "Representation of Sex-Pheromonal Information in the Insect Brain, in Døving, K. B. (ed.), *Proceedings of the 10th International Symposium on Olfaction and Taste. (pp. 142–150).* Oslo.

CHRISTENSEN, T. A., MUSTAPARTA, H., and HILDEBRAND, J. G. (1989). "Discrimination of Sex Pheromone Blends in the Olfactory System of the Moth," *Chem. Senses* **14**, 463–477.

CHRISTENSEN, T. A., MUSTAPARTA, H., and HILDEBRAND, J. G. (1991). "Chemical Communication in Heliothine Moths. II. Central Processing of Intra and Interspecific Olfactory Messages in the Male Corn Earworm Moth *Helicoverpa zea*," *J. Comp. Physiol.*, **169**, 259–274.

DAVID, C. T. (1986). "Mechanisms of Directional Flight in Wind, in Payne, T. L., Birch, M. C., and Kennedy, C. E. J. (eds.), *Mechanisms in Insect Olfaction* (pp. 49–57), Clarendon Press, Oxford.

DAVID, C. T., and KENNEDY, J. S. (1987). "The Steering of Zigzagging Flight by Male Gypsy Moths," *Naturwissenschaften.* **74**, 194–195.

DAVID, C. T., KENNEDY, J. S., and LUDLOW, A. R. (1983). "Finding of a Sex Pheromone Source by Gypsy Moths Released in the Field," *Nature* **303**, 804–806.

DAVID, C. T., KENNEDY, J. S., LUDLOW, A. R., PERRY, J. N., and WALL, C. (1982). "A Reappraisal of Insect Flight Towards a Distant, Point Source of Wind-Borne Odor," *J. Chem. Ecol.* **8**, 1207–1215.

DOMBROWSKI, U. (1991). "Untersuchungen zur funktionellen organisation des flugsystems von *Manduca sexta* (L.)," Doctoral dissertation, Universität zu Köln.

DOMBROWSKI, U., and WENDLER, G. (1989). "Dynamics of Visually Guided Flight Manoeuvers in *Manduca sexta*," in Elsner, N., and Singer, W. (eds.), *Dynamics and Plasticity in Neuronal Systems* (p. 47), *Proc. 17th Göttingen Neurobiology Conference.*

DØVING, K. B., WESTERBERG, H., and JOHNSEN, P. B."ROLE of Olfaction and Neuronal Responses of Atlantic Salmon, *Salmo salar*, to Hydrographic Stratification," *Can. J. Fish. Aquat. Sci.* **42**, 1658–1667.

DROST, Y. C., LEWIS, W. J., ZANEN, P. O., and KELLER, M. A. (1986). "Beneficial Arthropod Behavior Mediated by Air-Borne Semiochemicals. I. Flight Behavior and Influence of Pre-Flight Handling of *Microplitis croceipes* Cresson," *J. Chem. Ecol.* **12**, 1247–1262.

DUSENBERY, D. B., (1989a). "Ranging Strategies," *J. Theor. Biol.* **136**, 309–316.

DUSENBERY, D. B., (1989b). "Optimal Search Direction for an Animal Flying or Swimming in a Wind or Current," *J. Chem. Ecol.* **15**, 2511–2519.

DUSENBERY, D. B., (1990). "Upwind Searching for an Odor Plume is Sometimes Optimal," *J. Chem. Ecol.* **16**, 1971–1976.

ELKINTON, J. S., and CARDÉ, R. T. (1983). "Appetitive Flight Behavior of Male Gypsy Moths (Lepidoptera: *Lymantridae*)," *Environ. Entomol.* **12**, 1702–1707.

ELKINTON, J. S., CARDÉ, R. T., and MASON, C. J. (1984). "Evaluation of Time-Average Dispersion Models for Estimating Pheromone Concentration in a Deciduous Forest," *J. Chem. Ecol.* **10**, 1081–1108.

ELKINTON, J. S., SCHAL, C., ONO, T., and CARDÉ, R. T., (1987). "Pheromone Puff Trajectory and Upwind Flight of Gypsy Moths in a Forest,"*Physiol. Entomol.* **12**, 399–406.

EMANUEL, M. E., and DODSON, J. J. (1979). "Modification of the Rheotropic Behavior of the Male Rainbow Trout (*Salmo gairdneri*) by Ovarian Fluid," *J. Fish. Res. Board Can.* **36**, 63–68.

ERBER, J., HOMBERG, U., and GRONENBERG, W. (1987). "Functional Roles of the Mushroom Bodies in Insects," in Gupta, A. P. (ed.), *Arthropod Brain: Its Evolution, Development, Structure and Functions.* (pp. 485–511). Wiley, New York.

FARKAS, S. R., and SHOREY, H. H. (1974). "Mechanisms of Orientation to a Distant Pheromone Source," in Birch, M. C. (ed.), *Pheromones* (pp. 81-95). North-Holland, London.

GEWECKE, M. (1974). "The Antennae of Insects as Air-Current Sense Organs and Their Relationship to the Control of Flight," in Barton Browne, L. (ed.), *Experimental Analysis of Insect Behavior* (pp. 100–113) Springer-Verlag, Berlin.

GEWECKE, M. (1975). "The Influence of the Air Current Sense Organs on the Flight Behavior of *Locusta Migratoria*," *J. Comp. Physiol. (A)* **103**, 79–95.

GEWECKE, M., and NIEHAUS, M. (1981). "Flight and Flight Control by the Antennae in the Small Tortoiseshell (*Aglais urticae* L., Lepidoptera) I. Flight Balance experiments," *J. Comp. Physiol. (A)* **145**, 249–256.

GEWECKE, M., and WENDLER, G. (1985). *Insect Locomotion*. Paul Parey, Berlin.

GOLDSWORTHY, G. J., and WHEELER, C. H. (1989). *Insect Flight*. CRC Press, Boca Raton, FL.

GRUBB, T. C. (1972). "Smell and Foraging in Shearwaters and Petrels," *Nature,* **237**, 404–405.

GRUBB, T. C. (1974). "Olfactory Navigation to the Nesting Burrow in Leach's Petrel (*Oceanodroma leucorrhoa*)," *Anim. Behav.* **22**, 192–202.

GRUBB, T. C. (1979). "Olfactory Guidance of Leach's Storm Petrel to the Breeding Island," *Wilson Bull.* **91**, 141–143.

Chapter VIII   Organization of Goal-Oriented Locomotion 193

HANSSON, B. S., CHRISTENSEN, T. A., and HILDEBRAND, J. G., (1991). "Functionally Distinct Subdivisions of the Macroglomerular Complex in the Antennal Lobe of the Male Sphinx Moth *Manduca sexta*," *J. Comp. Neurol.* **312,** 264–278.

HASLER, A. D., and SCHOLZ, A. T. (1983). *Olfactory Imprinting and Homing in Salmon.* Spinger-Verlag, New York.

HAYNES, K. F., and BAKER, T. C. (1989). "An Analysis of Anemotactic Flight in Female Moths Stimulated by Host Odour and Comparison with the Males' Response to Sex Pheromone," *Physiol. Entomol.* **14,** 279–289.

HILDEBRAND, J. G., MATSUMOTO, S. G., CAMAZINE, S. M., TOLBERT, L. P., BLANK, S., FERGUSON, H., and ECKER, V. (1980). "Organization and Physiology of Antennal Centres in the Brain of the Moth *Manduca sexta*, in *Insect Neurobiology and Pesticide Action (Neurotox 79)* (pp. 375–382). Soc. Chem. Ind., London.

HILDEBRAND, J. G., CHRISTENSEN, T. A., ARBAS, E. A., HAYASHI, J. H., HOMBERG, U., KANZAKI, R., and STENGL, M. (1992). "Olfaction in *Manduca sexta*: Cellular Mechanisms of Responses to Sex Pheromone," in Duce, I. R. (ed.), *Neurotox '91 Molecular Basis of Drug and Pesticide Action.* Elsevier, London pp. 323–338.

HOMBERG, U., CHRISTENSEN, T. A., and HILDEBRAND, J. G. (1989). "Structure and Function of the Deutocerebrum in Insects," *Annu. Rev. Entomol.* **34,** 477–501.

HOMBERG, U., MONTAGUE, R. A., and HILDEBRAND, J. G. (1988). "Anatomy of Antenno-Cerebral Pathway in the Brain of the Sphinx Moth *Manduca sexta*," *Cell Tissue Res.* **254,** 255–281.

HOSKINS, S. G., HOMBERG, U., KINGAN, T. G., CHRISTENSEN, T. A., and HILDEBRAND, J. G. (1986). "Immunocytochemistry of GABA in the Antennal Lobes of the Sphinx Moth *Manduca sexta*," *Cell Tissue Res.* **244,** 243–252.

HUTCHISON, L. V., and WENZEL, B. M. (1980). "Olfactory Guidance in Procellariiforms," *Condor,* **82,** 314–319.

HUTCHISON, L. V., WENZEL, B. M., STAGER, K. E., and TEDFORD, B. L. (1982). "Further Evidence for Olfactory Foraging by Sooty Shearwaters and Northern Fulmars," in Nettleship, D. N., Sanger, G. A., and Springer, P. F. (eds.), *Marine Birds: Their Feeding Ecology and Commercial Fisheries Relationships* Proc. Pacific Seabird Group Symposium, (pp. 72–77), Seattle, WA.

JOHNSEN, P. B., (1981). "A Behavioral Control Model for Homestream Selection in Migratory Salmonids," in Brannon, E. L., and Salo, E. O. (eds.), Proc. of Salmon and Trout Migratory Behavior, Symposium (pp. 266–273).

JOHNSEN, P. B., and HASLER, A. D. (1980). "The Use of Chemical Cues in the Upstream Migration of Coho Salmon, *Oncorhynchus kisutch* Walbaum. *J. Fish. Biol.* **17,** 67–73.

JOHNSEN, P. B. and TEETER, J. H. (1985). "Behavioral Responses of Bonnethead Sharks (*Sphyrna tiburo*) to Controlled Olfactory Stimulation,"*Mar. Behav. Physiol.,* **11,** 283–291.

KAISSLING, K. E. (1986). "Temporal Characteristics of Pheromone Receptor Cell Responses in Relation to Orientation Behaviour of Moths," in Payne, T. L., Birch, M. C., and Kennedy, C. E. J. (eds.), *Mechanisms in Insect Olfaction* (pp.193–200). Clarendon Press, Oxford.

KAISSLING, K. E., and KRAMER, E. (1990). "Sensory Basis of Pheromone-Mediated Orientation in Moths," *Verh. Dtsch. Zool. Ges.* **83,** 109–131.

KAISSLING, K. E., HILDEBRAND, J. G., and TUMLINSON, J. H. (1989). "Pheromone Receptor Cells in the Male Moth *Manduca sexta*," *Arch. Insect Biochem. Physiol.* **10,** 273–279.

KAMMER A. E. (1971). "The Motor Output During Turning Flight in a Hawkmoth, *Manduca sexta*," *J. Insect Physiol.* **17**, 1073–1086.

KAMMER, A. E., and NACHTIGALL, W. (1973). "Changing Phase Relationships among Motor Units during Flight in a Saturniid Moth," *J. Comp. Physiol. (A)* **83**, 17–24.

KANZAKI, R. (1989). "Physiology and Morphology of Higher-Order Neurons in Olfactory Pathways of the Moth Brain: Pheromone-Processing Neurons in the Protocerebrum," in Erber, J., Menzel, R., Pflüger, H. J., and Todt, D. (eds.), *Neural Mechanisms of Behavior* (pp. 253–254). *Proc. 2nd Int. Congress of Neuroethology.* Georg Thieme Verlag, Stuttgart.

KANZAKI, R. and ARBAS, E. A. (1990). "Olfactory and Visual Influences on Fictive Flight in *Manduca sexta*," *Soc. Neurosci. Abstr.* **16**, 758.

KANZAKI, R. and SHIBUYA, T. (1986). "Descending Protocerebral Neurons Related to the Mating Dance of the Male Silkworm Moth," *Brain Res.* **377**, 378–382.

KANZAKI, R., ARBAS, E. A., and HILDEBRAND, J. G. (1991a). "Physiology and Morphology of Protocerebral Olfactory Neurons in the Male Moth *Manduca sexta*," *J. Comp. Physiol. (A)* **168**, 281–298.

KANZAKI, R., ARBAS, E. A., and HILDEBRAND, J. G. (1991b). "Physiology and Morphology of Descending Neurons in Pheromone-Processing Olfactory Pathways in the Male Moth *Manduca sexta*," *J. Comp. Physiol. (A)* **169**, 1–14.

KANZAKI, R., ARBAS, E. A., STRAUSFELD, N. J., and HILDEBRAND, J. G. (1989). "Physiology and Morphology of Projection Neurons in the Antennal Lobe of the Male Moth *Manduca sexta*," *J. Comp. Physiol. (A)* **165**, 427–453.

KEIL, T. A., (1989). "Fine Structure of the Pheromone-Sensitive Sensilla on the Antenna of the Hawkmoth, *Manduca sexta*," *Tissue & Cell* **21**, 139–151.

KENNEDY, J. S. (1939). The Visual Responses of Flying Mosquitoes," *Proc. Zool. Soc. Lond.* **109**, 221–242.

KENNEDY, J. S. (1977). "Olfactory Responses to Distant Plants and Other Odor Sources," in Shorey, H. H., and McKelvey, J. J. (eds.), *Chemical Control of Insect Behavior* (pp. 67–91). Wiley-Interscience, New York.

KENNEDY, J. S. (1983). "Zigzagging and Casting as a Programmed Response to Windborne Odour: A Review," *Physiol. Entomol.* **8**, 109–120.

KENNEDY, J. S. (1986). "Some Current Issues in Orientation to Odour Sources. in Payne, T. L., Birch, M. C., and Kennedy, C. E. J. (eds.), *Mechanisms in Insect Olfaction* (pp. 11–25, Clarendon Press, Oxford.

KENNEDY, J. S., and MARSH, D. (1974). "Pheromone-Regulated Anemotaxis in Flying Moths," *Science* **184**, 999–1001.

KENNEDY, J. S., LUDLOW, A. R., and SANDERS, A. R. (1980). "Guidance System Used in Moth Sex Attraction," *Nature* **288**, 474–477.

KENNEDY, J. S., LUDLOW, A. R., and SANDERS, A. R. (1981). "Guidance of Flying Male Moths by Wind-borne Sex Pheromone," *Physiol. Entomol.* **6**, 395–412.

KLAASSEN, L. W., and KAMMER, A. E. (1985). "Octopamine Enhances Neuromuscular Transmission in Developing and Adult Moths, *Manduca sexta*," *J. Neurobiol.* **16**, 227–243.

KONDOH, Y., and OBARA, Y. (1982). "Anatomy of Motoneurons Innervating Mesothoracic Indirect Flight Muscles in the Silkmoth *Bombyx mori*," *J. Exp. Biol.* **98**, 23–37.

Chapter VIII Organization of Goal-Oriented Locomotion

KUENEN, L. P. S. and BAKER, T. C. (1982). "The Effects of Pheromone Concentration on the Flight Behavior of the Oriental Fruit Moth, *Grapholitha molesta*," *Physiol. Entomol.* **7**, 423–434.

LEE, J. K., and STRAUSFELD, N. J. (1990). "Structure, Distribution and Number of Surface Sensilla and Their Receptor Cells on the Olfactory Appendage of the Male Moth *Manduca sexta*," *J. Neurocytol.* **19**, 519–538.

LEHMAN, H. K. (1990). "Circadian Control of *Manduca Sexta* Flight," *Soc. Neurosci. Abstr.* **16**, 1334.

LINN, C. E., and ROELOFS, W. L. (1989). Response Specificity of Male Moths to Multicomponent Pheromones," *Chem. Senses* **14**, 421–437.

LINN, C. E., CAMPBELL, M. G., and ROELOFS, W. L. (1986). "Male Moth Sensitivity to Multicomponent Pheromones: Critical Role of Female-Released Blend in Determining the Functional Role of Components and Active Space of the Pheromone," *J. Chem. Ecol.* **12**, 659–68.

LINN, C. E., CAMPBELL, M. G., and ROELOFS, W. L. (1987). "Pheromone Components and Active Spaces: What Do Moths Smell and Where Do They smell it?" *Science* **237**, 650–652.

LUDLOW, A. R. (1984). "Application of Computer Modelling to Behavioral Coordination," Ph.d. thesis, University of London.

LYNCH, G., and GRANGER, R. (1990). "Olfactory Rhythms and the Encoding of Memory," in Døving, K. B. (ed.), *Proc 10th International Symposium on Olfaction and Taste,* (pp. 246–264). Oslo.

MARSH, D., KENNEDY, J. S., and LUDLOW, A. R. (1978). "An Analysis of Anemotactic Zigzagging Flight in Male Moths Stimulated by Pheromone," *Physiol. Entomol.* **3**, 221–240.

MATHEWSON, R. F., and HODGSON, E. S., (1972). "Klinotaxis and Rheotaxis in Orientation of Sharks Toward Chemical Stimuli," *Comp. Biochem. Physiol.* **42A**, 79–84.

MATSUMOTO, S. G., and HILDEBRAND, J. G. (1981). "Olfactory Mechanisms in the Moth *Manduca sexta:* Response Characteristics and Morphology of Central Neurons in the Antennal Lobes," *Proc. R. Soc. Lond. (B)* **213**, 249–277.

MÖHL, B. (1989). "Sense Organs and the Control of Flight. in Goldsworthy, G. J., and Wheeler, C. H. (eds.) *Insect Flight* (pp. 75–92) CRC Press. Boca Raton, FL.

MOORE, P. A., and ATEMA, J. (1991). "Spatial Information in the Three-Dimensional Fine Structure of an Aquatic Odor Plume," *Biol. Bull.* **181**, 408–418.

MURLIS, J., and JONES, C. D., (1981). "Fine-Scale Structure of Odour Plumes in Relation to Insect Orientation to Distant Pheromone and Other Attractant Sources," *Physiol. Entomol.* **6**, 71–86.

MURLIS, J., BETTANY, B. W., KELLEY, J. and MARTIN, L. (1982). "The Analysis of Flight Paths of Male Egyptian Cotton Leafworm Moths, *Spodoptera littoralis,* to a Sex Pheromone Source in the Field," *Physiol. Entomol.* **7**, 435–441.

MURLIS, J., ELKINTON, J. S. and CARDÉ, R. T., (1992). "Odor Plumes and How Insects Use Them," *Annu. Rev. Entomol.* **37**, 505–532.

MURLIS, J., WILLIS, M. A., and CARDÉ, R. T., (1990). "Odour Signals: Patterns in Time and Space," in Døving, K. B. (ed.), *Proc. 10th International Symposium on Olfaction and Taste* (pp. 6–17). Oslo University.

NACHTIGALL, W. (1989). "Mechanics and Aerodynamics of Flight," in Goldsworthy, G. J., and Wheeler, C. H. (eds.), *Insect Flight* (pp. 1–29). CRC Press, Boca Raton, FL.

NAKAMURA, K., and KAWASAKI, F. (1977). "The Active Space of *Spodoptera litura* (F.) Sex Pheromone and the Pheromone Component Determining this Space," *Appl. Entomol. Zool.* **12,** 162–77.

NIEHAUS, M. (1981). "Flight and Flight Control by the Antennae in the Small Tortoiseshell (*Aglais urticae* L., Lepidoptera). II. Flight Mill and Free Flight Experiments," *J. Comp. Physiol. (A)* **145,** 257–264.

NIEHAUS, M. and GEWECKE, M. (1978). "The Antennal Movement Apparatus in the Small Tortoiseshell (*Aglais urticae* L., Insecta, Lepidoptera)," *Zoomorphology* **91,** 19–36.

OLBERG, R. M. (1983). "Pheromone-Triggered Flip-Flopping Interneurons in the Ventral Nerve Cord of the Silkworm Moth, *Bombyx mori*," *J. Comp. Physiol. (A)* **152,** 297–307.

OLBERG, R. M., and WILLIS, M. A. (1990). "Pheromone-Modulated Optomotor Response in Male Gypsy Moths, *Lymantria dispar* L.: Directionally Selective Visual Interneurons in the Ventral Nerve Cord," *J. Comp. Physiol. (A)* **167,** 707–714.

ORONA, E. and AGEE, H. (1987). "Thoracic Mechanoreceptors in the Wing Bases of *Heliothis zea* (Lepidoptera: noctuidae) and Their Central Projections," *J. Insect Physiol.* **33,** 713–721.

PAWSON, M. G. (1977). "The Responses of Cod *Gadus morhua* (L) to Chemical Attractants in Moving Water," *J. Int. Explor. Mer.* **37,** 316–318.

PAYNE, T. L., BIRCH, M. C. and KENNEDY, C. E. J. (1986). *Mechanisms in Insect Olfaction.* Clarendon Press, Oxford.

PREISS, R. (1991). "Separation of Translation and Rotation by Means of Eye-Region Specialization in Flying Gypsy Moths (Lepidoptera: *Lymantriidae)*," *J. Insect Behav.* **4,** 209–219.

PREISS, R., and FUTSCHEK, L. (1985). "Flight Stabilization by Pheromone-Enhanced Optomotor Responses" *Naturwissenschaften,* **72,** 435–436.

PREISS, R., and KRAMER, E. (1983). "Stabilization of Altitude and Speed in Tethered Flying Gypsy Moth Males: Influence of (+) and (-)-Disparlure," *Physiol. Entomol.* **8,** 55–68.

PREISS, R. and KRAMER, E., (1986a). "Mechanism of Pheromone Orientation in Flying Moths," *Naturwissenschaften* **73,** 555–557.

PREISS, R., and KRAMER, E., (1986b). "Pheromone-Induced Anemotaxis in Simulated Free Flight," in Payne, T. L., Birch, M. C., and Kennedy, C. E. J. (eds.), *Mechanisms in Insect Olfaction* (pp. 69-79). Clarendon Press, Oxford.

RAINEY, R. C. (1985). "Insect Flight: New Facts—and Old Fantasies?" in Gewecke, M. and Wendler, G. (eds.) *Insect Locomotion* (pp. 241–244). Paul Parey, Berlin.

RHEUBEN, M. B. (1985). "Quantitative Comparisons of the Structural Features of Slow and Fast Neuromuscular Junctions in *Manduca*," J. Neurosci. **5,** *1704–1716.*

RHEUBEN, M. B., and KAMMER, A. E. "Structure and Innervation of the Third Axillary Muscle of *Manduca* Relative to Its Role in Turning Flight," *J. Exp. Biol.* **131,** 373–402.

RIND, C. F. (1983). "The Organization of Flight Motoneurones in the Moth, *Manduca sexta,*" *J. Exp. Biol.* **102,** 239–251.

ROELOFS, W. L., and CARDÉ, R. T. (1974). "Sex Pheromones in the Reproductive Isolation of Lepidopterous Species," in Birch, M. C. (ed.), *Pheromones* (pp. 96–114). North-Holland, London.

ROWELL, C. H. F. (1988). "Mechanisms of Flight Steering in Locusts," *Experientia* **44,** 389–395.

Chapter VIII   Organization of Goal-Oriented Locomotion   197

Rumbo, E. R., and Kaissling, K. E. (1989). "Temporal Resolution of Odour Pulses by Three Types of Pheromone Receptor Cells in *Antheraea polyphemus*," *J. Comp. Physiol. (A)* **165**, 281–291.

Sabelis, M. W., and Schippers, P. (1984). "Variable Wind Directions and Anemotactic Strategies of Searching for an Odour Plume," *Oecologia.* **63**, 225–228.

Sanders, C. J. (1985). "Flight Speed of Male Spruce Budworm Moths in a Wind Tunnel at Different Wind Speeds and at Different Distances from a Pheromone Source," *Physiol. Entomol.* **10**, 83–88.

Sanes, J. R., and Hildebrand, J. G. (1976a). "Structure and Development of Antennae in a Moth, *Manduca sexta*," *Dev. Biol.* **51**, 282–299.

Sanes, J. R., and Hildebrand, J. G. (1976b). "Origin and Morphogenesis of Sensory Neurons in an Insect Antenna," *Dev. Biol.* **51**, 300–319.

Sasaki, M., and Riddiford, L. M. (1984). "Regulation of Reproductive Behaviour and Egg Maturation in the Tobacco Hawk Moth, *Manduca sexta*," *Physiol. Entomol.* **9**, 315–327.

Schmitt, B. C., and Ache, B. W. (1979). "Olfaction: Responses of a Decapod Crustacean Are Enhanced by Flicking," *Science* **205**, 204–206.

Schneiderman, A. M., Hildebrand, J. G., Brennan, M. M. and Tumlinson, J. H. (1984). "Trans-sexually Grafted Antennae Alter Pheromone-Directed Behavior in a Moth," *Nature* **323**, 801–803.

Schöne, H. (1984). *Spatial Orientation.* Princeton University Press, Princeton, NJ.

Schweitzer, E. S., Sanes, J. R., and Hildebrand, J. G. (1976). "Ontogeny of Electroantennogram Responses in the Moth, *Manduca sexta*," *J. Insect Physiol.* **22**, 955–960.

Stager, K. E. (1964). "The role of Olfaction in Food Location by the Turkey Vulture (*Cathartes aura*)," *Los Angeles County Museum, Contrib. Sci.* **81**, 3–63.

Stager, K. E. (1967). "Avian Olfaction," *Am. Zool.* **7**, 415–419.

Starratt, A. M., Dahm, K. H., Allen, N., Hildebrand, J. G., Payne, T. L., and Röller, H. , (1978). "Bombykal, a Sex Pheromone of the Sphinx Moth, *Manduca sexta*," *Z. Naturforsch.* **34C**, 9–12.

Stengl, M., Homberg, U., and Hildebrand, J. G. (1990). "Acetylcholinesterase Activity in Antennal Receptor Neurons of the Sphinx Moth *Manduca sexta*," *Cell Tissue Res.* **262**, 245–252.

Sutton, O. G. (1953). *Micrometeorology,* Mcgraw-Hill, New York.

Tsujimura, H. (1989). "Metamorphosis of Wing Motor System in the Silk Moth *Bombyx mori*: Origin of Wing Motor Neurons," *Devel. Growth Differ.* **31**, 331–339.

Tumlinson, J. H., Brennan, M. M., Doolittle, R. E., Mitchell, E. R., Brabham, A., Mazomenos, B. E., Baumhover, A. H., and Jackson, D. M. (1989). "Identification of a Pheromone Blend Attractive to *Manduca sexta* (L.) Males in a Wind Tunnel," *Arch. Insect Biochem. Physiol.* **10**, 255–271.

Vande Berg, J. (1971). "Fine Structural Studies of Johnston's Organ in the Tobacco Hornworm Moth, *Manduca sexta* (Johannson)," *J. Morphol.* **133**, 439–456.

Weis-Fogh, T. (1949). "An Aerodynamic Sense Organ Stimulating and Regulating Flight in Locusts," *Nature* **163**, 873–874.

Wendler, G. (1983). "The Interaction of Peripheral and Central Components in Insect Locomotion, in Huber, F., and Markl, H. (eds.), *Neuroethology and Behavioral Physiology* (pp. 42-53). Springer-Verlag, Berlin.

WESTERBERG, H. (1990). "Properties of Aquatic Odour Trails. Oslo, Døving, K. B. (ed.), *Proceedings of the 10th International Symposium on Olfaction and Taste* (pp. 45-54). Oslo University, Oslo.

WILLIS, M. A., and ARBAS, E. A. (1991a). "Odor-Modulated Upwind Flight of the Sphinx Moth, *Manduca sexta* L.," *J. Comp. Physiol. (A)* **169**, 427–440.

WILLIS, M. A., and ARBAS, E. A. (1991b). "Flight Muscle Activity Underlying Pheromone-Modulated Zigzagging Flight in Male Moths, *Manduca sexta*," *Soc. Neurosci. Abstr.* **17**, 1245.

WILLIS, M. A., and BAKER, T. C. (1984). "Effects of Intermittent and Continuous Pheromone Stimulation on the Flight Behavior of the Oriental Fruit Moth, *Grapholita molesta*," *Physiol. Entomol.* **9**, 341–358.

WILLIS, M. A., and BAKER, T. C. (1988). "Effects of Varying Sex Pheromone Component Ratios on the Zigzagging Flight Movements of the Oriental Fruit Moth, *Grapholita molesta*," *J. Insect Behav.* **1**, 357–371.

WILLIS, M. A., and CARDÉ, R. T. (1990). "Pheromone-Modulated Optomotor Response in Male Gypsy Moths, *Lymantria dispar* L.: Upwind Flight in a Pheromone Plume in Different Wind Velocities," *J. Comp. Physiol. (A)* **167**, 699–706.

WILLIS, M. A., MURLIS, J. and CARDÉ, R. T. (1991). "Pheromone-Mediated Upwind Flight of Male Gypsy Moths, *Lymantria dispar,* in a Forest," *Physiol. Entomol.* **16**, 507–521.

WILSON, D. M. (1968). "Inherent Asymmetry and Reflex Modulation of the Locust Flight Motor Pattern," *J. Exp. Biol.* **48**, 631–641.

WRIGHT, R. H. (1958). "The Olfactory Guidance of Flying Insects," *Can. Entomol.* **90**, 81–89.

ZANEN, P. O., LEWIS, W. J., CARDÉ, R. T., and MULLINIX, B. G. (1989). "Beneficial Arthropod Behavior Mediated by Airborne Semiochemicals VI. Flight Response of Female *Microplitis croceipes* (Cresson), a Braconid Endoparasitoid of *Heliothis* spp., to Varying Olfactory Stimulus Conditions Created with a Turbulent Jet," *J. Chem. Ecol.* **15**, 141–168.

Chapter IX   A New Role for the Insect
                                    Mushroom Bodies:
                                   Place Memory and
                                       Motor Control

MAKOTO MIZUNAMI, JOSETTE M. WEIBRECHT,
AND NICHOLAS J. STRAUSFELD
Arizona Research Laboratories Division of Neurobiology,
The University of Arizona,
Tucson, Arizona

Present address: Makoto Mizunami, Department of Biology, Faculty of Science, Kyushu University, Fukuoka 812, Japan.

I. Introduction

This chapter summarizes the results of behavioral and electrophysiological experiments that suggest a novel role for a pair of higher centers in the insect brain known as the mushroom bodies or corpora pedunculata. We have investigated the participation of the mushroom bodies in tasks involving place recognition and memory and have identified mushroom body activity before and during specific locomotory activity. The results suggest that in cockroaches mushroom bodies are higher centers that, among other functions, participate in place memory and in the planning and execution of appropriate locomotory actions. Preliminary results of this study (Weibrecht *et al.*, 1991; Mizunami and Strausfeld, 1991) have been extended and the results will be published in full elsewhere (Mizunami *et al.*, and Mizunami and Strausfeld, in preparation).

The nervous systems of insects have often been cited as examples of simple neural networks from which can be determined the organization of circuits that provide the relationship between a predictable behavior and the sensory stimuli evoking it. This is exemplified by a specific line of research on the insect visual system that abstracts a complex system comprising many thousands of neurons to its simplest form, in which two interdependent networks, one detecting panoramic motion, the other detecting differences of velocity between objects and their background, are claimed to be sufficient to explain a wide range of visually induced behaviors (Reichardt et al., 1983; Egelhaaf et al., 1988).

This useful but reductionist approach to neural integration contrasts with other studies that try to integrate physiological and behavioral studies with the abundant evidence that insect central nervous systems are numerically and structurally complex. At their most elaborate, many hundreds of thousand of neurons contribute to circuits that are as sophisticated as those found in mammals. Examples are the organization of sensorimotor circuits governing posture (e.g., Burrows, 1987; Laurent, 1987), the organization of olfactory neurons in antennal lobe glomeruli (Homberg et al., 1989a,b; Stocker et al., 1990), the parallel organization of color sensitive and color insensitive pathways in the visual system (Strausfeld and Lee, 1991), and, in the midbrain, the cellular intricacy of higher centers such as the central body (Hanesch et al., 1989) and paired mushroom bodies (Mobbs, 1982, 1984). Each of these examples provides the experimenter with a system whose challenging and intricate networks are uniquely accessible for investigating the relationships between complex behavior and circuits comprising real (rather than hypothetical) neurons.

A major focus in insect neurobiology has been that of motor control: understanding the significance of connections within an ensemble of uniquely identified neurons regulating a highly stereotypic motor output. Two examples are the control and modulation of the locust flight generator (Rowell, 1988, 1989) and escape behavior, such as midleg extension in *Drosophila* and in locusts (Tanouye and Wyman, 1980; Bacon and Strausfeld, 1986; Pearson, 1983). However, in addition to such predictable (and quantifiable) actions, insects express highly adaptive behaviors. These may involve learning by the individual, sometimes after a single trial (Balderrama, 1980), or the cooperative phenomenon of social interactions employing chemical and tactile communication in response to the perception of specific environmental events. Two examples are (1) the interaction between the queen and her subjects within an ant nest by means of chemical cues and (2) communication between individual worker honeybees about distance and direction of a food source, which is mediated by a complex motor routine (the waggle dance) encoding past events relating to a directional vector and time (see Hölldöbbler and Wilson, 1990; von Frisch, 1967).

An early attempt to match specific features of the insect brain to social behavior was that of Dujardin in 1850. He suggested that because of their prominence in social insects, a pair of mushroom-shaped brain centers (corpora pedunculata) were likely to be the seat of an insect's "intellect," society itself being then considered rooted in the capacity for intelligent thought. The greater number of neurons contributing to the mushroom body calyces in worker honeybees than in drones (Witthöft, 1967) or different elaborations of calyces of different species of ants

Chapter IX Place Memory and Motor Control

(Pandazis, 1930) provide some support for Dujardin's contention. So also do observations such as those by Alloway (1972) suggesting that behavior learned by the larva and carried through to the imago must be mediated by a brain region, such as the mushroom bodies, that persists through metamorphosis.

Von Frisch's (1967) classic studies on bee learning and memory, and the numerous research groups spawned by this field of research, encouraged neuroanatomical studies focused on hymenopteran brains (Vowles, 1955; Goll, 1967; Mobbs, 1982, 1984). Generally, these emphasize that the largest and most elaborate mushroom bodies occur in species that show the highest degree of social organization. However, as pointed out by Howse (1974), in the Hymenoptera it is the calyces that are especially complex. In many nonsocial or quasi-social insects, such as orthopteroids, the mushroom bodies can occupy as substantial a volume of the brain, having less elaborate calyces but larger pedunculi and lobes. For example, in adult *Periplaneta americana,* there are 403.6K intrinsic neurons (Kenyon cells) in the mushroom bodies (Neder, 1959) compared to the 340K in worker honeybees (Witthöft, 1967). A study on the elaboration of honeybee calyces (Buitkamp-Möbius, 1975) beautifully demonstrated clear dimorphisms between workers and drones and demonstrated from studies of gynandromorphs that both mushroom bodies must have worker-type morphology for the expression of worker-type behavior.

Attempts to demonstrate the role of the mushroom bodies have employed a wide variety of techniques. These range from extirpation (van du Kloot and Williams, 1953; Howse, 1974), electrical (Huber, 1960) and chemical (Mercer and Menzel, 1982; Menzel *et al.,* 1988) stimulation, through immunocytochemistry of the putative transmitters (Schäfer and Bicker, 1986; Schäfer and Rehder, 1989) and neuropeptides (Schäfer *et al.,* 1988; Schürmann and Erber, 1990), to structural and molecular genetics (Heisenberg *et al.,* 1985; Nighorn *et al.,* 1991).

The first technically sophisticated experiments demonstrating the possible roles of higher brain centers in controlling sequences of motor output were by Huber (1959, 1960), who electrically stimulated the mushroom bodies and their surrounding neuropil to elicit extended episodes of courtship behavior in field crickets. These experiments were later repeated using focal stimulation through glass electrodes, followed by attempts to identify stimulated neuropils using cobalt iontophoresis (Wahdepuhl, 1983). Although the marked areas were usually quite large, the results demonstrated the accuracy with which the stimulating electrode could be placed. The importance of mushroom bodies in courtship behavior has also been ascertained genetically, using histochemical identification of male tissue in gynandromorph mosaics in which neuropils associated with the mushroom bodies were required to be male for the individual to demonstrate male behavior (Hall, 1979).

The role of the mushroom bodies in associative learning has been tested experimentally using the stereotypic proboscis extension reflex first described by

Kuwabara in 1957. In honeybees, proboscis extension in response to gustatory stimuli can easily be conditioned using a specific odor or color (Menzel et al., 1974). To test which neuropils were involved in retaining this association, small cryoprobes were applied to the antennal lobes or mushroom body α-lobes at different times after the conditioning stimulus (Erber et al., 1980). Cooling the antennal lobes and both α-lobes 1 min after antennal stimulation blocked memory formation, whereas cooling after 6 min did not. This suggests a role for the α-lobes in short-term memory acquisition. Erber et al. (1980) demonstrated that cooling the antennal lobes up to 3 min after stimulation also blocked information storage, suggesting that the antennal lobes may also be involved in consolidating stored information (Menzel et al., 1974).

The possible role of the mushroom body in learning and memory is even more strongly suggested by studies on *Drosophila*. In wild-type *Drosophila*, the number of axons intrinsic to the mushroom bodies (Kenyon cells) depends on both age and behavioral experience (Technau, 1984). Behavioral studies (Dudai et al., 1976; Byers, 1980) have isolated a number of mutants that are defective in associative learning. Certain mutations that give rise to structural defects of the mushroom bodies (Heisenberg et al., 1985) also involve defects in olfactory conditioning. The best-known behavioral mutant is *dunce (dnc)*, which is deficient in conditioning (Dudai et al., 1976; Tully and Quinn, 1985), habituation (Duerr and Quinn, 1982), and learning-dependent courtship (Hall, 1986). All these defects are the result of lesions in a gene encoding cAMP-specific phosphodiesterase (PDE) (Davis and Kauvar, 1984). Antibodies raised against *dnc* PDE reveal intense staining in wild-type *Drosophila* specifically within the lobes and calyces of the mushroom bodies (Nighorn et al., 1991). *In situ* hybridization using *dnc* RNA demonstrates intense enrichment in the cell bodies of neurons intrinsic to the mushroom bodies (Kenyon cells).

Although the studies reviewed above provide compelling evidence that the mushroom bodies are involved in learning and memory, their specific roles in controlling behavior are still conjectural. In this chapter, we attempt to elucidate one of these roles from experiments that demonstrate that bilateral microlesions in the pedunculi of mushroom bodies of cockroaches (*P. americana*) abolish the ability to acquire spatial memory but do not disrupt visually directed orientation behavior. We also report experiments that demonstrate neural activity in and connections from areas in which lesions cause behavioral defects. Recordings from freely moving cockroaches show specific neural activity preceding and then accompanying specific motor actions. We demonstrate that copper released from the recording electrode is taken up by efferent neurons at the recording site and is transported to their terminals. Many of these are situated in neuropils giving rise to descending premotor neurons that reach the thoracic ganglia and supply motor circuits governing flight and walking (Milde and Strausfeld, 1990; Gronenberg and Strausfeld,

1990). We conclude with the hypothesis that one function of the mushroom bodies is to acquire place memory and to mediate appropriate motor programs.

II. Methods

A. *Microablations*

Lesions were made by implanting microscopic aluminum foil blades into the brain. Blades approximately 50–100 μm wide, 100–250 μm long, and 15–20 μm thick were fashioned under the dissection microscope. Cockroaches were anesthetized by cooling. Each individual was mounted in a special plastic holder surrounded by packed ice to effectively stop its heart during dissection so as to minimize the loss of hemolymph. A small flap of the frontal head capsule was cut and folded back to expose the dorsal brain surface. Care was taken not to disturb the tracheae except at the site of implant. Minute uni- or bilateral incisions were made on the brain's surface with a sharpened Minuten pin. This permitted the insertion of a blade, holding one end of it with watchmaker's forceps. The flap was then closed and sealed with bee's wax. Operations targeted the pedunculata/β-lobe junction, the β-lobes, or the α-lobes. Lesions in areas superficial to the lobes or lateral to the calyces served as control operations. Animals fully recovered a day or two after surgery and their spontaneous behavior was indistinguishable from that of unoperated animals.

B. *Behavioral Tests*

1. Hidden Targets. For reasons that will become clear, we have termed the arrangement used to test spatial learning the "Tennessee Williams" paradigm. The paradigm was designed to imitate as closely as possible the swimming-rat paradigm used by researchers testing spatial memory in rats. The apparatus consists of an arena 30 cm in diameter. Its floor was of uniform texture and a nonreflecting black. It was maintained noxiously hot, at 50–55°C except at a single location. This was a circular area (6 cm in diameter; referred to as the goal or "target"), which was maintained at a temperature of 20–22°C. The target was visibly and texturally indistinguishable from the surrounding floor. The wall of the arena was colored gray. In earlier experiments the wall was maintained at room temperature, but because the cockroaches tended to run along the walls during the first few trials, in later experiments the wall was also heated. The wall was decorated with four visual cues, which had a specific geometric relationship to the cool area. The cues were a black rectangle, a white rectangle, and two rectangles patterned with either vertical or horizontal gratings. In an experiment, the target was initially situated either between the vertical gratings and black rectangle or directly in front of the vertical gratings. The floor of the arena was covered by a thin transparent carpet, which could be rotated or replaced between trials to test whether chemical cues released by the subjects themselves influenced the outcome of the experiment.

2. Visible Targets. A "visible target paradigm" tested goal-directed behavior to the target when this itself was colored white against the black arena. The walls of the arena were undecorated. Temperature conditions were the same as in the preceding paradigm.

3. Training Procedures. Two different training procedures were used. The first employed trials in which the cockroach (subject) was released from an opaque container into the arena. The subject had no experience of the arena, and the release site differed in each trial. Trials were repeated 12 times with intervals of 3–4 min. During the interval, the cockroach was allowed to remain on the target, where it occasionally antennated the hot surface surrounding it. The second procedure used pretraining sensitization. The subject was placed on the target beneath a 12-cm-diameter glass beaker that covered the cool area and a concentric surround of the hot surface. During this pretraining, subjects were observed to leave the target often but immediately return to it. Normal trials commenced after 5 min of pretraining with the subject entering the arena at different locations on each trial. In the control experiment, the arena walls were bare, but the target itself was visible ("visible target" paradigm) as a white disc on the arena surface. Behavioral observations were made by one of the authors (J.M.W), who was not informed what surgeries had been made on operated subjects. Possible effects of cues outside the arena were controlled by rotating the whole arena between trials of unoperated subjects. This did not affect the learning curves.

C. Histology

After completing the behavioral trials, each operated subject was killed by injection of 3% formaldehyde carried in cockroach saline. After 20 min, the head capsule was opened and the brain, with its implanted blades, dissected out and postfixed for 20 min in artificially aged Bouin (Gregory, 1980). It was then dehydrated and embedded in hard Araldite to preserve the original location of the blades in the surrounding tissue during sectioning. Serial 30-µm sections were cut with a steel knife on a sliding microtome and mounted on albumin-subbed slides. Sections showed no displacement of the blades (now also sectioned). Slides were stained with toluidine blue or ethyl gallate and the locations of the blades relative to the brain regions were recorded using dark-field microscopy.

D. Statistics

The time to reach the target (run time) was used to evaluate spatial learning in unoperated and operated subjects. Chi-square and a Mann-Whitney U-test were employed for evaluating the data.

E. Recording and Stimulation

Electrodes were implanted into cooled, restrained animals. A small flap of cuticle was opened to reveal the paired mushroom bodies, which have characteristic spatial relationships with respect to the entry of the ocellar nerve, the optic pedunculi, and pars intercerebralis. Calyces can be seen under the binocular microscope as paired, relatively opaque tori. A small incision was made on the brain surface and a 2 × 2 (or 2 × 3) array of 20-μm-diameter Teflon-insulated copper wires was inserted into the brain. The wires were waxed to the head capsule and to an insulated 60-μm-diameter copper wire, which served as the ground electrode into the thorax and a tether limiting locomotion within the field of view of a video camera. Recording wires were connected to differential AC amplifiers.

Experiments were run in disposable X-shaped cardboard arenas and commenced 30–60 min after electrode implantation. Single-unit or multiunit recordings were obtained with any of the four to six electrodes in the array. The electrode showing the fewest units and the best signal-to-noise ratio was selected for recording for the duration of the experiment. The recorded signals were stored on line on videotape sound track during video recording of the experiment. Analysis was made using a Mackintosh SE/Plus computer with a WPI 121 window discriminator and MacLab software for data analysis.

Behavioral observation was confined to ambulation (forward, backward movement, turns), head movements, antennation, and grooming. A gentle air puff or a slight touch with a probe to trigger movement was applied to one of the following: an antenna, the head, the dorsal thorax, dorsal abdomen, tarsus, or a cercus. Episodes of behavior recorded on videotape were drawn from the video monitor. After recording, copper ions were released into the tissue by passing a 1–5-μA current for 30 sec. After 5–20 min diffusion, copper ions were precipitated with 5% ammonium sulfide in cockroach saline. The brain was fixed in Bouin, removed from the head, washed in 70% alcohol, dehydrated, and permeabilized with propylene oxide. Rehydrated brains were silver intensified (Strausfeld and Obermayer, 1976), dehydrated, embedded in Araldite, and sectioned at 30 μm. Sections were serially reconstructed using a camera lucida at an initial magnification of 250×. A library of Golgi-impregnated brains was used to match copper sulfide–silver stained elements to single impregnated neurons.

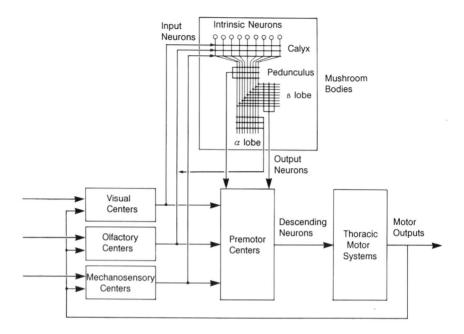

FIGURE 1. Summary of sensorimotor pathways in the insect brain. Primary afferents supply sensory neuropils (left), which provide efferents that converge to premotor integration centers. From these arise descending neurons to thoracic motor systems. This is referred to as the "direct" sensorimotor pathway. Outputs from sensory neuropils also diverge to an "indirect" pathway, supplying inputs (afferents) whose terminals are distributed within the calyx and overlap other sensory modalities there. Afferents synapse onto Kenyon cells, whose axons supply the pedunculi, α-lobes, and β-lobes. Kenyon cell axons are transected by the dendritic trees of output neurons (efferents). Efferent terminals from the pedunculus and part of the β-lobe supply premotor centers in the brain, and hence the descending output.

III. Results

A. Introductory Review: Organization of the Mushroom Bodies

The anatomical organization of the mushroom bodies is schematized in Fig. 1, which also serves as a summary diagram of the "direct" and "indirect" sensorimotor pathways considered in the Discussion.

The earliest description of the mushroom bodies was that by Kenyon (1896), and much of it basically holds good today. Anatomically, the mushroom body consists of two distinct entities: the cups, or calyces, and the lobes. Calyces are supplied by the terminals of projection neurons from the antennal lobes, regions of the midbrain, and the optic lobes. As described by Mobbs (1982), sensory projection neurons provide extensive branching within calycal neuropil. Anatomical observations have revealed that spatial relationships among the dendritic trees of nerve cells in neuropils supplying the calyces are not represented by their terminals in the calyces themselves. For example, retinotopically organized neurons in the optic lobes supplying the calyces do not give rise to retinotopic terminals there, as demonstrated by Gronenberg (1986) using intracellular cobalt diffusion. Instead, the sensory endings of retinotopic neurons overlap extensively within their termination area of the calyx. The same has been demonstrated for terminals of antennal lobe projection neurons (Lepidoptera: Kanzaki *et al.*, 1990; Diptera: Stocker *et al.*, 1990): spatial relationships between neighboring projection neurons in the olfactory glomeruli are apparently lost among the widely overlapping terminals in the calyces.

Situated within the calycal cups and around their lips, many thousands of small perikarya give rise to regularly arranged dendritic trees within the calycal neuropil (Weiss, 1974, 1981). These intrinsic neurons (termed Kenyon cells, after their discoverer) are the most densely packed neurons in any brain area (including mammalian cerebellum; see Strausfeld, 1976). Typically, their dendrites are postsynaptic to the terminals of sensory projection neurons (review; Schürmann, 1987) and send axons into the pedunculi. The axons then branch into at least two lobes, the α- and β-lobes. In many species, though not in cockroaches or honeybees, there are additional lobes (e.g., γ-lobes in the Lepidoptera; Pearson, 1971), which are derived from additional axon branches from Kenyon cells or from special subsets of Kenyon cells. In social Hymenoptera, the calyces show specific and complex subdivisions of their neuropils, in which certain sensory projections provide local systems of terminals. In cockroaches, the calyces have a uniformly stratified appearance, with sensory axons of different modalities overlapping each other and thus providing a substrate for associative convergence among visual, olfactory, and mechanosensory inputs (Strausfeld *et al.*, in preparation).

The axons of Kenyon cells project in parallel through the lobes and intersect dendritic trees of "extrinsic" neurons. These are the mushroom body's efferent neurons, the axons of which extend into surrounding neuropil or project recurrently back into the calyces. Extrinsic neurons comprise a great variety of morphological cell types, each being defined by the appearance of its dendritic tree and the geometric relationships between its dendritic branches across or parallel with Kenyon cell axons (Mobbs, 1982, 1984; Schürmann, 1987).

B. Role of Mushroom Bodies in Spatial Orientation

At the start of trials, both pretrained "sensitized" and naive cockroaches have the initial tendency to walk around the wall of the arena. This behavior, which possibly reflects negative phototaxis or sporadic attempts to escape, persists for two or three trials, during which, in normal unoperated animals, the time to reach the target decreases markedly. Fig. 2 illustrates a learning curve using two spatial configura-

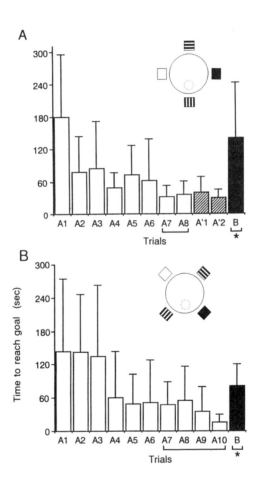

FIGURE 2. Results of successive trials by four groups of subjects ($n=9$). The run time (time to reach the target) is plotted in seconds (ordinate), with each trial indicated along the abscissa (A1–B*). (A) Trials A1–A8 demonstrate the learning curve. In A'1 and A'2 the plastic carpet was rotated to test the role of olfactory self-cues. There was no significant change in run times (compare trials A7, A8, with A'1, A'2). In trial B*, arena decorations were rotated by 90°. Run times are significantly longer (compare trials A7, A8 with B*; $p<.05$). (B) Trials with visual cues ($n=8$). A1–A10 demonstrate the learning curve. After A10, patterns were rotated through 90°, thus changing geometric relationships between the target and decorations. Statistical comparison between A7–A10 and B* (Mann-Whitney U-test) showed a significant ($p<.01$) increase in the run time of trial B*.

tions between the visual spatial cues and the hidden target. In Fig. 2A the target was situated directly in front of the vertical grating. In Fig. 2B the target was located midway between the vertical grating and the black rectangle. In both tests, the subjects showed similar learning curves. The possibility that subjects themselves laid down their own olfactory cues was tested by rotating the transparent floor sheet indicated as A'1–2 in Fig. 2A. Rotating the carpet did not increase the run time. In contrast, however, subjects required significantly longer to reach the target when, after learning the first configuration between the target and visible cues, the cues were rotated to a second configuration (in Fig. 2A compare trials A7–8 and B*; Mann-Whitney U-test, $p<.05$). A second group of individuals were first tested with the hidden target situated between the vertical grating and black rectangle (Fig. 2B). After learning, the configuration was again changed, resulting in significantly longer run times (compare trials A7–A10 with trial B*; Mann-Whitney U-test, $p<.01$). These results suggest that subjects locate the hidden target by matching the organization of the learned visual world relative to their current location in the arena, as has been suggested for flying honeybees (Collett and Cartwright, 1983). In the absence of such cues, there was no significant improvement in target location over time. Tests using olfactory cues (Mizunami *et al.*, in preparation) have provided comparable results.

We next tested the effects of microsurgical lesions (the implantation of small aluminum foil blades) on place memory. To distinguish deficits in place memory from deficits of other neural functions required for target location (such as primary visual perception and motor output), tests first employed the Tennessee Williams paradigm, as above, and then the visible target paradigm. Operations were performed on cockroaches that had already demonstrated place memory, as described above. Before each set of postsurgical trials, subjects were pretrained on the target as described in the Methods section.

Control surgery (Fig. 3A) lesioned superficial brain areas but left intact both mushroom bodies and their sensory supply. The results of control and bi–pedunculus–lesioned subjects are summarized in Fig. 3A–D. *Postmortem* histology established that the mushroom bodies of subjects that provided data for Fig. 3C and D had suffered bilateral lesions that completely eliminated connections between afferent Kenyon cell supply to the pedunculi and the extrinsic neurons arising from them.

Comparison of Fig. 3A and Fig. 2B (trials A4 etc.) demonstrates that the results from control operated subjects differ little from those obtained from naive unoperated subjects. Also, the performance of control operated subjects in the Tennessee Williams paradigm (Fig. 3A) differs little from their performance using the visible target paradigm (Fig 3B). However, the results for nine subjects shown *postmortem* to have suffered bilateral pedunculus lesions demonstrate that, although they were unable to learn the location of a hidden target by its association with the visual cues around it (Fig. 3C), they did learn the location of the visible target (Fig. 3D).

In summary, cockroaches whose pedunculi are bilaterally lesioned reach the visible target as quickly as controls, but take significantly longer than control subjects to reach the hidden target (Mann-Whitney U-test, $p<.05$). Fig. 4A compares the times to reach the hidden target after trial 3 using the Tennessee Williams paradigm, that is, when a normal unoperated animal makes the association between landmarks to locate an otherwise covert target. Fig. 4B summarizes the results of the visible target paradigm, tested on the same individuals. In one series of experiments (Mizunami *et al.*, in preparation), we have shown that neither unilateral β-lobe lesions nor bilateral lesions of the α-lobes significantly affect the time to reach either the hidden or the visible target. The significance of this is considered in the Discussion.

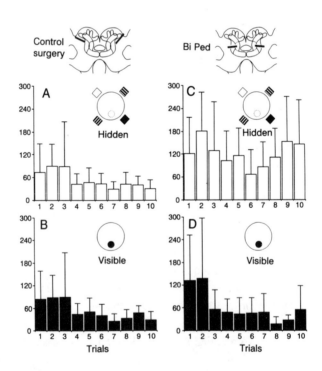

FIGURE 3. Comparison of performance (after pretraining) of control operated and bilateral pedunculotomized subjects. Learning of control operated subjects ($n=9$) (A) in the Tennessee Williams paradigm and (B) in the visible target paradigm. (C, D). Results of bilateral pedunculus lesions ($n=9$) showing no learning of hidden target (C) but learning of visible target (D).

Chapter IX Place Memory and Motor Control

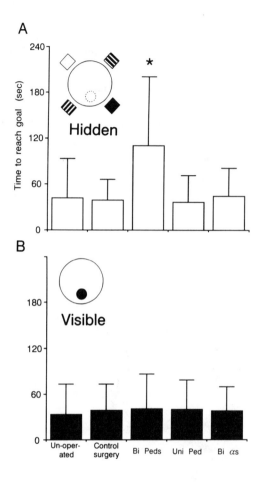

FIGURE 4. Comparison of run times of subjects tested with the Tennessee Williams paradigm (A) and subjects tested with a visible target paradigm (B). Numbers of subjects tested are $n=15$ (unoperated), $n=11$ (control operated), $n=9$ (bilateral pedunculus lesions), $n=9$ (unilateral pedunculus lesions), and $n=10$ (bilateral α-lobe lesions). Run times are averages measured after the third trial of each series. In (A) only subjects with bilateral pedunculus section showed no significant decrease in run times, compared with trial 3 of unoperated subjects. However, all subjects showed improved run times in (B) testing with a visible target.

C. Electrical Activity in the Mushroom Body Pedunculi

1. Motor-Associated Units. We have recorded and stained units whose activity in areas of the mushroom bodies (1) accompanies sensory input, (2) reflects reafference copy of motor activity, and (3) precedes and accompanies sequential motor actions, such as grooming or walking (Mizunami and Strausfeld, in preparation). In this account we shall exemplify two units associated with movement and the behavior with which they are associated. These units typically begin to fire before a motor action and continue during a specific motor repertoire. Such "motor preparatory units" can be triggered by sensory stimulation, or they can fire spontaneously in the apparent absence of environmental change.

In Figs. 5 and 6, the behavioral episodes are lettered a–c in A, with individual frames numbered. The corresponding frames are indicated above the electrophysiological traces. In Fig. 5, the mushroom body units responded to gentle tactile stimulation of either hind leg tarsus (Fig. 5Aa and 5Ba) or stimulation of the thorax (Fig. 5Ab,c and 5Bb,c). The discharge pattern differed according to the behavioral response. Tactile stimulation that failed to elicit a behavioral response gave rise to relatively short-lasting phasitonic activity of about 1 sec duration, fol-

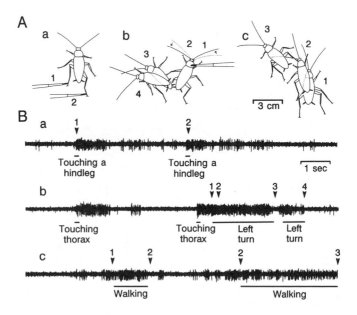

FIGURE 5. Sensorimotor units recorded from the pedunculus. 2–3 units respond tonically to tactile stimulation, (Aa 1, 2) with sporadic activity persisting 1–3 sec. after stimulation (Ba, 1, 2). Vigorous discharges, however, accompany movement and persist during locomotion (Ab 1–4, Bb 1–4, Ac 1–3, Bc 1–3), ceasing during breaks in movement (Bb 3) when locomotion itself ceases.

Chapter IX Place Memory and Motor Control

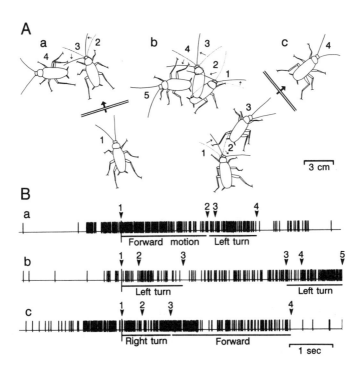

FIGURE 6 Recording from the pedunculus–β-lobe junction showing a motor preparatory unit. Videoframe drawings (A) show the sequences of events during recordings (B). The corresponding episodes of behavior (frame 1, 2, etc. in A) are indicated for each recording sweep. Recordings are from one individual. They demonstrate the onset of unit response 0.8 –0.3 sec before the onset of a motor sequence (event 1 in each trace a,b,c). The response of the unit persists during locomotion. The firing rate increases during accelerated forward motion (trace a, events 1 to 2) and during turns (trace a, events 3–4, b 4–5; c, 2–3). Cessation of locomotion results in a return to the resting discharge frequency.

lowed by occasional burst-like discharges (Fig. 5Ba, 1,2). However, when tactile stimulation elicited locomotion (such as a turning movement away from the direction of stimulus as in Fig. 5Ab, episodes 1–4), unit activity persisted for the duration of the motor action although the source of stimulation was long removed (Fig. 5Bb, events 1–4). Interestingly, this unit also showed bursts of spontaneous activity (Fig. 5Bc), which preceded an episode of behavior (walking) and lasted some seconds. The unit maintained its activity during the motor sequence (Fig.5Ac, episodes 1–3, and Fig. 5Bc).

In the second example (Fig. 6), the onset of unit activity typically preceded walking without being triggered by any apparent sensory input (Fig. 6Aa, episodes 1–4, recorded in Fig. 6Ba). Start of motion (Fig. 6Aa, frame 1) was preceded by 300–800 msec of sustained activity by the recorded unit (Fig. 6Ba, to the left of event 1), which continued to fire during forward motion as well as during turns to the left or right. Interestingly, the unit sustained its firing rate during a pause in walking, as filmed in Fig. 6A, episode b, at frame 3 (during which only the antennae moved from right to left) and revealed in the recording of Fig. 6Bb between events 3 (left) and 3 (right). There are two interpretations of this recording: (1) activity in the unit associated with a specific motor sequence (walking) was maintained during an interval of a predefined motor program or (2) activity in the unit contributed to two discrete actions, walking and antennal movement in the direction in which the subject would subsequently turn.

2. **Connections of the Pedunculus to Premotor Brain Areas.** Copper was released from electrodes that recorded motor-associated activity in the pedunculus. Cobalt sulfide–silver staining demonstrated neurons arising from two clusters of as many as 30 cell bodies (Fig. 7). One cluster is situated posterior to the pedunculus β-lobe, in rind flanking the inner margin of the dorsal deutocerebrum (Fig. 7B). The other is situated anterior to the β-lobe, near the ipsilateral margin of the pars intercerebralis (Fig. 7D). The dendrites of the marked neurons invade the pedunculus, with some branches also giving rise to one or two small dendritic fields within the β-lobes. Of the 11 successful stainings, only one showed branches extending into the base of the α-lobe. Two preparations contained stained axons projecting to the contralateral dorsal deutocerebrum. Three preparations revealed recurrent axons to the ipsilateral calyces (Fig. 8), and in one preparation axons arising from the pedunculus projected to the ipsilateral protocerebrum, medial to the α-lobes and lateral to the calyces. In all the preparations, however, efferent axons extended ipsilaterally to terminate in neuropil lateral to the pedunculus–β-lobe junction (the dorsolateral deutocerebrum) and in the dorsal deutocerebrum. A summary of these pathways (Fig. 8), derived from nine copper electrode recordings, shows the main projection routes (black), their dendritic domains in the pedunculus/β-lobes (striped), and the target neuropils, most of which are dendritic areas of descending neurons (stippled). As described for other insect orders (Diptera: Gronenberg and Strausfeld, 1990; Hymenoptera: Ibbotson 1991), the dorsolateral and dorsal deutocerebrum are integration neuropils receiving converging projection neurons from unimodal sensory neuropil.

Chapter IX Place Memory and Motor Control

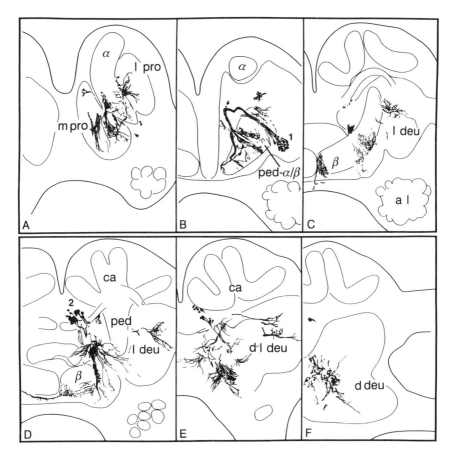

FIGURE 7. Serial reconstructions of the right supraesophageal hemisphere showing the projections of copper–silver intensified neurons stained from an electrode recording activity of motor preparatory units. Two cell body clusters (1, 2 in B and D), each comprising 25–30 neurons, send neurites that provide extensive dendritic trees with the confluence of the pedunculus– α-,β-lobes, pedunculus, and a small patch of the β-lobe medially. The neurons send axons that branch in the lateral (l pro) and medial protocerebrum (m pro), around the α-lobes (α) and in the dorsolateral (d l deu) and dorsal deutocerebrum (p deu).

IV. Discussion

We will first consider whether the results of the behavioral experiments suggest a role for the mushroom bodies in place memory. We will then consider the significance of the recordings and cell tracings with respect to the relationship between the mushroom bodies and descending premotor pathways. We conclude by proposing that the mushroom bodies acquire information about a specific sensory surround and orchestrate motor programs allowing them to reestablish that relationship. We emphasize that we are not proposing that cockroaches use cognitive maps to recognize and update any momentary position in the environment. The present experiments are not designed to test this, nor could they be so interpreted. Rather, our observations are compatible with those on honeybees that suggest template matching between a learned image of the environment and behavior that achieves a match between it and the current image (Collett and Cartwright, 1983; Cartwright and Collett, 1987).

We have demonstrated that two learning paradigms lead to rapid acquisition of an adaptive behavior: escape from heat. Behavioral experiments demonstrate that unoperated adult cockroaches learn to locate a covert target as efficiently as a visible one. The position of the former is defined only by its geometric relationship with distant visual cues. One objection to the observed learning curves (Fig. 2) might be that the subject randomly encounters the target and then uses its own chemical trail to relocate it. Using a carpet that could be removed or turned between trials eliminated this. Changing the configuration between the target and visual cues required relearning a new configuration, except that during the first one to three trials in the new stimulus configuration the subject briefly visited the previously learned target location, which was now indistinguishable from the ambient temperature of the arena. Another objection could be that the subject associated the approximate location of the target with light intensity gradients in the laboratory. This was also eliminated by rotating the entire experimental apparatus between trials.

The lesion experiments have used a novel procedure, which is to implant the lesioning agent—a thin blade of aluminum—into the brain and to leave it there throughout the experiment and subsequent histology. By this means it is possible to determine the precise depth and width of the injury and to identify any secondary pathological effects. These included, in four of the nine subjects, hemocytes lining the lesion and, in three ethyl gallate preparations, unusual darkening of some axon profiles leading from the site of lesion. This suggested that they were in the process of degeneration at the time of fixation. The results of lesion experiments demonstrate that the ability to learn a hidden target depends on the integrity of at least one mushroom body's pedunculus. Interestingly, bilateral pedunculus lesions did not diminish the ability of the subjects to locate a visible target, suggesting that the mushroom bodies mediate associative spatial memory. Another noteworthy result was the observation that bilateral α-lobe lesions did not significantly impair the acquisition of place memory, suggesting that different parts of the mushroom bodies, supplied by the same ensembles of Kenyon cells, might be involved in different and possibly independent tasks. For example, olfactory memory may be the domain of the α-lobe, as suggested by studies on proboscis reflex conditioning in honeybees (Erber *et al.*, 1980). It will be of great interest to determine whether place memory in bees is mediated by their pedunculi and β-lobes. Although such functional subdivisions cannot be confirmed by conventional neuroanatomical studies, a modified Golgi impregnation that reveals glial cells demonstrates that the pedunculi and α- and β-lobes in *Periplaneta* are complexly subdivided by glial partitions and barriers (Strausfeld et al., in preparation).

The results of extracellular recordings and subsequent copper sulfide–silver impregnation of neurons in the cockroach mushroom bodies confirm other neuroanatomical studies. These show that the calyces are the primary site of convergence among sensory projection neurons from the olfactory lobes (Orthoptera: Ernst and Boeckh, 1977; Burrows et al., 1982; Schildberger, 1983; Diptera: Strausfeld, 1976; Lepidoptera: Kanzaki et al., 1990), visual neuropils (Hymenoptera: Jawlowski, 1958; Mobbs, 1982), and regions of the deutocerebrum receiving primary afferents, or receiving terminals of ascending neurons from the thoracic–abdominal ganglia (Strausfeld, 1976; Mobbs, 1982). The same sensory neuropils (except the antennal lobes) send other projection neurons to the deutocerebrum, where they converge onto the dendrites of multimodal descending neurons. It is these, comprising some 200 elements in the supraesophageal ganglia, that contribute synaptic inputs to thoracic motor neuron–interneuron circuits controlling locomotion.

Although quite a lot is known about the sensory interneuron supply to the calyces, there is relatively little information about the functional properties of the pedunculi and lobes. Exceptions are observations on honeybees (Homberg, 1984) and field crickets (Schildberger, 1983, 1984), which have identified efferent neurons that are activated by multimodal sensory inputs. Certain of these efferent neurons also show long-lasting afterdischarges or exhibit modulations of their sensitivity to certain stimuli (Schildberger, 1984). This suggests that these neurons are involved in circuits that maintain, for short periods, information about a specific combination of sensory cues after these elapse. Studies of recurrent efferents between the pedunculus and calyces of honeybees (Gronenberg, 1987) also demonstrate stimulus-dependent modifications of the neurons' sensitivity to multimodal stimuli. Perhaps the most accurate indication of plastic changes within an identified neuron comes from studies by Mauelshagen (1990), who demonstrated specific alterations in the firing patterns and thresholds of an identified α-lobe efferent neuron during conditioning of fictive proboscis extension reflexes.

Despite this background, the position of the mushroom bodies with respect to descending pathways has been conjectural (Strausfeld, 1989). Yet, it is clear that if the mushroom bodies are involved in learning and memory, they must also mediate adaptive behavior. If orchestrated motor actions are at least in part controlled by the brain, then it is to be expected that mushroom bodies should influence activity among descending pathways. As pointed out by Boeckh and Ernst (1987), apart from rare exceptions (see below), most anatomical observations have so far failed to support this. Intracellular dye filling (Gronenberg, 1987), as well as neuroanatomical studies on bees (Mobbs, 1982; Strausfeld and Hurley, in preparation), Orthoptera (Schildberger, 1983, 1984), and Diptera (Hurley *et al.*, 1991), suggest that one major output from the mushroom bodies comprises recurrent axons that extend back into the calyces that supplied their original input. Other studies (Mobbs, 1982, 1984; Mauelshagen, 1990) suggest that α-lobe efferents send axons into concentric areas of synaptic neuropil encircling the lobes.

A few observations on Orthoptera (Schildberger, 1984) and calliphorids (Strausfeld *et al.*, 1984) have, however, described efferents from the mushroom bodies projecting to neuropils that contain the dendritic trees of descending neurons (Strausfeld *et al.*, 1984). The present results show convincing evidence for such pathways (Mizunami and Strausfeld, in preparation) by means of the novel method of releasing copper from extracellular electrodes to stain neurons that extend to, or from, the recording site. Recordings from the pedunculus demonstrate groups of neurons reaching descending neuron neuropils in the deutocere-

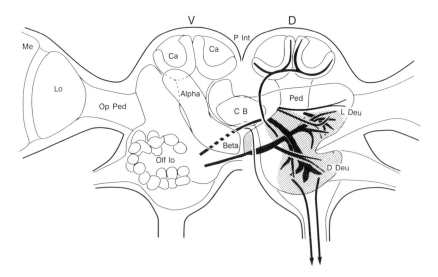

FIGURE 8. Summary diagram showing projections of units recorded from the pedunculi and β-lobes (motor preparatory and motor units). Striped areas indicate dendritic domains in the mushroom bodies. Axons are indicated in black, and deutocerebral neuropils containing the dendrites of descending neurons are stippled. Other areas are: Me, medulla; Lo, lobula; Op Ped, optic peduncle; Olf lo, olfactory (antennal) lobe; CB, central body; Ca, calyx; Ped, pedunculus; alpha, α-lobe; beta, β-lobe; L Deu and D Deu, dorsolateral and dorsal deutocerebrum; P Int, pars intercerebralis; V, ventral view; D, dorsal view.

Chapter IX Place Memory and Motor Control

brum. A parsimonious explanation for this organization is that the pedunculus and possibly part of the β-lobe (Mizunami and Strausfeld, in preparation) provide a direct input onto descending pathways (Fig. 8).

2. *The Relationship of the Mushroom Bodies to Sensorimotor Pathways*

Studies on the mushroom bodies of Hymenoptera, Diptera, and Orthoptera support a relatively straightforward morphological context for mushroom body operation (Fig. 1). Sensory interneurons converge in the calyx onto many thousands of Kenyon cell dendrites from which arise axons that project in parallel through the pedunculi and lobes. Characteristic ensembles of Kenyon cell axons are intersected and sampled at specific positions through the lobes by extrinsic (efferent) neuron dendrites. As suggested above, it is likely that the number of such extrinsic

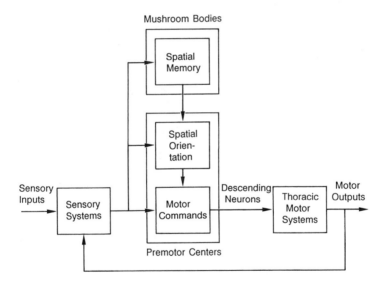

FIGURE 9. Suggested hierarchy among integration pathways from sensory systems to the motor output. Two separate "higher" control systems are hypothesized. One involves spatial (place) memory (testable with the Tennessee Williams paradigm); the other involves nonspatial memory revealed by the visible target paradigm. In the scheme shown, interruption between the pedunculi and subordinate centers abolishes place memory but leaves target orientation intact. Escape responses by cockroaches in which the protocerebrum has been largely destroyed suggest a lower-level organization reminiscent of that of ascending pathways from the terminal ganglia.

neurons has been greatly underestimated, although neuroanatomical studies (Mobbs, 1982, 1984) illustrate that outgoing bundles consist of scores of axons.

Where do extrinsic neurons project? The evidence so far is that they target (1) the calyces, (2) concentric protocerebral neuropils around the pedunculi and α-lobes, and (3) the sensory integration neuropils in the dorsolateral and dorsal deutocerebrum. These deutocerebral areas are the site of sensory convergence onto premotor descending neurons and thus represent a "direct" pathway between sensory neuropils and thoracic–abdominal motor circuits. The results of this chapter support the hypothesis that mushroom body efferents to deutocerebral neuropils provide the output from a second "indirect" pathway that involves the most intricate and densest of any known neuropil, including those of vertebrates. Although intracellular recordings demonstrate that descending neurons carry multimodal information about events leading to rapid behavioral responses (some in the millisecond range; Land and Collett, 1974; Gronenberg and Strausfeld, 1991), we suggest that the orchestration of descending neurons and possibly their sensitivity to specific combinations of sensory inputs are governed by mushroom body pedunculus efferents whose activities are themselves modified by sensory experience (e.g., place memory). Without this efferent supply, cockroaches are still able to locate a visible target and learn its position (spatial orientation). Even in experiments that lesion across the protocerebrum and destroy both mushroom bodies entirely, sensory stimuli still elicit simple motor commands (Mizunami *et al.*, in preparation). In conclusion, we suggest that this range of sensorimotor performance is governed by a hierarchy of integration pathways in the brain (Fig. 9), the most sophisticated of which involves the mushroom bodies.

Acknowledgments

We have profited from discussions with Drs. Cole Gilbert, Klaus Schildberger, Randolf Menzel, and Roy Ritzmann. We thank Zhiqiang Yang, M.S. for technical and photographic assistance. This study was supported by an NIH Fogarty International Fellowship F05 TW 04390 to M.M., a Howard Hughes Undergraduate Biology Research Program Scholarship to J.M.W., and a BRSG award to N.J.S. from the University of Arizona.

References

ALLOWAY, B. W. (1972). "Retention of Learning through Metamorphosis in the Grain Beetle (*Tenebrio molitor*)," *Am. Zool.* **12**, 471–477.

BACON, J. P., and STRAUSFELD, N. J. (1986). "The Dipteran "Giant Fibre" Pathway: Neurons and Signal," *J. Comp. Physiol. A* **150**, 439–452.
BALDERRAMA, N. (1980). "One Trial Learning in the American Cockroach, *Periplaneta americana*," *J. Insect Physiol.* **26**, 499–504.
BOECKH, J., and ERNST, K.-D. (1987). "Contribution of Single Unit Analysis in Insects to an Understanding of Olfactory Function," *J. Comp. Physiol. A.* **161**, 567–582.
BUITKAMP-MÖBIUS, K. (1975). "Strukturuntersuchungen an den Pilzkörpern im Oberschlundganglion von *Apis mellifica-Gynandromorphen* unter Berücksichtigung ihres Verhaltens," Ph.D. Dissertation, Universität Bonn.
BURROWS, M. (1987). "Inhibitory Interactions between Spiking and Nonspiking Local Interneurons in the Locust," *J. Neurosci.* **7**, 3282–3292.
BURROWS, M., BOECKH, J., and ESSLEN, J. (1982). "Physiological and Morphological Properties of Interneurons in the Deutocerebrum of Male Cockroaches Which Respond to Female Pheromone," *J. Comp. Physiol. A* **145**, 447–457.
BYERS, D. (1980). "A Review of the Behavior and Biochemistry of *dunce*, a Mutation of Learning in *Drosophila*," in: Siddiqi, O., Babu, P., Hall, L. M., and Hall, J. C. (eds.), *Development and Neurobiology of Drosophila*. (pp. 467–474.) Plenum, New York.
CARTWRIGHT, B. A., and COLLETT, T. S. (1987). "Landmark Maps for Honeybees," *Biol. Cybern.* **57**, 85–93.
COLLETT, T. S., and CARTWRIGHT, B. A. (1983). "Eidetic Images in Insects: Their Role in Navigation," *Trends Neurosci.* **6**, 101–105.
DAVIS, R. L., and KAUVAR, L. (1984). "Drosophila Cyclic Nucleotide Phosphodiesterases," in Strada, S., and Thompson, W. (eds.), *Advances in Nucleotide Research.* (pp. 393–402). New York, Raven Press.
DUDAI, Y., JAN, T. N., BYERS, D., QUINN, W. G., and BENZER, S. (1976). "*Dunce*, a Mutant of *Drosophila* Deficient in Learning," *Proc. Natl. Acad. Sci. USA* **73**, 1684.
DUERR, J. S., and QUINN, W. G. (1982). "Three *Drosophila* Mutations That Block Associative Learning and Also Affect Habituation and Sensitization," *Proc. Natl. Acad. Sci. USA* **79**, 3646–3650.
DUJARDIN, F. (1850). "Memoire sur le systeme nerveux des insects," *Ann. Sci. Nat. Zool.* **14**, 195–206.
EGELHAAF, M., HAUSEN, K., REICHARDT, W., and WEHRHAHN, C. (1988). "Visual Course Control of Flies Relies on Neural Computation of Object and Background Motion," *Trends Neurosci.* **11**, 351–356.
ERBER, J., MASUHR, T. and MENZEL, R. (1980). "Localization of Short-Term Memory in the Brain of the Bee, *Apis mellifera*," *Physiol. Entomol.* **5**, 343–358.
ERNST, K. D., and BOECKH, J. (1977). "A Neuroanatomical Study of the Central Antennal Pathways in Insects. III. Neuroanatomical Characterization of Physiologically Defined Response Types of Deutocerebral Neurons in *Periplaneta americana*," *Cell Tissue Res.* **229**, 1–22.
GOLL, W. (1967). "Strukturuntersuchungen am Gehirn von *Formica*," *Z. Morph. Ökol. Tiere* **59**, 143–210.
GREGORY, G. E. (1980). "The Bodian Protargol Technique," in Strausfeld, N. J. (ed.), *Neuroanatomical Techniques: Insect Nervous System.* (pp. 77–97). Springer-Verlag, New York.
GRONENBERG, W. (1986). "Physiological and Anatomical Properties of Optical Input-Fibers to the Mushroom Body of the Bee Brain," *J. Insect Physiol.* **32**, 695–704.

GRONENBERG, W. (1987). "Anatomical and Physiological Properties of Feedback Neurons of the Mushroom Bodies in the Bee Brain," *Exp. Biol.* **46**, 115–125.

GRONENBERG, W., and STRAUSFELD, N. J. (1990). "Descending Neurons Supplying the Neck and Flight Motor of Diptera: Physiological and Anatomical Characteristics," *J. Comp. Neurol.* **32**, 973–991.

GRONENBERG, W., and STRAUSFELD, N. J. (1991). "Descending Pathways Connecting the Male-Specific Visual System of Flies to the Neck and Flight Motor," *J. Comp. Physiol. A* **169**, 413–426.

HALL, J. C. (1979). "Control of Male Reproductive Behavior by the Central Nervous System of *Drosophila;* Dissection of a Courtship Pathway by Genetic Mosaics," *Genetics* **92**, 437–457.

HALL, J. C. (1986). "Learning and Rhythms in Courting Mutant *Drosophila,*" *Trends Neurosci.* **9**, 414–418.

HANESCH, U., FISCHBACH, K. F., and HEISENBERG, M. (1989). "Neuronal Architecture of the Central Complex in *Drosophila melanogaster,*" *Cell Tissue Res.* **257**, 343–366.

HEISENBERG, M., BORST, A., WAGNER, S., and BYERS, D. (1985)."Drosophila Mushroom Body Mutants Are Deficient in Olfactory Learning," *J. Neurogenet.* **2**, 1–30.

HÖLLDÖBBLER, B., and WILSON, E. O. (1990). *The Ants.* Belknap, Harvard.

HOMBERG, U. (1984). "Processing of Antennal Information in Extrinsic Mushroom Body Neurons of the Bee Brain," *J. Comp. Physiol. A* **154**, 825–836.

HOMBERG, U., CHRISTENSEN, T. A., and HILDEBRAND, J. G. (1989a). "Structure and Function of the Deutocerebrum in Insects," *Annu. Rev. Entomol.* **34**, 477–501.

HOMBERG, U., MONTAGUE, R. A., and HILDEBRAND, J. G. (1989b). "Anatomy of Antenno-Cerebral Pathways in the Brain of the Sphinx Moth *Manduca sexta,*" *Cell Tissue Res.* **254**, 255–281.

HOWSE, P. E. (1974). "Design and Function in the Insect Brain," in Browne, L. B. (ed.), *Experimental Analysis of Insect Behavior.* (pp. 180–195). Springer-Verlag, New York.

HUBER, F. (1959). "Auslösung von Bewegungsmustern durch elektrische Reizung des Oberschlundganglions bei Orthopteren (Saltatoria: Gryllidae, Acridiidae)," *Z. Vergl. Physiol.* **43**, 359–391.

HUBER, F. (1960). "Untersuchungen über die Funktion des Zentralnervensystems und insbesondere des Gehirns bei der Fortbewegung und der Lauterzeugung der Grillen," *Z. Vergl. Physiol.* **44**, 60–132.

HURLEY, B. A., GRONENBERG, W., STRAUSFELD, N. J. (1991). "Structure of Output Neurons of the Mushroom Bodies in Honeybee Brains," *Soc. Neurosci. Abstr.* **17**, 1228.

IBBOTSON, M. R. (1990). "Wide-Field Motion-Sensitive Neurons Tuned to Horizontal Movement in the Honeybee, *Apis mellifera,*" *J. Comp. Physiol. A.* **168**, 91–102.

JAWLOWSKI, H. (1958). "Nerve Tracts in the Bee (*Apis mellifica*) Running from the Sight and Antennal Organs to the Brain," *Ann. Univ. M. Curie-Sklodowska. C* **12**, 265–268.

KANZAKI, R., ARBAS, E. A., STRAUSFELD, N. J., and HILDEBRAND, J. G. (1990). "Physiology and Morphology of Projection Neurons in the Antennal Lobe of the Male Moth *Manduca sexta,*" *J. Comp. Physiol. A* **165**, 427–453.

KENYON, F. C. (1896). "The Brain of the Bee. A Preliminary Contribution to the Morphology of the Nervous System of the Arthropoda," *J. Comp. Neurol.* **6**, 133–210.

KUWABARA, M. (1957). "Bildung des bedingten Reflexes von Pavlovs Typus bei der Hoigbiene (Apis *mellifica*)," *J. Fac. Sci. Hokkaido Univ.* **13**, 458–467.

Chapter IX Place Memory and Motor Control

LAURENT, G. (1987). "Parallel Effects of Joint Receptors on Motor Neurons and Intersegmental Interneurons in the Locust," *J. Comp. Physiol. A* **160,** 341–353.

LAND, M. F., and COLLETT, T. S. (1974). "Chasing Behavior of Houseflies (*Fannia canicularis*)," *J. Comp. Physiol. A* **89,** 331–357.

MAUELSHAGEN, J. (1990). "An Identified Neuron in the Honeybee Brain —A Correlate for Olfactory Learning? in Elsner, N., and Roth, G (eds.), *Brain, Perception, Cognition.* Theime, Stuttgart.

MENZEL, R., ERBER, J., and MASUHR, T. (1974). "Learning and Memory in the Honeybee," in Barton-Browne, L (ed.) *Experimental Analysis of Insect Behavior.* (pp. 218–227). Springer-Verlag, New York.

MENZEL, R., MICHELSEN, B., RUFFER, P., AND SUGAWA, M. (1988). "Neuropharmacology of Learning and Memory in Honeybees," in Herting, G., and Spatz, H-C. (eds.), *Modulation of Synaptic Transmission and Plasticity.* NATO ASI Series. (pp. 333–350). Springer-Verlag, New York.

MERCER, A. R., and MENZEL, R. (1982). "The Effects of Biogenic Amines on Conditioned and Unconditioned Responses to Olfactory Stimuli in the Honeybee *Apis mellifera*," *J. Comp. Physiol.* **145,** 363–368.

MILDE, J. J., and STRAUSFELD, N. J. (1990). "Cluster Organization and Response Characteristics of the Giant Fiber Pathway of the Blowfly, *Calliphora erythrocephala*," *J. Comp. Neurol.* **294,** 59–75.

MIZUNAMI, M., and STRAUSFELD, N. J. (1991). "Neural Activities of the Mushroom Bodies of the Cockroach During Locomotory Behavior," *Soc. Neurosci. Abstr.* **17,** 1228.

MOBBS, P. G. (1982). "The Brain of the Honeybee *Apis mellifera.* 1. The Connections and Spatial Organization of the Mushroom Bodies," *Philos. Trans. R. Soc. London Ser. B* **298,** 309–354.

MOBBS, P. G. (1984). "Neural Networks in the Mushroom Bodies of the Honeybee," *J. Insect Physiol.* **30,** 43–58.

NEDER, R. (1959). "Allometrisches Wachstum von Hirnteilen bei drei verschiedenen größen Schabenarten," *Zool. Jahrb. Anat.* **4,** 411–464.

NIGHORN, A., HEALY, M. J., and DAVIS, R. L. (1991). "The Cyclic AMP Phosphodiesterase Encoded by the *Drosophila dunce* Gene is Concentrated in the Mushroom Body Neuropil," *Neuron* **6,** 455–467.

PANDAZIS, G. (1930). "Über die relative Ausbildung der Gehirnzentren bei biologisch verschiedenen Ameisenarten," *Z. Morph. Ökol. Tiere* **18,** 114–169.

PEARSON, K. G. (1983). "Neural Circuits for Jumping in the Locust," *J. Physiol.* **78,** 765–771.

PEARSON, L. (1971). "The Corpora Pedunculata of *Sphinx ligustri* (L) and Other Lepidoptera: An Anatomical Study," *Philos. Trans. R. Soc. London Ser. B* **259,** 477–516.

REICHARDT, W., POGGIO, T. and HAUSEN, K. (1983). "Figure-Ground Discrimination by Relative Movement in the Visual System of the Fly. Part II: Towards the Neural Circuitry," *Biol. Cybern.* **46,** 1–30.

ROWELL, C. H. F. (1988) "Mechanism of Flight Steering in Locusts," *Experentia* **44,** 389–395.

ROWELL, C. H. F. (1989). "Descending Interneurones of the Locust Reporting Deviation from Flight Course. What Is Their Role in Steering? *J. Exp. Biol.* **146,** 177–194.

SCHÄFER, S., and BICKER, G. (1986). "Distribution of GABA-Like Immunoreactivity in the Brain of the Honeybee," *J. Comp. Neurol.* **246,** 287–300.

SCHÄFER, S., and REHDER, V. (1989). "Dopamine-Like Immunoreactivity in the Brain and Suboesophageal Ganglion of the Honeybee," *J. Comp. Neurol.* **280**, 43–58.

SCHÄFER, S., BICKER, G., OTTERSEN, O. P., and STORM-MATHISEN, J. (1988). "Taurine-Like Immunoreactivity in the Brain of the Honeybee," *J. Comp. Neurol.* **265**, 60–70.

SCHILDBERGER, K. (1983). "Local Interneurons Associated with the Mushroom Bodies and the Central Body in the Brain of *Acheta domesticus*," *Cell Tissue Res.* **230**, 573–586.

SCHILDBERGER, K. (1984). "Multimodal Interneurons in the Cricket-Brain: Properties of Identified Extrinsic Mushroom Body Cells," *J. Comp. Physiol. A* **154**, 71–79.

SCHÜRMANN, F. W. (1987). "The Architecture of the Mushroom Bodies and Related Neuropils in the Insect Brain," in Gupta, A. P. (ed.). *Arthropod Brain. (pp.* 231–264). Wiley Interscience, New York.

SCHÜRMANN, F. W., and ERBER, J. (1990). "FMRF Amide-Like Immunoreactivity in the Brain of the Honeybee (*Apis mellifera*). A Light and Electron Microscopical Study," *Neuroscience* **38**, 797–807.

STOCKER, R. F., LIENHARD, M. C., BORST, A., and FISCHBACH, K. F. (1990). "Neuronal Architecture of the Antennal Lobe in *Drosophila melanogaster*,". *Cell. Tissue Res.* **262**, 9–34.

STRAUSFELD, N. J. (1976). *Atlas of an Insect Brain.* Springer-Verlag, Berlin.

STRAUSFELD, N. J. (1989). "Insect Vision and Olfaction: Common Design Principles of Neuronal Organization," in Singh, R. N., Strausfeld, N. J. (eds.), *Neurobiology of Sensory Systems.* (pp. 319–354). Plenum, New York.

STRAUSFELD, N. J., BASSEMIR, U., SINGH, R. N., and BACON, J. P. (1984). "Organizational Principles of Outputs from Dipteran Brains," *J. Insect. Physiol.* **30**, 73–93.

STRAUSFELD, N. J., and LEE, J. K. (1991). "Neuronal Basis for Parallel Visual Processing in the Fly," *Vis. Neurosci.* **7**, 13–33.

STRAUSFELD, N. J., and OBERMAYER, M. (1976). "Resolution of Intraneuronal and Transsynaptic Migration of Cobalt in the Insect Visual and Central Nervous System," *J. Comp. Physiol. A.* **110**, 1–12.

TANOUYE, M. A., and WYMAN, R. J. (1980). "Motor Outputs of the Giant Nerve Fiber in *Drosophila*," *J. Neurophysiol.* **44**, 405–421.

TECHNAU, G. (1984). "Fiber Number in the Mushroom Bodies of Adult *Drosophila* Depends on Age, Sex and Experience," *J. Neurogenet.* **1**, 113–126.

TULLY, T., and QUINN, W. G. (1985). "Classical Conditioning and Retention in Normal and Mutant *Drosophila melanogaster*," *J. Comp. Physiol. A* **157**, 263–277.

VAN DER KLOOT, W. G., and WILLIAMS, C. M. (1953). "Cocoon Construction by the *Crecopia* silkworm," *Behavior* **5**, 141–174.

VON FRISCH, K. (1967). *The Dance Language and Orientation of Bees.* Belknap, Harvard.

VOWLES, D. M. (1955). "The Structure and Connections of the Corpora Pedunculata in Bees and Ants," *Q. J. Microsc. Sci.* **96**, 239–255.

WAHDEPUHL, M. (1983). "Control of Grasshopper Singing Behavior by the Brain: Responses to Electrical Stimulation," *Z. Tierpsychol.* **63**, 173–200.

WEIBRECHT, J. M., MIZUNAMI, M., and STRAUSFELD, N. J. (1991). "The Mushroom Body of Cockroach Brain Participates in Spatial Memory Processing," *Soc. Neurosci. Abstr.* **17**, 1228.

WEISS, M. J. (1974). "Neuronal Connections and the Function of the Corpora Pedunculata in the Brain of the American Cockroach, *Periplaneta american* (L),". *J. Morphol.* **142**, 21–70.

WEISS, M. J. (1981). "Structural Patterns in the Corpora Pedunculata of Orthoptera: A Reduced Silver Analysis," *J. Comp. Neurol.* **203,** 515–525.
WITTHÖFT, W. (1967). "Absolute Anzahl und Verteilung der Zellen im Hirn der Honigbiene," *Z. Morph. Tiere.* **61,** 160–184.

PART III

COMPUTER MODELING

Chapter X
Modeling a Reprogrammable Central Pattern Generating Network

A. I. SELVERSTON, P. ROWAT, AND M. E. T. BOYLE

Department of Biology
University of California, San Diego
La Jolla, California

I. Strategies for the Incorporation of Biological Data into Robotic Models

Biological systems have evolved a "technology" to perform a wide variety of movements. One of the most complex of these is rhythmic locomotion. The biological technology underlying insect walking, for example, has been around for millions of years. But in order to incorporate some of this biotechnology into an artificial insect or similar device capable of walking through and interacting with the environment, there must be a match between the biology (central nervous system, nerves, muscles) and the engineering (actuators, control systems, sensors, etc.). It is probably within present capabilities to obtain a transfer function between bursts of impulses in a motor nerve and the required input to a robotic actuator. Such a transfer function would enable artificial movements to be commanded by neural-like bursts instead of their usual input signal. A more difficult problem, however, is determining how central nervous systems are able to generate bursts of impulses at the proper phase and frequency to produce locomotion in the first place. It has been suggested that understanding the principles governing the operation of the nervous system would mean that a technological analog could be made. But there is a problem here: real neurons and their synaptic relationships are governed by principles completely different from the principles underlying modern electronic and robotic theory. There is not a direct path from biology to hardware implementation, but there are biological mechanisms that can suggest ways to design artificial systems that may be more robust and have more desirable features than systems derived from engineering principles alone. Here we demonstrate how a detailed knowledge of an extremely simple locomotor pattern generator may be useful in this respect.

II. Invertebrate Central Pattern Generators

The basic neural circuit underlying animal locomotion is the central pattern generator or CPG (Grillner, 1991). A CPG is a group of interconnected neurons that, when isolated from the rest of the central nervous system and deprived of sensory feedback, can be activated to generate a "fictive" motor pattern, i.e., spatiotemporal activity in the motor nerves similar to the patterned activity observed in intact animals. In fact, the *in vivo* pattern (in the intact animal) is different because of sensory feedback, neuromodulation, and other forms of input to the CPG neurons. The CPG concept can serve as a starting point for the analysis of locomotion, with the other factors, particularly sensory feedback, being added in as the analysis proceeds.

Invertebrate CPGs are especially useful in the analysis of cyclic patterns like locomotion, in that the number of cells producing the rhythm can be quite small and their circuitry determined in some detail by paired intracellular recording (Selverston and Moulins, 1985). The spatiotemporal output patterns for these small systems can be quantified for different states of activity, and questions of how the different states are turned on as well as how they are controlled by sensory feedback have been well studied (Barnes and Gladden, 1985).

III. Why the Stomatogastric System Can Inform Us about Locomotion Controllers

Electromyographic recordings taken from animals during locomotion consist of bursts of impulses to the various muscles moving the legs. Critical components of this spatiotemporal activity are the phase relationships between activity patterns in different muscles, the length and intensity of the bursts, and the overall burst frequency. The whole pattern changes significantly during different locomotory gaits and sensory feedback provides cycle-by-cycle adjustment to the pattern as the animal interacts with its environment. Although this peripheral activity has been well studied in both *in vivo* and *in vitro* preparations, in most cases virtually nothing is

Chapter X   A Reprogrammable Central Pattern Generating Network

known about the detailed makeup of the CPGs involved. Only in the lamprey are data available which suggest some of the relevant cellular mechanisms involved (Grillner *et al.*, 1991).

Although the CPGs in the lobster stomatogastric ganglion have nothing to do with locomotion, they do produce two different motor patterns, one of which is very similar to a locomotion pattern. The pattern generated by the gastric mill CPG controls three teeth in the dorsal region of the stomach. All of the muscles involved are striated; i.e., they are exactly like the leg muscles, requiring patterns of impulses in their motor nerves for contractions to occur. This may seem like a strange system with which to establish bridges between biology and engineering, but an examination of the motor patterns in vertebrate appendages during locomotion and stomatogastric patterns reveals striking similarities (Selverston, 1989).

The advantage of the stomatogastric system is that *all* of the neurons in the CPG have been identified (i.e., can be repeatedly studied from preparation to preparation) and their synaptic connectivity is well known (Selverston and Moulins, 1985). Furthermore, a fair amount of data on sensory feedback to the CPGs are also available (Katz *et al.*, 1989; Simmers and Moulins, 1988).

Our thesis is that the output of the stomatogastric CPGs is in principle no different than the output of more complex vertebrate CPGs and that by knowing how to model the stomatogastric system, particularly the gastric mill CPG, one can use this information to control artificial forms of rhythmic behavior.

IV.   Central Pattern Generators and Peripheral Feedback

The concept of the central pattern generator was experimentally proved when it was demonstrated that a motor pattern similar to that observed *in vivo* could be produced by the central nervous system in the absence of sensory feedback (Delcomyn, 1980). However, the role of sensory feedback in the intact animal is vital for adapting the rhythmic output to the environment and for fine-tuning the output even under constant environmental conditions. A recurring question has been whether or not to consider the sensory apparatus *part* of the CPG. The question is raised because in some cases the pattern can be significantly altered when sensory feedback is removed (Altman, 1982). The importance of sensory feedback biologically is related to the speed of locomotion. When animals are moving slowly, feeling their way around the environment, sensory input to the central nervous system obviously plays a vital role. During more rapid, ballistic-type movements, sensory feedback plays a lesser role.

V. Neurophysiology of the Lobster Stomatogastric System

There are four principal regions of the lobster foregut, each with its own unique movement pattern and central pattern generator: the gastric mill, the pyloric, the esophagus, and the cardiac sac (Fig. 1). When the nervous system is removed from the wall of the stomach all of the patterns can be observed by recording "fictive" activity from motor nerves innervating these regions. The esophageal rhythm is the slowest and most irregular of all the patterns, occurring at intervals of 2 to 20 s. The cardiac sac pattern has a period of approximately 7 s but is also somewhat irregular. The pyloric and gastric mill patterns have been studied in the most detail and have patterns with frequencies of approximately 2 and 0.33 Hz, respectively. The movements for each part of the foregut with the exception of the gastric mill are rather simple mechanically. In general, contraction of intrinsic muscles narrows the lumen and propels food in a caudal direction by sequential contractions. The two lateral ossicles and single medial ossicle of the gastric mill, which serve as teeth, are driven by a more complex six-phase pattern that produces a chewing

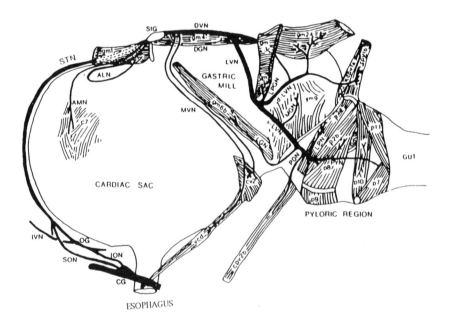

FIGURE 1. Lateral view of the lobster stomach indicating the nerves and muscles involved in both the gastric and pyloric rhythms. The major regions of the stomach are the esophagus, cardiac sac, gastric mill, and pyloric. The four ganglia involved are the stomatogastric (STG), the esophageal (OG), and the two commissural (CG). Adapted by permission of Selverston and Moulins.

Chapter X A Reprogrammable Central Pattern Generating Network

rhythm (Heinzel, 1988). Not all of the motor neurons to the cardiac sac and esophagus are known at present, and since they may be distributed among several ganglia, there is little likelihood that all of them will be found.

The detailed firing pattern of motor neurons in the pyloric and gastric mill circuits is shown in Fig. 2. These patterns are from combined *in vitro* preparations (stomatogastric, commissural and esophageal ganglia) and can be thought of as the basic output of the system. The pyloric pattern consists of three phases. The first phase consists of PD firing, which activates extrinsic muscles arranged in a way that dilates the pyloric chamber. This is followed by two phases of intrinsic muscle contraction, which produces a wave of contraction from the front of the chamber rearward.

The gastric mill has two sets of antagonistic movements. The medial tooth is pulled forward by the four GM motor neurons and reset by the DG neuron. The lateral teeth are pulled together by the LG and MG neurons and opened by the two LPG neurons. The phase relationship between all of the cells is an emergent prop-

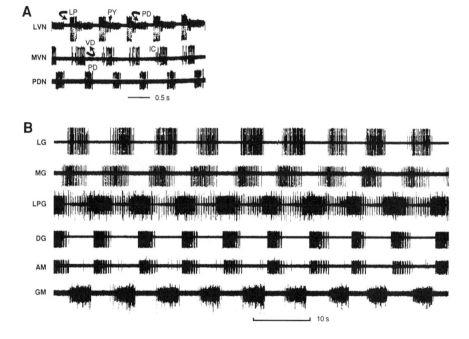

FIGURE 2. Burst patterns recorded extracellularly from motor nerves of the stomatogastric ganglion. (A) Pyloric pattern recorded from three nerves but containing all of the motor neurons. (B) Gastric mill pattern. Top three traces control the lateral teeth; bottom three traces control the medial tooth.

erty of the stomatogastric circuits shown in Fig. 3. Note that only 11 neurons are involved in the production of the basic gastric pattern and 14 in the pyloric.

VI. Modulation of the Circuits

Both the pyloric and gastric mill circuits are modulated by chemical substances released from specific input fibers or carried to the stomatogastric ganglion in the blood. The modulatory substances bind to receptors on some of the gastric or pyloric neurons, activating biochemical cascades that alter the biophysical properties of the cells and synapses in specific ways. For example, modulators can induce oscillatory properties, turn on plateau potentials, or change synaptic strength. Such changes fundamentally alter the circuit, resulting in modification of the state of the

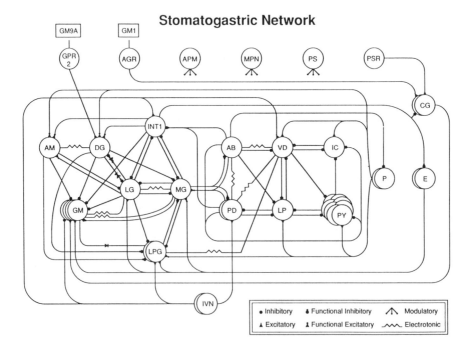

FIGURE 3. Neural circuits present in the stomatogastric ganglion and some of the extrinsic inputs to the system. The gastric mill is composed of the neurons on the left, AM, DG, Int1, LG, MG, GM, and LPG. The pyloric CPG is made up of the AB, PD, VD, LP, IC, and PY neurons. Note that the two systems are coupled via several synaptic pathways. GPR and AGR are sensory neurons and APM, MPN, PS, and PSR are known neuromodulatory inputs.

Chapter X A Reprogrammable Central Pattern Generating Network

output pattern for as long as the substance is present. This change in the functional architecture is different from the pattern changes normally associated with conventional synaptic input. There are many other types of input to the pattern-generating circuit. Sensory feedback, command fibers, coordinating fibers, etc., all act to modify the normal rhythmic output pattern, but differ from the neuromodulatory input in that their action is on ligand-gated channels and terminates immediately. We know that the neuromodulator-induced changes can persist both *in vivo* and *in vitro* for many hours.

In addition to "rewiring" any single circuit present in the stomatogastric system, studies have shown that neurons can move from one circuit to another (Hooper and Moulins, 1989; Weimann *et al.,* 1991), can form blended circuits (Dickinson *et al.,* 1990), and in some cases form entirely new circuits (Meyrand *et al.,* 1991). All of these changes can be produced with the appropriate neuromodulatory substance or by stimulating a neuron containing such a substance. These results point strongly to the conclusion that there are not separate "hardwired" circuits for particular behaviors but that CPGs are extremely flexible in carving out the circuits for different behaviors (Grillner, 1991). It is now clear that both the individual neurons and the circuit as a whole are in a fluid state governed by both fast synaptic inputs and chemical modulation. The circuits can be thought of as reprogrammable by analogy with a computer CPU, in which each set of instructions, or program, causes a fixed input–output behavior. In the biological case, a combination of modulators programs a specific, repeatable, spatiotemporal motor pattern by functionally rewiring the circuit, i.e., changing cellular properties and synaptic strengths.

There are typically three distinct kinds of movements when one observes the gastric mill directly with an endoscope (Fig. 4). One set of movements can be referred to as the "squeeze" mode because the three cusps of all three teeth converge onto a common point, squeezing any food caught between them. In the rest position, when the gastric mill is inactive, the lateral teeth are spread apart and the medial tooth is held in the dorsal–caudal position. When the squeeze mode starts, there is a simultaneous movement of the teeth toward each other. The teeth then move apart simultaneously, so this mode consists of only two phases, one when the teeth come together and the second when they move apart. In this mode note that only the cusps of the teeth come in contact with one another.

A second mode is called "cut and grind." In the cut and grind mode, the lateral teeth come together simultaneously and then move slightly backward. There is a delay in the forward, or power stroke movement of the medial tooth, but when the movement starts, the serrated edges of the medial tooth grind along the file of the lateral teeth until all three of the cusps have come together. The initial pulling

236　　　　　　　　　　　　　　　　　　　　　　　　Part III　Computer Modeling

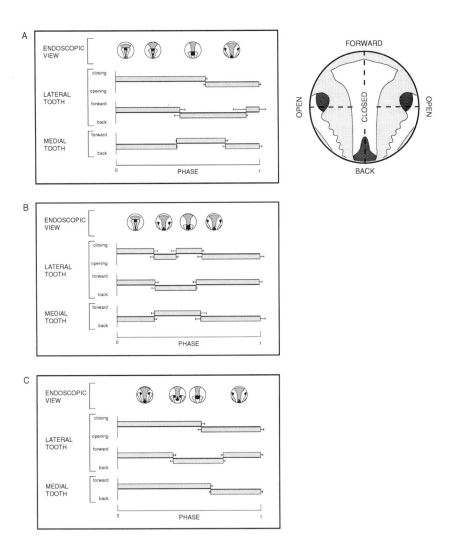

FIGURE 4. Chewing modes observed with an endoscope. Top shows diagrammatic view of teeth in their open resting position (lateral teeth are open and medial tooth is back) and their principal movement directions. (A) Cut and grind mode of chewing starts from the open resting position. The cut phase begins as the lateral teeth close until the serrated edges are together and the cusps are as far forward as possible. The grind phase starts while the lateral teeth remain in the closed position and move back as they grind along the file of the medial tooth. This phase ends when the cusps of the lateral and medial teeth come together. The teeth then return to the original resting position. (B) The cut and squeeze mode starts from the open resting position. The cut phase is identical to that just described. The squeeze has two phases. First,

together of the lateral teeth has the effect of cutting, whereas the medial tooth movement has the effect of grinding. Note that there are four separate phases to the cut and grind movement, making it considerably more complex than the squeeze mode and more similar to the locomotor sequences of appendages.

The third and least studied mode is termed "cut and squeeze." Here, the lateral teeth come together but then open and all three cusps then produce a squeeze movement. This mode is sometimes observed spontaneously, but is most often seen following injection of the mammalian peptide cholecystokinin.

VII. Mechanisms of Proctolinergic Modulation

Using two neuropeptides, proctolin and cholecystokinin (CCK), we have examined the neural mechanisms underlying different chewing modes exhibited by the gastric mill. Proctolin is a small peptide first isolated from the gut of cockroaches (Brown, 1967). The substance was subsequently purified and sequenced (Brown and Staratt, 1975) and commercially made antibodies are available. Proctolin has been studied extensively in insects and has been shown to have both central and peripheral modulatory effects (O'Shea and Schaffer, 1985). Authentic proctolin has also been located in cells of the commissural ganglia and fibers leading to the neuropil of the stomatogastric ganglion (Marder *et al.*, 1986).

Injection of proctolin into the animal's bloodstream had two different effects, depending on the concentration. Low blood concentrations of approximately 10^{-9} M led to squeeze movements, and higher concentrations (10^{-4} to 10^{-5} M) elicited cut and grind movements. When proctolin was bath applied to a combined *in vitro* preparation, dose-dependent excitatory effects could be seen superimposed on the spontaneous gastric mill pattern as shown in Fig. 5A and B, The effects were characterized by strongly increased burst durations as well as an increased spike rate in all motor neurons except the two LPGs. The effects were strongest in the DG and LG neurons and were accompanied by a phase change of the DG burst. DG continued firing during a large part of the LG burst period and into GM time even though the GMs are antagonists of DG and would not be expected to fire coincidentally with them. If we examine the movement trajectories (Fig. 5C and D), the

FIGURE 4. (cont.) the lateral teeth open and move back while the medial tooth starts to come forward. Second, all three cusps come together to squeeze food between them. (C) The squeeze mode of chewing starts from the open resting position. The lateral teeth cusps close and move forward, followed by a forward movement of the medial tooth until all cusps are together. The teeth then return to the resting position. Note that at no time do the lateral teeth edges come together, nor is there contact between the lateral teeth and the medial tooth until the cusps meet.

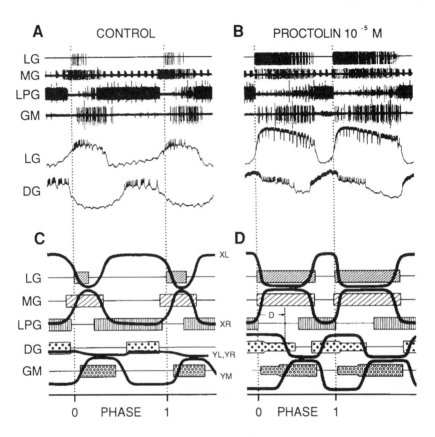

FIGURE 5. Differences in motor patterns produced by the peptide proctolin. (A) Normal control pattern. (B) Pattern produced after bath application of proctolin to the stomatogastric ganglion. (C) Block diagram of the burst pattern superimposed on the movement profile of the teeth during control conditions. (D) Pattern changes that occur after injection of proctolin. The letter D indicates a delay that is introduced and that alters the trajectory of the tooth movements. Reprinted from "Lobster Gastric Mill Oscillator" by A. I. Selverston, in *Neuronal and Cellular Oscillators* (1989), edited by J. W. Jacklet, p. 348, by courtesy of Marcel Dekker Inc.

control condition is consistent with the squeeze mode of chewing as would be triggered by low doses of proctolin. It is characterized by relatively short closings of the lateral teeth, consistent with the short LG and MG bursts and the simultaneous short gap in the firing of the LPG opener motor neurons. Under the influence of higher doses of proctolin, however, the time for the closing phase of the lateral teeth is much longer and this is reflected in the long burst durations of LG and MG. There is also a delay in the forward movement of the medial tooth and backward movement of the lateral teeth. This would be expected from the physiological data, since the antagonistic motor neurons DG and GM fire together for a long time and the resulting cocontraction would hold back the medial tooth until the muscles driven by the GM neurons overcome the ongoing contraction of their antagonist. The predicted movement is therefore entirely consistent with chewing in the cut and grind mode elicited by high doses of proctolin. This mode not only is characterized by different coordination of the motor neurons and different trajectories of the tips of the teeth but also shows the use of functionally different parts of the three teeth as compared to the squeeze mode.

It is important to note that although comparison of the *in vitro* recordings with the behavioral activity is suggestive, the data were obtained from different animals under drastically different conditions. What is required is simultaneous monitoring of the motor activity and behavior with different doses of proctolin. We have performed such experiments using a different neuromodulatory substance, a cholecystokinin peptide. A crustacean cholecystokinin-like peptide has been shown to be present in the lobster stomatogastric system and released into the stomatogastric ganglion following stimulation of the stomatogastric nerve, the sole input pathway to the ganglion (Turrigiano and Selverston, 1989). This peptide has also been shown to be important in the normal activity of the animal by correlating its feeding behavior (monitored by electromyograms of the GM muscles) and blood levels of a crustacean CCK-like peptide. When mammalian CCK-8 is injected into the blood of a lobster, its gastric mill is turned on and the activity parallels the rise and fall of the crustacean CCK-like peptide in the blood (Turrigiano and Selverston, 1990). A similar rise in blood levels with concomitant activation of the gastric mill occurs after feeding, and both natural and artificial activation of the mill can be terminated by injection of the specific CCK blocker proglumide (Turrigiano and Selverston, 1990).

Preliminary examination of gastric mill movements following injection of CCK-8 shows teeth movements that appear initially to have a cut and squeeze trajectory, i.e., movements entirely different from those produced by proctolin. This represents an entirely new pattern, and the cellular mechanisms will have to be examined *in vitro* as they were in the case of proctolin. We also have to be able to account for the fact that cut and squeeze is sometimes seen following injection of proctolin and cut and grind late after injection of CCK.

The lobster has many neuromodulatory substances whose effects have been studied at the cellular level but whose behavioral effects are as yet unexamined. If the pattern of present findings continues, then the possibility of producing a wide variety of stable behavioral outputs with this very simple circuit exists.

VIII. A Model for Simulating the Different Spatiotemporal Patterns Produced by Multipotent Neural Circuits

Given that considerable physiological data from invertebrate CPGs are available, a theoretical model of such systems is a good starting point for determining if they can be used as robotic controllers. As with all computer simulations, however, it is crucial to define a model that is simple enough to be computationally tractable while being realistic enough to generate all of the properties of the biological system. At one end of the modeling spectrum are abstract models that represent neurons as computational units with very simplified input–output properties. The units are connected to each other with "synapses" that can be strengthened or weakened according to some learning rules. These adjust the synaptic weights in a way that allows the units to retain a representation of the patterned input which they have been taught. This includes dynamic as well as static patterns, so in principle they can be used to model rhythmic motor outputs.

At the other end of the spectrum are neuron-based models incorporating the full menu of biophysical parameters. Such models are often too cumbersome to be of much value because rarely are values for all of the parameters known. Often they are as complicated as the neural systems they hope to simulate and so have little explanatory value in trying to deduce general principles. Any model of the gastric mill, for example, must be able not only to replicate the major features of the basic pattern but also to account for the different patterns produced by particular neuromodulators. Among the most important aspects of motor patterns are the phase relationships that exist between bursts of activity in different channels. Synchrony has to occur between bursts that fire agonists and disynchrony between those that fire antagonistic muscles. There are, in addition, bursts that occur at intermediate frequencies. A fundamental question that is still unanswered biologically is how the bursts are generated and how the phase relationships are established. Models that are able to represent some of the ionic currents, the basis for most of the cellular biophysical properties, appear to be most useful in simulating small CPGs.

We have found that overly simplified models of single gastric neurons were inadequate for simulating the most basic features of the gastric network when they were used in connectionist-type algorithms. By incorporating rather idealized representations of ionic currents into single-compartment neurons and synaptic connections with both feedforward and feedback loops constrained exactly as they are in the biological system, much better fits to the real data could be obtained. Before attempting to model the whole gastric mill network, we initially examined a small subset of the circuit that is found not only throughout the entire stomatogastric system but also in most other invertebrate motor circuits—the reciprocal inhibitory pair (Selverston and Moulins, 1985).

Each model neuron contains a fast and a slow conductance. By changing a single parameter (σ_f), the I–V characteristic curve of the fast conductance could be given a region of negative resistance that has the effect of making the cell oscillate. Changes in the strength (σ_s) or the time constant (τ_s) of the slow conductance alter the frequency of bursting. The idealized fast current corresponds biologically to the sum of a fixed leakage current and a fast noninactivating sodium current. The idealized slow current is similar to a combination of a slow hyperpolarization-activated ("sag") current plus a Ca^{2+}-activated K^+ current.

The current equation for a single cell is

$$\tau_m \frac{dV}{dt} + I_{fast} + I_{slow} + I_{syn} + I_{gap} = 0$$

where V is the membrane potential, and τ_m the membrane time constant. I_{fast} is the fast current, given by $I_{fast} = V - A^f \tanh[(\sigma^f/A^f)V]$. The expression for I_{fast} allows the parameter σ^f to be adjusted independently of the distance between the asymptotic values for I_{fast}, $V \pm A^f$. When $\sigma^f > 1$, I_{fast} has a negative slope characteristic (negative resistance), a property of all neurons that show endogenous bursting. $I_{slow} = q$ is the slow current, whose activation is $\tau_s \, dq/dt = -q + \sigma_s V$, with time constant τ_s, steady-state value $\sigma_s V$, and slow conductance σ_s. I_{syn} is the sum of all postsynaptic currents, where the current S from one synapse with maximum conductance W is given by $S = W_f(V_{pre})(V - E_{post})$. Here $f(V) = (1 + e^{-2V})^{-1}$ and E_{post} is the synaptic reversal potential. The postsynaptic current is inhibitory when $V > E_{post}$ and excitatory when $V > E_{post}$. Thus, for an inhibitory synapse the reversal potential E_{post} must be less than the lowest value of the postsynaptic membrane potential during slow wave oscillations. W is also called the synaptic weight or strength. I_{gap} is the sum of all gap junction currents, where the current due to the gap conductance G is $G(V_{pre}-V)$.

Spikes were not modeled because it has been shown in this system that synaptic interactions can occur without unitary postsynaptic potentials; i.e., transmission occurs in a graded fashion in addition to that following spikes (Graubard et al., 1980).

IX. Modeling Reciprocal Inhibitory Chains and Matrices

The most common form of connection in the gastric mill circuit is reciprocal inhibition. If two neurons with endogenous oscillatory properties are connected with weak reciprocal inhibitory synapses, it can be shown that the neurons will oscillate out of phase with one another. Because when one neuron fires it will inhibit the other and the firing period for each neuron is self-terminating, inhibition will be periodically removed and the network will produce alternating activity. Now if a third neuron with reciprocal connections to one of the first two is added, the total cell number is odd and the connectivity pattern ensures that when the middle cell fires it inhibits the end cells and when the end cells fire they inhibit the middle cell. As a result, the two end cells become synchronized and the middle cell becomes unsynchronized. This, like the two-cell pair, is intuitive because both end cells will be inhibited at the same time. Not intuitive, however, is that removing the endogenous bursting capability of the middle neuron has no effect on the synchronized bursting of the end cells. That is, weak inhibition between the oscillatory end cells and the middle passive neuron is able to force the end cells to entrain each other.

This phenomenon could be extended to chains with large numbers of neurons, the only stipulation being that the total number of neurons be odd. As shown in Fig. 6, the end neurons in a chain of seven neurons could be brought into synchrony after only a few cycles. The frequencies in the two end cells had to be reasonably similar for synchronization to occur, and the extent of frequency difference could be compensated for by changing the synaptic strengths within the chain. The stronger the synapses, the greater could be the frequency disparity between the end cells. If, for example, the frequency of one end cell is 25% greater than the frequency of the other end cell, entrainment to a common frequency occurs but the two end cells have a phase difference of over 30%.

These findings were extended to a 5 by 5 matrix of reciprocally coupled neurons (Fig. 7a). Cells in opposite corners were made to oscillate by giving them a region of negative resistance at different starting times, and they came into synchronization almost immediately. However, if the other two corner cells were made to oscillate together but at a different frequency from the first pair, all four cells would be entrained at a blended frequency but with *each pair firing at a different phase* from the other. This phase was not 0.5 but some intermediate value (Fig. 7b). This result may be of key importance in understanding motor pattern generation where neurons fire not only out of phase with one another but at intermediate phases as well. This suggests that a central pattern generator with only 11 neurons, but with a number of nested reciprocal inhibitory pairs, could produce the complex patterns seen in the gastric mill. If oscillatory neurons had the same frequency, they would fire either in phase or out of phase, depending on whether the chain was even or odd. But adding oscillatory neurons with different frequencies would ensure that some would fire at intermediate phases and a common overall frequency, a replica of the biological condition.

Chapter X A Reprogrammable Central Pattern Generating Network

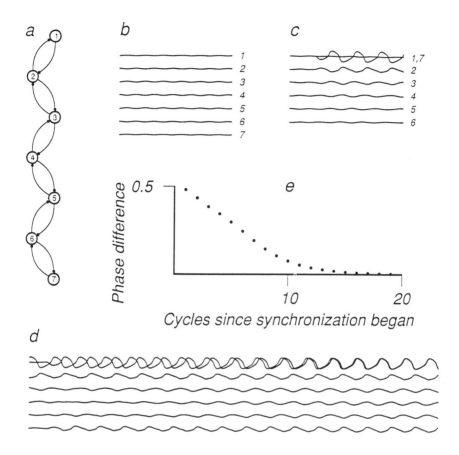

FIGURE 6. Synchronization of end cells of a seven-member chain. (a) Network diagram for a chain of seven cells connected by reciprocal inhibition. The odd-numbered cells synchronize out of phase with the even-numbered cells. (b) When the fast currents of all seven cells are not endogenous bursters, there are low-amplitude oscillations due to the synaptic currents that are generated spontaneously. (c) The traces of cells 1 and 7 are superimposed and cell 7 is made to oscillate. (d) Subsequently, cell 1 is turned into an oscillator and the initial phase difference of 0.5 begins to decrease immediately and quickly disappears. The synchronization process is summarized in (e), which shows the phase difference between the end cells plotted against the number of cycles until synchronization occurs.

X. Modeling the Gastric Mill CPG

We have started to construct a minimal model of the gastric mill having the cell and synaptic properties just described. The aim is not only to duplicate the important characteristics of the gastric motor pattern and perturbations to it but also to be

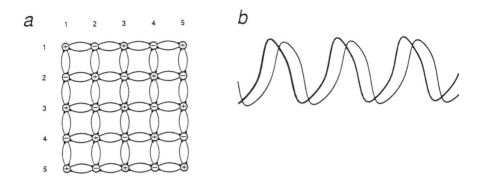

FIGURE 7. Synchronization in a 5 by 5 matrix of reciprocal inhibitory connections. When cells in opposite corners, cell 1: 1 and cell 5: 5, are made to oscillate, they become synchronized. Similarly, cells 5: 1 and 1: 5 can become synchronized. However, if the latter two cells are also given a different frequency, both pairs become entrained to a common frequency but at an intermediate phase.

able to shift the pattern to states equivalent to those produced by modulatory action. The gastric mill pattern generator is very robust, so that when the membrane potentials of single cells are perturbed by current injection, the pattern very quickly resumes its normal operation. When the overall frequency changes, the phase relationships between the cells remain constant. Even when individual cells are removed from the network, the remaining cells continue to generate a rhythm. The pattern-generating mechanism appears to be distributed throughout the gastric mill network (Selverston, 1987) and requires nonspecific modulatory excitation from higher centers in order to be active.

A model built with cells based on relaxation oscillator principles (Pavlidis, 1973) and using graded synaptic transmission captures many experimental characteristics of the gastric motor pattern. The cells of the biological network (Fig. 3) can be divided into two subsets, the medial, containing the DG, AM, and GM neurons, which control the medial tooth, and the lateral subset, LG, MG, and LPGs, controlling the two lateral teeth. An interneuron, Int1, oscillates in phase with the lateral subset and helps coordinate the two subsets. There is a phase delay between the two subsets that has been difficult to model in the past; however, when the slow synapse between Int1 and the DG/AM (Selverston and Mulloney, 1974) is incorporated into a reduced model, this phase delay can be accounted for. The reduced model also can account for the coordinating role played by Int1, including the effects of experimentally removing it from the circuit.

XI. Modeling the Complete Network

When the entire gastric mill is simulated and constrained by the known physiological connections, a reasonably good fit to the biological output can be obtained if all of the cells are made into endogenous oscillators (Fig. 8A). The approximately correct patterns could be obtained through a wide range of gap junction and synaptic strength parameters. For this type of approach, the problem is not so much one of using learning algorithms to find particular sets of parameter values as one of understanding the phase-locking properties of networks of nonlinear oscillators.

When all of the cells are set so as not to be endogenous oscillators, the model again produces approximately correct patterns, the major discrepancy being the LPGs. With the inhibitory input to the LPGs reduced so that LG and MG were the dominant inputs, the LPGs fired in the correct phase (Fig. 8B). In addition, when membrane potentials were perturbed, the correct pattern resumed in less than a cycle. Not only did phase relationships remain constant over a wide range of frequencies, but also the pattern was not sensitive to changes in parameter values of up to 10%. The network could continue to generate a pattern even when several component cells were removed; in fact, as long as a single inhibitory pair remains in the network, some form of oscillatory activity continues. When the model's output is matched against the smoothed activity of the *in vitro* preparation, it can be seen that the model output can be phase matched to the lateral subset or to the medial subset, but the phase delay between the subsets requires a more simplified model.

XII. A Reduced Model of the Gastric Mill

To better examine the mechanism underlying the phase delay between the lateral and medial subsets and to simulate the results of Int1 killing, a four-cell collapsed model of the GM circuit was constructed. DG and AM are normally phase locked because of the strong electrotonic coupling between them and were modeled as a single cell. Similarly, LG and MG fire at approximately the same time, are electrically coupled, and could be lumped together. The four GM cells, which act mainly as followers to DG and AM and have weak feedback to the rest of the circuit, were ignored (Fig. 3). The model was significantly altered, however, by incorporating a slow synapse, known to exist physiologically (Selverston and Mulloney, 1974). This synapse, thought to be mediated by an A-current, could act as a delay line between Int1 and the DG/AM cells and thus be the source of the lateral–medial phase delay. With this change, the model reproduces the delay quite well (Fig. 9A).

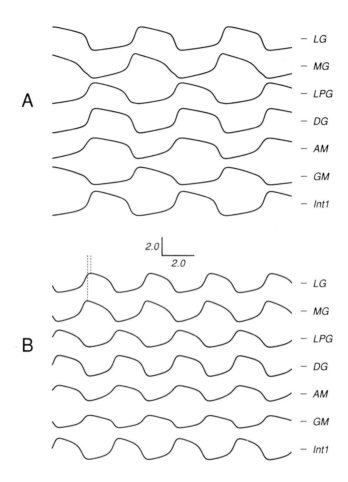

FIGURE 8. Pattern generation in the model network, with all known biological connections present. (A) Each cell is an endogenous oscillator. (B) No cell is an endogenous oscillator. In both, the pattern has approximately correct phase relationships (compare with Fig. 10A). Note the increased pattern frequency in B. The only difference in parameter values between A and B is that the common σ_f value for all the cells has been reduced, so that in B each cell would be quiescent if isolated. The phase of MG has been advanced by increasing the gain of the slow current, σ_s, so that the frequency of MG, if isolated and made to oscillate by increasing σ_f, would be higher. This is shown by the dotted vertical line in B. The horizontal line between each trace and its label indicates the zero level of the trace. All the excitatory weights had the same value, and with two exceptions all the inhibitory weights had the same, slightly larger value. The exceptions were for the LPG cell: the inhibitory weights from DG and AM were reduced by an order of magnitude. (Otherwise LPG was receiving inhibitory input at all phases of its cycle, and its trace would have reduced amplitude and sometimes exhibit phase walk-through.) All the gap junctions had the same conductance.

Chapter X   A Reprogrammable Central Pattern Generating Network   247

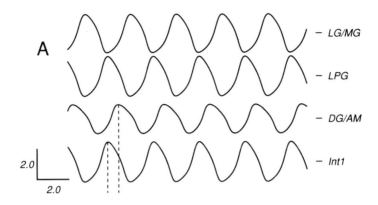

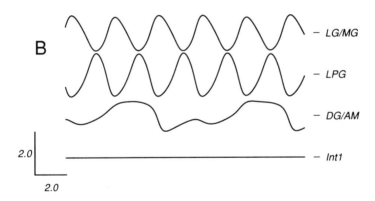

FIGURE 9. Pattern generation in the reduced four-cell model network and the effect of removing a cell. (A) The phase delay between LG/MG, LPG, Int1, and DG/AM, shown by the vertical dashed lines, was obtained by making a slow excitatory synapse from Int1 and DG/AM and by giving DG/AM a reduced gain (σ_s) for its slow current. The σ_f parameter for DG/AM was set so that DG/AM was oscillatory in isolation, with a very low frequency. (B) Int1 has been removed from the network by setting all its connections to zero. DG/AM is now entrained (3:1) to the LG/MG and LPG cells. In A, the DG/AM phase delay is in rough agreement with the DG/AM delay in Fig. 10A. In B, the frequency ratio (3:1) between the LG/MG, LPG pair, and DG/AM corresponds roughly to the frequency ratio (4:1) seen in the traces in Fig. 10B. The prolonged peak in the model DG/AM trace corresponds to the long bursts in DG and AM in Fig. 10B, and the small peak corresponds to one of the two small bursts in the AM trace.

Accounting for the pattern changes following Int1 kills proved to be a more difficult problem. For example before killing the Int1 cell shown in Fig. 10A, all of the cells are phase locked to a common frequency. After Int1 is killed (Fig. 10B), DG/AM and GM are phase locked and LG/MG and LPG are phase locked. DG/AM and GM also have a much lower frequency than do LG/MG and LPG; without Int1, the lateral and medial subsets appear to oscillate independently at different frequencies, whereas when Int1 is present, they oscillate at a common frequency. Since there is reciprocal inhibition between DG/AM and LG/MG, the output of the reduced model shown in Fig. 9A suggests that when Int1 is present, the reciprocal inhibition is sufficiently strong to entrain the lateral and medial subsets, but when Int1 is absent, the reciprocal inhibition is too weak to cause 1:1 entrainment (Fig. 9B).

XIII. Output of the Gastric Mill CPG and Behavior

The fictive output of the gastric mill CPG is useful for studying the cellular mechanisms that underlie neuromodulation, and endoscope studies can graphically

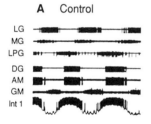

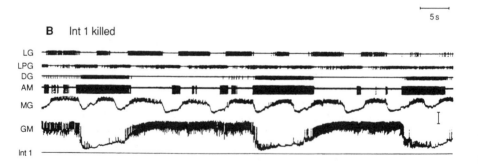

FIGURE 10. The effect of killing a cell on biological pattern generation. (A) Control intracellular recordings in combined preparation. (B) Same pattern after the Int1 cell was killed by photoinactivation. In A, all cells have the same frequency. In B, the LG, MG, and LPG have one frequency, whereas DG, AM, and GM are phase locked at a much lower frequency.

demonstrate the behavioral changes that occur in the tooth movements. However, the gap between these two levels of analysis is large. In order to convert the motor patterns to tooth movements it is necessary to know the relationship between the impulse pattern and muscle tension. And in order to convert the tension development to behavior, one must understand the mechanics of the muscle–ossicle system. Ideally, one would like to record from the CPG neurons while observing the teeth, and although this is difficult technically, work in this direction has already begun. A more experimentally tractable system is recordings from the muscles (electromyograms), which provide a good monitor of the centrally generated activity while simultaneously videotaping the movements of the three teeth. We have developed a system with which to do this that displays a cartoon of the three teeth on a computer terminal along with the centrally generated patterns. At present, an artificial transfer function moves the teeth in response to artificial impulse patterns in the motor nerves, and our principal effort is to obtain data that refine this transfer function to biologically realistic values. We are also obtaining more data about the relationship between the muscles and the actual movements.

As yet, we have not incorporated sensory feedback into the model; by doing so and making this a closed-loop system, the gastric mill CPG can serve as realistic biological inspiration for artificial legged robots. Such an approach can circumvent some of the problems encountered in treating the "central nervous system" of these models as a black box.

Acknowledgments

The work reported here is supported by ONR grant N00014-91-J-1720, and NIH grants 09322 and 5PO1 NS25916. The authors thank I-Teh Hsieh for expert technical assistance.

References

ALTMAN, J. (1982). "The Role of Sensory Inputs in Insect Flight Motor Pattern Generation" *Trends Neurosci.* **5,** 257–260.

BARNES, W. J. P., and GLADDEN. M. H. (1985). *Feedback and Motor Control in Invertebrates and Vertebrates.* Croon Helm, London and Dover, NH.

BROWN, B. E. (1967). "Occurrence of Proctolin in Six Orders of Insect," *J. Insect Physiol.* **23,** 861–864.

BROWN, B. E., and STARATT. A. M. (1975). "Isolation of Proctolin, a Myotropic Peptide from *Periplaneta Americana,*" *J. Insect Physiol.* **21,** 1879–1881.

DELCOMYN, F. (1980). "Neural Basis of Rhythmic Behavior in Animals," *Science* **210**, 492–498.
DICKINSON, P. S., MECSAS, C., and MARDER, E. (1990). "Neuropeptide Fusion of Two Motor-Pattern Generator Circuits," *Nature* **344**, 155–158.
GRAUBARD, K., RAPER, J. A., and HARTLINE, D. K. (1980). "Graded Synaptic Transmission between Spiking Neurons," *Proc. Natl. Acad. Sci. USA* **77**, 3733–3735.
GRILLNER, S. (1991). "Recombination of Motor Pattern Generators," *Current Opinion Neurobiol.* **1(4)**, 231–233.
GRILLNER, S., WALLEN, P., BRODIN, L., and LANSER, A. (1991). "Neuronal Network Generating Locomotor Behavior in Lamprey—Circuitry, Transmitters, Membrane Properties and Simulation," *Annu. Rev. of Neurosci.* **14**, 169–199.
HEINZEL, H. G. (1988). "Gastric Mill Activity in the Lobster. I. Spontaneous Modes of Chewing," *J. Neurophysiol.* **34**, 528–550.
HOOPER, S. L., and MOULINS, M. (1989). "Switching of a Neuron from One Network to Another by Sensory-Induced Changes in Membrane Properties," *Science* **244**, 1587–1589.
KATZ, P. S., EIGG, M. H., and HARRIS-WARRICK, R. M. (1989). "Serotonergic/Cholinergic Muscle Receptor Cells in the Crab Stomatogastric Nervous System: I. Identification and Characterization of the Gastropyloric Receptor Cells," *J. Neurophysiol.* **62**, 558–570.
MARDER, E. E., HOOPER, S. L., and SIWICKI, K. K. (1986). "Modulatory Action and Distribution of the Neuropeptide Proctolin in the Crustacean Stomatogastric Nervous System," *J. Comp. Neurol.* **243**, 454–467.
MEYRAND, P., SIMMERS, J., and MOULINS, M. (1991). "Construction of a Pattern Generating Circuit with Neurons of Different Networks," *Nature* **351**, 60–63.
O'SHEA, M., and SCHAFFER, M. (1985). "Neuropeptide Function: The Invertebrate Contribution," *Annu. Rev. Neurosci.* **8**, 171–198.
PAVLIDIS, T. (1973). *Biological Oscillators: Their Mathematical Analysis.* Academic Press, New York.
SELVERSTON, A. I. (1987). "Gastric Mill Mechanisms," in Selverston, A. I., and Moulins, M. (eds.), *The Crustacean Stomatogastric System* (pp. 147–180). Springer-Verlag, Heidelberg.
SELVERSTON, A. I. (1989). "Lobster Gastric Mill Oscillator," in *Neuronal and Cellular Oscillators,* J. W. Jacklet (ed.), Marcel Dekker, New York (339–370).
SELVERSTON, A. I., and MOULINS, M. (1985). *The Crustacean Stomatogastric Nervous System.* Springer-Verlag, New York.
SELVERSTON, A. I., and MULLONEY, B. M. (1974). "Organization of the Stomatogastric Ganglion of the Spiny Lobster: II. Neurons Driving the Medial Tooth," *J. Comp. Physiol.* **33–51**.
SIMMERS, J., and MOULINS, M. (1988). "A Disynaptic Sensorimotor Pathway in the Lobster Stomatogastric System," *J. Neurophysiol.* **59**, 740–756.
TURRIGIANO, G. G., and SELVERSTON, A. I. (1989). "A Cholecystokinin-like Peptide is a Modulator of a Crustacean Central Pattern Generator," *J. Neurosci.* **9**, 2486–2501.
TURRIGIANO, G. G., and SELVERSTON, A. I. (1990). "A Cholecystokinin-like Hormone Activates a Feeding-Related Neural Circuit in Lobster." *Nature* **344**, 866–868.
WEIMANN, J. M., MEYRAND, P. and MARDER, E. (1991). "Neurons That Form Multiple Pattern Generators: Identification and Multiple Activity Patterns of Gastric/Pyloric Neurons in the Crab Stomatogastric System," *J. Neurophysiol.* **65**, 111–122.

Chapter XI

Voyages through Weight Space: Network Models of an Escape Reflex in the Leech

SHAWN R. LOCKERY AND TERRENCE J. SEJNOWSKI

*Computational Neurobiology Laboratory, Salk Institute for Biological Studies,
La Jolla, California,
and The Howard Hughes Medical Institute*

I. Introduction

Whether the interest is in discovering the neural basis of behavior or in reverse engineering the nervous system to reveal its secrets of computation and control, modeling and simulation play a central role in the process of discovery. Many interesting behaviors are subserved by large, nonlinear, and highly interconnected neural networks that are too complicated to grasp intuitively. Modeling of complex networks could be used to gain insight into their biological counterparts. However, such models typically contain many free parameters that cannot be set by the available physiological or anatomical data. One approach to these difficulties is to choose values for these parameters using an optimization algorithm constrained by biological data. This chapter illustrates several different applications of one such algorithm called backpropagation (Rumelhart *et al.,* 1986), a widely used gradient descent technique, to the well-defined neural circuit of the local bending reflex of the leech. After introductory remarks on optimization in network modeling, we review our use of optimized network models to demonstrate the plausibility of distributed processing in the local reflex. We next show how varying the assumptions of the model led to unexpected local bending networks involving dedicated rather than distributed processing mechanisms. A final section demonstrates the use of optimization to study how the memory for nonassociative conditioning can be stored in distributed processing networks. These examples show how optimization techniques can be used to survey the range of possible network solutions subserving animal behaviors.

II. Parameter Space and Optimization

In modeling a neural system, one goal is to fit the performance of the model to the performance of the biological network. The first task is to define the architecture of the network and to identify the relevant parameters that govern the performance of the model. Ideally, each of these parameters is measured experimentally, as in the well-known Hodgkin–Huxley model of the action potential in the squid axon (Hodgkin and Huxley, 1952). In practice, however, most models contain free parameters that have not been measured. This is especially true in models of neural networks, since they may contain hundreds of synaptic connections and each connection strength (or weight) is a parameter. Models with many free parameters are hypotheses whose confirmation awaits experimental determination of the actual parameter values. They nevertheless have many important uses, including formulation of quantitative hypotheses, generation of experimental strategies and tests, and construction of biologically inspired artificial neural networks.

In mathematical terms, fitting a model network with free parameters is equivalent to searching parameter space for locations where the performance of the model matches the biological network. The axes in such a space are the parameters to be fit (Fig. 1). In many model networks these are the weights of connections between pairs of neurons and the space is a weight space. However, parameter space can include additional variables such as input resistances, time constants, and maximum ionic conductances.

A performance measure is associated with each point in parameter space. In a simple feedforward network, the performance measure is the network's input–output function, or the set of relationships between patterns of activity in the input neurons and patterns of activity in the output neurons. In feedback and other dynamical networks, the performance measure includes the patterns of activity of output neurons as functions of time. Experience with neural network models has shown that many different points in parameter space can have equivalent performance functions (Sejnowski and Rosenberg, 1987). Thus, the task of fitting a model to data is to find one of the many points in parameter space whose performance measure matches the biological one. Conclusions from the model should be robust across the many possible sets of parameters consistent with the data.

Many methods exist for searching parameter space. When the number of free parameters is small, all possible combinations can be tested. However, neural networks usually contain many hundreds of free parameters, and the search time

Chapter XI Voyages through Weight Space

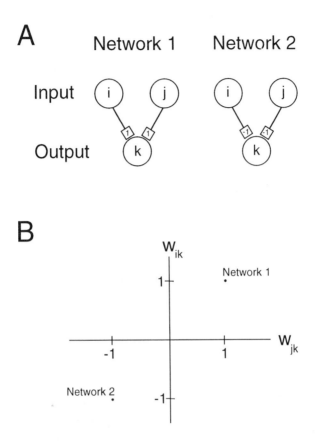

FIGURE 1. The concept of parameter space. A pair of networks with two input neurons and one output neuron is shown in A. The values in the boxes are the strengths of the synaptic connections in relative units. Each network can be located in a two-dimensional space in which the strength of the connection from neuron j to neuron k is plotted on the abscissa and the connection from neuron i to k is plotted on the ordinate as shown in B. Fitting a model network to data is equivalent to searching parameter space for locations where the performance of the model matches that of the biological network.

scales exponentially with the number of parameters for an exhaustive search. In this case, computational optimization methods are preferred, of which two widely used types are random (Monte Carlo) methods and gradient descent. In a simple random method, new points in parameter space are selected using a random number generator to pick parameter values. If the new point has better performance, the new parameters are adopted; if not, another random set is chosen (Miall and Wolpert, 1990). This procedure is repeated for many iterations until a satisfactory

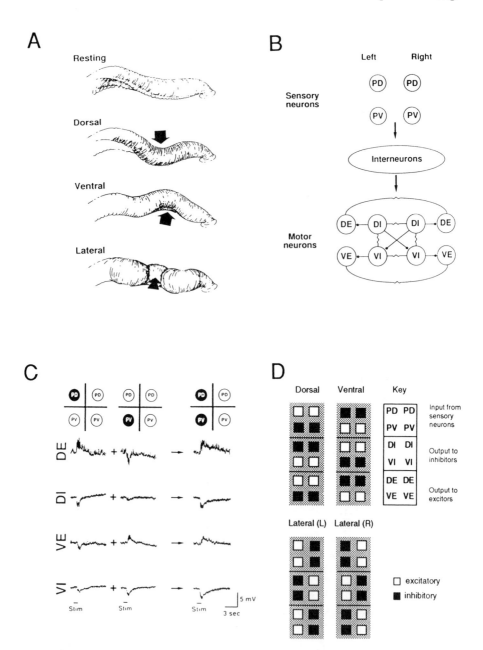

FIGURE 2. Network model of the local bending reflex of the leech. (A) Behavior: dorsal, ventral, and lateral stimuli produce local U-shaped bends. (B) Simplified neural circuit: the main input to the reflex is provided by the dorsal and ventral P cells (PD and PV). Control

Chapter XI Voyages through Weight Space

fit to the biological data is obtained. More efficient random search strategies include simulated annealing (Kirkpatrick *et al.,* 1983) and genetic algorithms (Goldberg, 1989; Beer and Chiel, Chapter XII, this volume). In gradient descent methods, the effect of each parameter on the fit of the model to the data is determined and parameters are changed in the direction that improves the fit, i.e., reduces the error of the model. In numerical differentiation, a simple gradient descent method, parameters in the model are increased one at a time by a small fraction. If this reduces the error, the change is retained; if not, the opposite change is made. As in random methods, this procedure is repeated until a satisfactory fit is obtained. Backpropagation is a more efficient procedure for computing the derivatives in which the changes required for all the parameters are calculated simultaneously. Once the derivatives have been calculated, they can be used to update the parameters as in numerical differentiation. Many variations of gradient descent are available, such as conjugate gradient (Battiti, 1992) and methods using second derivatives (Parker, 1987).

III. The Local Bending Reflex

We use backpropagation as means of searching the parameter space associated with a model of the local bending reflex in the leech. In response to a moderate mechanical stimulus, the leech withdraws from the site of contact (Fig. 2A). This is accomplished by contracting longitudinal muscles beneath the stimulus and relaxing longitudinal muscles on the opposite side of the body, resulting in a local U-shaped bend (Kristan, 1982). Major input to the local bending reflex is provided by dorsal or ventral pressure-sensitive mechanoreceptors or P cells (Fig. 2B, PD and PV; Nicholls and Baylor, 1968). Contraction and relaxation of longitudinal muscles are controlled by a total of eight types of motor neurons, an excitatory

FIGURE 2. (cont.) of local bending movements is largely provided by motor neurons whose projective field is restricted to one quadrant (left or right, dorsal or ventral) of the body. Dorsal and ventral quadrants are innervated by both excitatory (DE and VE) and inhibitory (DI and VI) motor neurons. Inhibitors inhibit excitors of the same body quadrant, and dorsal inhibitors inhibit contralateral ventral inhibitors (filled terminals). (C) Physiological input–output function: intracellular recordings from the four motor neurons in response to stimulation of one or two P cells (filled circles). The motor neurons shown have projective fields ipsilateral to the stimulated P cell(s). Similar recordings were obtained with other patterns of P cell stimulation and from contralateral motor neurons (not shown). (D) Hypothetical local bending interneurons dedicated to the detection of dorsal, ventral, and left (L) and right (R) lateral stimulus locations. Each interneuron has effects on motor output that are consistent with withdrawal from the stimulated site. White boxes represent excitatory connections; black boxes represent inhibitory connections. The presynaptic or postsynaptic neuron for each connection is given in the key.

(DE or VE) and inhibitory (DI or VI) type for the dorsal and ventral quadrants on the left and right side of each body segment (Fig. 2B; Stuart, 1970; Ort et al., 1974). The task of the interneurons in the local bending reflex is to compute a behavioral input–output function: the mapping relation between patterns of P cell activation and patterns of motor neuron excitation and inhibition sufficient for the animal to withdraw from the stimulus. The input–output function has been studied experimentally by making intracellular recordings from each of the eight types of motor neuron in response to P cells stimulated singly or in dorsal, ventral, or lateral pairs (Fig. 2C) (Lockery and Kristan, 1990a).

In a simple model of the local bending reflex, dorsal, ventral, and lateral bends are produced by types of interneurons specific for each form of the response (Fig. 2D). To determine how the interneurons in the reflex computed the local bending input–output function, a subpopulation of local bending interneurons contributing to dorsal local bending was identified using physiological and morphological criteria. Nine types of dorsal bending interneuron, which have excitatory connections to DE and receive excitatory connections from PD (Lockery and Kristan, 1990b), were found. Interestingly, several aspects of the other connections made by this subpopulation (Fig. 3A) are inconsistent with a commitment to only dorsal local bending and thus with the simple model (Fig. 2D). First, all but one type of dorsal

FIGURE 3. (opposite) (A) Average connection strengths of identified local bending interneurons (Lockery and Kristan, 1990b). Interneurons are numbered according to their location on the standard map of the leech midbody ganglion (Muller et al., 1981). The left (L) member of each left–right pair of interneurons is shown, except for cell 218, which is unpaired. Symbols as in Fig. 2D, except that box area is proportional to synaptic strength determined from pairwise intracellular recordings. White plus signs indicate excitatory connections of unknown strength determined from extracellular recordings of DE. Blank spaces indicate connections that have not been determined because the presynaptic neuron lies on the ventral surface of the ganglion while the postsynaptic neuron lies on the dorsal surface. The connections are not consistent with the dedicated interneuron model (Fig. 2D). (B) Connection strengths of interneurons 1L to 9L in a 40-interneuron model after optimization. Like all other interneurons in the model, these are excited by ventral as well as dorsal stimuli and have connections to most motor neurons. Thus the connections of model interneurons are qualitatively similar to the connections of identified local bending interneurons. (C) The 36-interneuron model. Interneurons 1L–9L (dorsal bending interneurons) were constrained to receive four excitatory P cell inputs and have outputs to excitatory motor neurons consistent with dorsal bending. Interneurons 10L-18L (unconstrained interneurons) were constrained only to receive 4 excitatory P cell inputs; no constraints were placed on the sign or amplitude of their output connections. After optimization to the local bending data set, most of the unconstrained interneurons had developed connections to the excitatory motor neurons that were consistent with ventral bending. (D) The 4-interneuron model. Both members of each left–right pair are shown. Networks with fewer interneurons could not be optimized to produce local bending motor output.

Chapter XI Voyages through Weight Space 257

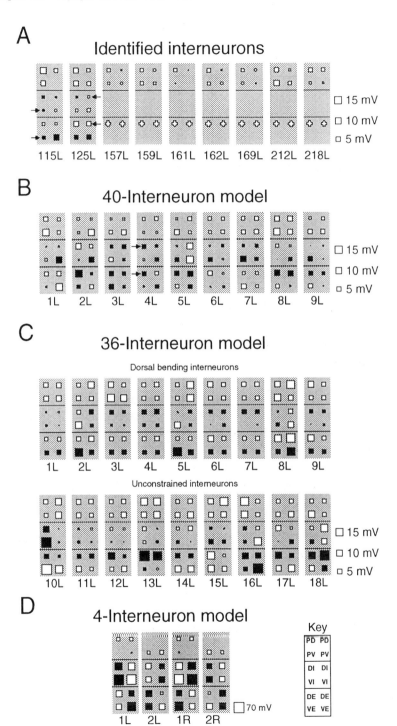

bending interneuron receive substantial excitatory input from PV, indicating that these neurons are also active in ventral and lateral local bends. Second, the connections from an interneuron to the inhibitory motor neurons are not always opposite in sign to its outputs to the excitatory motor neurons controlling the same body quadrant (Fig. 3A, interneuron 125, arrows). Thus, the connections of the subpopulation of local bending interneurons suggest a distributed processing strategy in which each interneuron is active in some or all forms of local bending; motor neuron excitation and inhibition would thus result from balanced combinations of appropriate and inappropriate inputs from many interneurons acting in concert.

IV. The Distributed Model of Local Bending

Modeling the reflex was prompted by the need to demonstrate that the distributed processing hypothesis is consistent with the responses of the interneurons and with the physiological details of the input–output function of the reflex. The possibility remained that another type of interneuron, as yet undiscovered, is required to produce accurately the known set of local bending input–output relations. The model has 4 sensory neurons, 8 motor neurons, and 40 interneurons and thus 480 connections, representing the actual local bending circuit (Fig. 2B) (Lockery and Sejnowski, 1992). This is referred to as the 40-interneuron model. The number of interneurons was based on an estimate of the number of local bending interneurons that remain to be identified in the biological network. Each neuron in the model is represented as a single electrical compartment (Segev *et al.*, 1989) with a physiologically determined input resistance and a time constant. The membrane potential is updated as a function of time and depends only on the synaptic current injected by chemical and electrical synapses.

The large number of connections in the model entails a parameter space too large to search by hand. Therefore, the backpropagation algorithm was used to adjust the connections. To make the model more realistic, backpropagation was forced to operate within additional physiological constraints. First, only excitatory connections were allowed from sensory neurons to interneurons in the model, because only excitatory connections have so far been found between sensory neurons and interneurons in the biological network (Lockery and Kristan, 1990b). Second, the sigmoidal function for interneurons and motor neurons was shifted so that the output of a unit that receives no net input is zero, as in leech neurons (Granzow *et al.*, 1985). Third, each interneuron on the left was paired with one on the right to maintain homologous input and output connections, in accordance with the overall bilateral symmetry of connections in the leech nervous system. Fourth, no connections between interneurons were allowed. Fifth, the model included all the chemical and electrical connections between the motor neurons. Synaptic weights from sensory neurons to interneurons and from interneurons to motor neurons were adjusted by the algorithm until the amplitudes and time courses of synaptic potentials recorded in the model motor neurons in response to each pattern of sensory input matched the smoothed and scaled replicas of the physiological synaptic potentials (Fig. 2C).

After training, the input and output connections of hidden units (Fig. 3B) in the model network qualitatively resemble the connections of identified local bending interneurons (Fig. 3A). In particular, interneurons receive inputs from ventral as well as dorsal input units, most have connections to all motor neurons, and the connections to the inhibitory motor neurons are not always opposite in sign to those onto the excitatory motor neurons controlling the same body quadrant (Fig. 3B, interneuron 4L, arrows). The similarity between model hidden units and interneurons in the biological network shows that the local bending input–output function can be achieved with interneurons similar to those identified physiologically. Additional interneurons with receptive and projective fields (output targets) that differ radically from the subpopulation of identified interneurons are not required. In hundreds of optimization runs from different initial positions in weight space, a different final point in weight space was reached each time. Thus, there are many different points in weight space that produce a physiologically accurate local bending input–output function utilizing a distributed processing strategy for computing the input–output relations.

V. Distributed Models of Local Bending with Functionally Specific Interneuronal Subpopulations

The correspondence between the identified interneurons and the interneurons in the 40-interneuron network establishes the possibility of using a distributed processing model to account for the input–output function of the reflex. However, several aspects of the output connections of the identified interneurons suggest that the actual network may use a strategy intermediate between the fully distributed solution of the 40-interneuron network and the dedicated interneuron solution of Fig. 2D. Consistent with the dedicated solution, all nine types of interneuron excite the DE motor neurons, and interneurons 115 and 125 also inhibit the VE motor neurons, effects that quite possibly are shared by the identified interneurons whose output connections have not yet been measured. On the other hand, both the inputs and the outputs to the inhibitory interneurons are consistent with a distributed solution, as noted above. Thus the identified interneurons are likely to constitute a subpopulation that operates in a partly dedicated and partly distributed fashion. This suggests a new version of the model having two subpopulations of interneurons. The first, like the identified interneurons, is partially dedicated to dorsal bending. The second subpopulation might be partially dedicated to ventral or lateral bending.

To determine whether such a model can account for the input–output function of the reflex, a population of 36 model interneurons was divided into two separate subpopulations of 18 interneurons (Lockery and Sejnowski, 1992). We used 18 interneurons in this subpopulation to reflect the fact that nine types of interneurons have been identified that are partly dedicated to dorsal bending and that all but one of these comprises a pair of bilaterally symmetrical interneurons. During optimization, interneurons in the first subpopulation, referred to as dorsal bending interneurons, were constrained to excite DE and inhibit VE. Interneurons in the second subpopulation had no such constraints and were referred to as unconstrained interneurons. In both subpopulations, the constraints on the input connections and symmetry were the same as in the 40-interneuron model. After optimization, the performance of the 36-interneuron network was identical to that of the 40-interneuron model. Inspection of the connections of the subpopulation of model dorsal bending interneurons showed that they were like the identified dorsal bending interneurons, in accordance with the additional constraints placed on this subpopulation (Fig. 3C). Inspection of the unconstrained interneurons showed that most (61%) had output connections to DE and VE that were consistent with ventral bending and had mixed effects on the inhibitory motor neurons. The other major type of interneuron either excited or inhibited all four excitatory motor neurons. These results show that the local bending input–output function can be computed by networks with a separate subpopulation that corresponds closely to the identified interneurons. The unconstrained interneurons complement the effects of the dorsal bending interneurons and suggest possible connectivities of as-yet-unidentified interneurons in the biological network. In repeating this simulation many times, none of the networks contained interneurons with outputs consistent with lateral bending. Thus, this type of interneuron is not necessary for computing the input–output function.

VI. Minimal Local Bending Networks

To determine whether 36 interneurons are required for local bending or whether a smaller number would suffice, we used optimization to seek solutions having fewer interneurons (Lockery and Sejnowski, 1992). To increase the likelihood that a solution would be found, the requirement that each interneuron have an input

Chapter XI Voyages through Weight Space

from all four sensory neurons was removed. The symmetry constraint was retained, but no constraints were placed on the output connections. Networks with fewer than four interneurons could not be optimized to produce recognizable local bending motor output patterns. Therefore, the minimum number of interneurons appeared to be four (Fig. 3D). However, we cannot rigorously exclude the possibility that in the networks with fewer than four interneurons the optimization procedure became trapped in a local minimum. That the local bending input–output function can be produced by as few as four interneurons indicates a high degree of redundancy may be present in the biological network.

The four-interneuron networks require two pairs of interneurons to accommodate three basic types of local bending: dorsal, ventral, and lateral. Enforcing this requirement led to hitherto unexpected mechanisms for computing the input–output function. While some of the four-interneuron networks used variations of the distributed processing network, some of the networks had interneurons that were specific for particular patterns of sensory input and motor output. The interneurons in the network shown in Fig. 4 were specific for dorsal or ventral inputs. Surprisingly, the same interneurons had exactly the motor outputs expected of lateral bending interneurons. Thus, the local bending input–output function can be produced by a novel solution involving dedicated interneurons in which there is a dissociation between the sensory and motor specificities of the two types of interneurons.

VII. Possible Engrams in Nonassociative Conditioning of the Local Bending Reflex

At the level of individual reflexes, learning can be defined as a change produced by experience in the input–output function of the underlying neural network. Learning in many systems is thought to be the result of changes in synaptic strength. Thus, when a reflex is conditioned, the network is moved to a point in weight space associated with a new input–output function. A major objective in the cellular analysis of learning and memory is to identify the sites of synaptic plasticity underlying the change in input–output function, a task that has been referred to as a search for the engram (Squire, 1987). Ideally, from an experimental point of view, the new point in weight space will be far from the original point so that the engram will comprise many large, hence easily detectable, changes. However, one might imagine that the same change in input–output function could be achieved by moving a much shorter distance in weight space. If so, then learn-

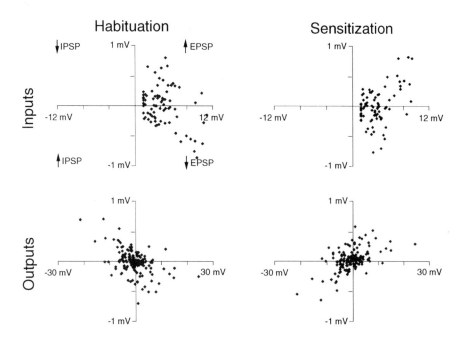

FIGURE 4. Scatter plots of the changes in synaptic strength produced by reoptimization. In each panel, the strength of a connection before reoptimization is plotted on the abscissa. The ordinate gives the change in that connection (before reoptimization − after reoptimization). Thus, connections whose strength increased fall in the upper right and lower left quadrants, while connections whose strength decreased fall in the lower right and upper left. Input connections were constrained to be positive in the model, hence there are no points on the left in the two upper panels. These plots show that habituation and sensitization were produced by the combined effect of increases and decreases in synaptic strength.

ing would be the result of some number of small changes in synaptic strength and the engram could thus be difficult to detect.

The local bending reflex exhibits several forms of nonassociative learning, including sensitization, warm-up, and habituation (Lockery and Kristan, 1991, and unpublished results). Little is known about the engrams for nonassociative learning in distributed processing systems. We therefore sought to examine the characteristics of engrams produced by using backpropagation to reoptimize the connections in a normal local bending network to produce habituated or sensitized local bending responses. Because backpropagation makes small changes in weights at each iteration, this approach was expected to yield solutions in which the final differences in synaptic strengths were small, if such solutions exist for the local bending network.

Chapter XI   Voyages through Weight Space

Backpropagation was thus used as a means of searching weight space for habituated or sensitized networks that were close to the original network, not as a model for the underlying mechanisms of synaptic plasticity whereby the learning is induced or retained. This approach could provide a worst-case scenario: if learning is distributed as widely as possible among the interneurons, how small are the changes in synaptic strength likely to be?

In these simulations we assumed that habituation entails a 50% reduction in the peak amplitude of the motor neuron synaptic potentials in each output pattern in the training set and that sensitization entails a 50% increase. Starting with the 40-interneuron network optimized for normal local bending, we reoptimized this network to the habituated or sensitized state. The change in synaptic strength at each synapse was determined by measuring the difference in the the peak of the simulated synaptic potential in response to a standard stimulus in the presynaptic neuron before and after reoptimization.

In reoptimizing six different 40-interneuron networks for habituation, the average change in synaptic strength (absolute value) was 0.22 mV; for sensitization it was 0.20 mV. The changes in synaptic strength were visualized in scatter plots where the change in synaptic strength was plotted against the strength of the connection before reoptimization (Fig. 4). In such a plot, the connections that increased in strength fall into the upper right and lower left quadrants, those that decreased in strength fall into the upper left and lower right, and unchanged connections lie along the abscissa. The scatter plots show that the engram produced by backpropagation was widely distributed, since almost every input and output connection in the network changed. A simple model of nonassociative learning predicts that habituation is due to decreases in synaptic strength and sensitization to increases in synaptic strength. For habituation, the scatter plots revealed that while most of the changes were consistent with the simple model, many increases in synaptic strength also occurred, in both the input and output connections of the interneurons. A similar effect was noted in sensitization, where many decreases in synaptic strength occurred. Taken together, these results show that, for each normal local bending network model, there exist nearby positions in weight space associated with habituated or sensitized motor output. Moreover, the nearby solutions involve a mixture of increases and decreases in synaptic strength, regardless of whether motor output increases or decreases in the learning.

The existence of habituated and sensitized networks involving many small changes raises the question of whether such changes would be detectable in practical physiological experiments. This was addressed by asking how much of the change in motor output could be accounted for by all the changes that were larger than a given sensitivity threshold. In a quantal analysis of a central synapse in the leech, Nicholls and Wallace (1978) were able to resolve differences in synaptic potentials as small as 0.25 mV. At this level of resolution, approximately 40% of the learning encoded by the distributed engrams produce by backpropagation would be detectable. This sets an approximate lower bound on the detectability of nonassociative learning in the local bending reflex.

VIII. Conclusion

We have used backpropagation as an optimization algorithm to explore the weight space associated with a model of the distributed processing of sensory information in the local bending reflex of the leech. In optimizing the 40-interneuron model we found, as in other networks, that there are many different points in weight space that produce a physiologically accurate input–output function. In restricting the algorithm to smaller regions of weight space by limiting the value of interneuron output weights to observed ranges, as in the 36-interneuron networks, we found that qualitatively different networks with populations of dorsal ventral bending interneurons are also possible. A further restriction to the still smaller region of weight space defined by a network with only four interneurons showed that this was the minimum number of interneurons necessary and revealed unexpected types of dedicated interneurons. Finally, in reoptimizing networks to produce habituated or sensitized local bending responses, we found that the memory for nonassociative learning in distributed processing networks can involve many small changes at almost every weight in the network, a situation that could be hard to uncover in practical physiological experiments.

Whether the local bending reflex operates as any of these models suggests will require identification of the as-yet-undiscovered local bending interneurons and measurement of their input and output connections strengths. Whether memory is encoded as reoptimization suggests will require identifying the sites of synaptic plasticity underlying nonassociative learning in the reflex. Whatever the results, these prior explorations of the local bending weight space provide a framework in which to place the actual biological solutions and thus deepen our understanding of the solutions nature has chosen. Far from being limited to well-defined invertebrate networks, this approach is a general one that can be applied to any neural system for which the input–output function is known or can reasonably be assumed. It should therefore be useful in a great variety of modeling studies (Zipser and Andersen, 1988; Lehky and Sejnowski, 1988; Anastasio and Robinson, 1990; Fetz *et al.,* 1990; Servan-Schreiber *et al.,* 1990; Tsung *et al.,* 1990; Zipser, 1991; Krauzlis and Lisberger, 1991; Pouget *et al.,* 1992).

Acknowledgments

Supported by NIH and HHMI postdoctoral fellowships.

References

ANASTASIO, T. J., and ROBINSON, D. A. (1990). *Biol. Cybern.* **63**, 161–167.
BATTITI, R. *Neural Comp.* **4**, 141–166.
FETZ, E. E., SHUPE, L.E., and VENKATESH, N. M. (1990). *International Joint Conference on Neural Networks.* San Diego, CA, 675–679.
GOLDBERG, D. E. (1989). *Genetic Algorithms in Search, Optimization and Machine Learning.* Addison-Wesley, Reading, MA.
GRANZOW, B., FRIESEN, W. O., and KRISTAN, W. B., JR., (1985). *J. Neurosci.* **5**, 2035–2050.
HODGKIN, A. L., and HUXLEY, A. F. (1952). *J. Physiol.* **117**, 500–544.
KIRKPATRICK, S. C., GELATT, D.J., and VECCHI, M. P. (1983). *Science* **220**, 671–680.
KRAUZLIS, R. J., and LISBERGER, S. G. (1991). *Science* **253**, 568–571.
KRISTAN, W. B., JR. (1982). *J. Exp. Biol.* **96**, 161–180.
LEHKY, S. R., and SEJNOWSKI, T. J. (1988). *Nature* **333**, 452–454.
LOCKERY, S. R., and KRISTAN, W. B., JR. (1990A). *J. Neurosci.* **10**, 1811–1815.
LOCKERY, S. R., and KRISTAN W. B., JR. (1990B). *J. Neurosci.* **10**, 1816–1829.
LOCKERY, S. R., and KRISTAN, W. B., JR. (1991). *J. Comp. Physiol. A* **168**, 165–177.
LOCKERY, S. R., and SEJNOWSKI, T. J. (1992). *J. Neurosci.* (to appear).
MIALL, C., and WOLPERT, D. (1990). In Selverston, A.I. (ed.), *Neural Computation* (pp. 77–88). Society for Neuroscience, Washington, DC.
MULLER, K. J., NICHOLLS, J.G., and STENT, G. S. (1981). *Neurobiology of the Leech,* Cold Spring Harbor Laboratory, Cold Spring Harbor, NY.
NICHOLLS, J., and WALLACE, B. G. (1978). *J. Physiol. (Lond.)* **201**, 157–170.
NICHOLLS, J. G., and BAYLOR, D. A. (1968). *J. Neurophysiol.* **31**, 740–756.
ORT, C A., KRISTAN, W.B., JR., and STENT, G.S. (1974). *J. Comp. Physiol. A* **94**, 121–154.
PARKER, D. B. (1987). *First International Conference on Neural Networks II,* San Diego, CA, pp. 593–600.
POUGET, A., FISHER, S.A., and SEJNOWSKI, T. J. (1992). In Moody, J.E., Hansons S. J., and Lippmann, R. P. (eds.), *Advances in Neural Information Processing Systems,* Vol. 4, Morgan Kaufmann Publishers, San Mateo, CA, (pp. 412–419).
RUMELHART, D. E., HINTON G. E., and WILLIAMS, R. J. (1986). *Nature* **323**, 533–536.
SEGEV, I., FLESHMAN, J.W., and BURKE, R. E. (1989). In Koch, C., and Seger, I. (eds.), *Compartmental Models of Complex Neurons* (pp. 63–96). MIT Press, Cambridge, MA.
SEJNOWSKI, T. J., and ROSENBERG, C. R. (1987) *Complex Syst.* **1**, 145–168.
SERVAN-SCHREIBER, D., PRINTZ, H., and COHEN, J. D. (1990). *Science* **249**, 892–895.

SQUIRE, L. R. (1987). *Memory and Brain.* Oxford University Press, New York.
STUART, A. E. (1970). *J. Physiol.* **209,** 627–646.
TSUNG, F. S., COTTRELL, G.W., and SELVERSTON, A. I. (1990). *International Joint Conference on Neural Networks I,* San Diego, CA, pp. 169–174.
ZIPSER, D. (1991). *Neural Comp.* **3,** 179–193.
ZIPSER, D. and ANDERSEN, R. A. (1988). *Nature* **331,** 679–684.

Chapter XII
Simulations of Cockroach Locomotion and Escape

RANDALL D. BEER
Department of Computer Engineering and Science and Department of Biology
Case Western Reserve University
Cleveland, OH

HILLEL J. CHIEL
Department of Biology and Department of Neuroscience
Case Western Reserve University
Cleveland, OH

I. Introduction

A basic premise of this volume is that the fields of neuroethology and robotics would benefit a great deal from a closer relationship. The twin problems of understanding the mechanisms underlying natural animal behavior and of designing controllers for autonomous mobile robots are not nearly so different as they might at first appear. In this chapter, we would like to argue that computer simulation has much to offer both of these endeavors and that it can often serve to facilitate their interaction.

Simulating the neural basis of natural animal behavior has been termed *computational neuroethology* (Beer, 1990; Achacoso and Yamamoto, 1990; Cliff, 1991). In order to bridge the gap between neural circuitry and behavior, computational neuroethology simulates the generation of behavior through the interaction between models of an animal's environment, its body, and its nervous system. Such simulations can be carried out for two related but distinct reasons: (1) they can deepen our understanding of the neural control of behavior; (2) they can serve as the basis for abstracting biological control principles for application in other contexts.

While these two endeavors have much in common, the goals of a biological model are not identical to those of an engineering model. A biological model should be judged by the extent to which it accounts for and illuminates the observed behavioral and neurobiological data, as well as the extent to which it generates testable experimental hypotheses. On the other hand, an engineering model should be judged by the extent to which it clarifies the feasibility of a proposed solution to some problem. Despite the fact that an engineering model might have very interesting biological *implications* (and vice versa), one should be clear about the primary intent of a given simulation.

In this chapter we describe two examples of work in computational neuroethology. In Section II we review a simulation of the neural basis of cockroach locomotion. Although the primary intent of this model was to demonstrate the practical applicability of biological control principles to robotics, we show that it has interesting neuroethological implications as well. Section III describes a simulation of the cockroach escape response. Although the primary goal of this model is to elucidate the neural basis of this behavior, we argue that it may also eventually have important engineering implications.

II. A Computer Model of Cockroach Locomotion

There is currently a great deal of interest in the construction of legged robots. Legs possess obvious advantages for locomotion over complex and varied terrain. However, controlling the locomotion of legged robots has turned out to be a difficult and computationally demanding task. In contrast, insects are capable of robustly solving the complex coordination problems raised by legged locomotion in real time for a wide variety of terrains (Graham, 1985). What can we learn from biology?

A. Background

During normal walking, insects exhibit a variety of statically stable gaits. Slowly walking insects often utilize gaits consisting of distinct metachronal waves, in which each leg begins its swing immediately following the swing of the leg behind it. Fast-walking insects, on the other hand, typically utilize the tripod gait, in which the front and back legs on each side of the body swing in phase with the middle leg on the opposite side. Wilson (1966) suggested that the entire range of observed insect gaits could be explained by assuming that fixed, antiphasic metachronal waves on each side of the body increasingly overlap as walking speed increases.

The overall neural organization of the walking system of the American cockroach *(Periplaneta americana)* has been studied by Pearson and his colleagues (Pearson *et al.*, 1973; Pearson, 1976). This work led to the development of the *flexor burst-generator* model for cockroach locomotion. In this model, the basic swing and stance movements of each leg are generated by a central pattern gener-

Chapter XII Simulations of Cockroach Locomotion and Escape

ator whose operation is modulated by feedback from sensory structures in the insect's legs. The overall speed of walking is set by descending influences from higher command centers. Interleg coordination is achieved by inhibitory interactions between adjacent pattern generators.

B. Model

Inspired by Pearson's flexor burst-generator model, we designed a distributed neural network architecture for hexapod locomotion (Beer *et al.*, 1989; Beer, 1990). This circuit was used to control the walking of a simulated insect, whose body model is illustrated at the top of Fig. 2. Each of the model's six legs can swing forward or backward and each leg possesses a foot that can be either up or down. A leg whose foot is up can swing freely. When its foot is down, any forces generated by a leg are applied to the body. The body can move only when it is statically stable (i.e., the center of mass falls within the polygon of support formed by the stancing legs).

Our locomotion circuit is shown in Fig. 1. Each of the model neurons is capable of capacitatively integrating the weighted sum of its inputs and passing this sum through a saturating linear threshold function. At the center of each leg controller is a pacemaker neuron whose output rhythmically oscillates due to the presence of a simplified voltage-dependent intrinsic current (for details see Beer, 1990). The interval between bursts depends linearly on the tonic level of excitation that the pacemaker receives, with excitation decreasing this interburst interval and inhibition increasing it. In addition, a strong excitatory pulse between bursts or a strong inhibitory pulse within a burst can reset a pacemaker's burst rhythm. These pacemaker neurons implement the flexor burst-generator in Pearson's model.

A pacemaker burst initiates a swing by inhibiting the foot and backward swing motor neurons and exciting the forward swing motor neurons, causing the foot to lift and the leg to swing forward. Between pacemaker bursts, the foot is down and tonic excitation from the command neuron moves the leg backward. The output of the central pattern generator is fine-tuned by feedback from two sensors that signal when a leg is nearing its extreme forward or backward position. A forward angle sensor encourages a pacemaker to terminate a burst by inhibiting it, whereas a backward angle sensor encourages a pacemaker to initiate a burst by exciting it. The forward angle sensor also makes direct connections to the motor neurons, modeling leg reflex pathways described in Pearson's model.

In order to generate statically stable gaits, the swings of the individual legs must be coordinated in some way. Following Pearson's model, we inserted mutually inhibitory connections between the pacemaker neurons of adjacent legs. At high frequencies of stepping, this architecture produces the commonly observed tripod gait. At slower stepping frequencies, however, the network is underconstrained. Statically stable gaits are not reliably generated. In particular, no metachronal waves are produced. Thus, we augmented Pearson's model with a mechanism for generating metachronal waves. The basic idea was to phase-lock the pattern gener-

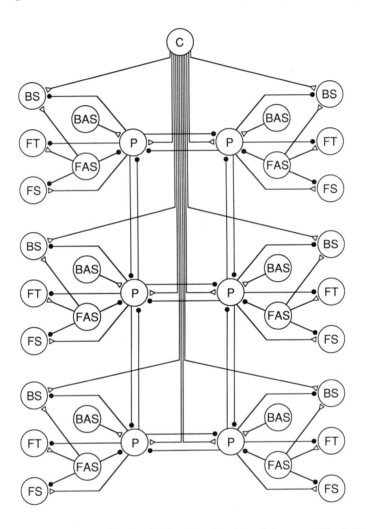

FIGURE 1. Locomotion controller circuit. The following abbreviations are utilized: C, command neuron; FT, foot motor neurons; FS and BS, forward swing and backward swing motor neurons; P, pacemaker neurons; and FAS and BAS, forward and backward angle sensors. Excitatory connections are shown as open triangles and inhibitory connections are shown as filled circles.

ators on each side of the body into a metachronal relationship by lowering the burst frequency of the rear pattern generators (Graham, 1977). We implemented this entrainment by slightly increasing the angle ranges of the rear legs, thereby lowering their stepping frequency relative to the middle and front legs. We found that this phase locking could also be accomplished by lowering the strengths of the connections from the command neuron to the rear pacemakers, thus lowering their natural frequency relative to the middle and front pacemakers for any given command neuron setting (Chiel and Beer, 1989; Beer, 1990).

C. Results and Discussion

When embedded within the body model described above, this locomotion circuit is capable of producing a continuous range of statically stable insect-like gaits (Fig. 2). These gaits range from the tripod gait at high stepping frequencies to slower metachronal gaits at lower stepping frequencies. All of these gaits are produced by the same circuit simply by varying the tonic level of activity of the command neuron. In addition, smooth transitions between these gaits can be generated by continuously varying the command neuron activity.

In order to explore the robustness of this locomotion circuit, we performed a series of lesion studies on it (Chiel and Beer, 1989; Beer, 1990). Specifically, we examined the controller's response to such perturbations as the removal of selected sensors or connections. We found that the ability of this circuit to generate statically stable hexapod gaits was quite robust to such perturbations. For example, lesioning any single sensor or interpacemaker connection did not generally disrupt locomotion. Indeed, the controller could often tolerate the random removal of up to half of the interpacemaker connections.

In order to demonstrate the practical utility of our locomotion circuit, we embedded it in an actual hexapod robot (Beer *et al.,* 1992; Quinn and Espenschied, Chapter XVI, this volume). This physical implementation allowed us to explore whether the gaits that the circuit generates in simulation persist in the presence of such physical effects as noise, friction, inertia, and delays. Not only did the robot exhibit the same range of gaits observed in simulation, but it was similarly robust to perturbations (Chiel *et al.,* 1992). These results demonstrate that the observed gaits are not simply artifacts of the many physical simplifications of the simulated body but are in fact robust properties of the locomotion circuit. Of course, it is important to point out that this controller addresses only the problem of gait control. Equally important problems in legged locomotion include postural control, directional control, negotiation of complex terrain, and compensation for leg damage or loss.

272　　　　　　　　　　　　　　　　　　　Part III　Computer Modeling

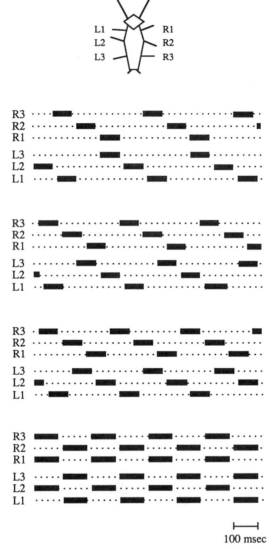

FIGURE 2. Gaits generated by the model as the command neuron activity is varied from lowest (top) to highest (bottom). The gaits shown correspond to command neuron outputs of 0.25, 0.3, 0.35, and 1, respectively. Black bars denote the swing phase of a leg. The body model and leg labeling conventions are shown at top.

Chapter XII Simulations of Cockroach Locomotion and Escape

It is interesting to compare our locomotion controller to one that has been proposed by Brooks (1989; Chapter XV, this volume). The two controllers share a common emphasis on distributed approaches to legged locomotion with low computational demands. In Brooks' controller, each leg is driven by a chain of peripheral reflexes. For example, whenever a leg is lifted it is swung forward, and whenever it reaches an extreme forward position it is put down. The movements of the individual legs are coordinated by a centralized mechanism that tells each leg when to lift. Different gaits are produced by explicitly modifying this mechanism. Such a controller is inherently less robust than ours because damage to a single sensor in a reflex chain or to the centralized gait sequencer may compromise the robot's locomotion.

Our locomotion circuit was part of a larger project to construct a simulated insect whose behavior was controlled by an artificial nervous system (Beer, 1990). The goal of this project was to demonstrate that neurobiological control principles could be effectively applied to artificial agents. The complete simulated insect could walk, wander, follow the edges of obstacles, and seek out and consume food. These behaviors were organized into a simple behavioral hierarchy. Like the locomotion controller, the neural circuit responsible for the artificial insect's consummatory behavior was directly inspired by a biological system, in this case the feeding circuitry of the marine mollusc *Aplysia*. Circuits controlling the remaining behaviors were designed to be as consistent as possible with known neurobiological principles but were not directly based on any existing circuit.

Although the primary intent of our locomotion model was to demonstrate the practical utility of biological control principles, we believe that it may have some interesting neuroethological implications as well. Perhaps the most obvious is that our simulation demonstrates the sufficiency of Pearson's flexor burst generator for generating the tripod gait. In addition, the simulation suggests a plausible entrainment mechanism that can account for a much larger range of insect gaits.

Interestingly, Delcomyn (Chapter II, this volume) has argued that, contrary to the impression left by Wilson's early analysis of insect gaits (Wilson, 1966), cockroaches and many other insects almost always use the tripod gait independently of their speed of progression. In this regard, it is worth noting that we have observed a nonlinear relationship between gait and command neuron activity in our model. The locomotion controller generates a tripod gait over two thirds of the command neuron's range of activity, with all of the remaining gaits occurring in the lowest third of this range (see Fig. 2). Because metachronal gaits have been reported in many insects (including cockroaches), it is possible that many different insects utilize a control architecture similar to that shown in Fig. 1, but in different operating ranges.

The lesion studies that we have performed on the locomotion model raise some additional issues of potential neuroethological interest. For example, we have found that the role of sensory information differs for the different gaits (Chiel and Beer, 1989; Beer, 1990). Sensory feedback is crucial for the establishment and maintenance of the slower metachronal gaits, but it appears to play no significant role in the tripod gait. Although the mechanisms are probably quite different, this result is intriguingly similar to conclusions that have been drawn for the American cockroach (Zill, 1985; Zill, Chapter III, this volume). In general, the interactions between the central and peripheral components of our locomotion circuit appear to be remarkably complex. This may shed some light on the difficulty of resolving the current controversy regarding the relative roles of central neural circuitry and sensory feedback in pattern generation (Pearson, 1985).

Given the results summarized in this section, it appears that we have succeeded in abstracting from the American cockroach a locomotion controller that offers a number of potential advantages over conventional locomotion controllers. First, whereas conventional controllers are often centralized (i.e., optimal control decisions are made by a single processor after simultaneously considering all sensory inputs and all performance requirements), our controller is fully distributed. No single component is uniquely responsible for the observed gaits. Rather, these gaits are actively constructed from the dynamics of interaction between the coupled pattern generators and the sensory feedback that they receive. Second, whereas centralized control approaches are often brittle, the lesion studies have demonstrated that our controller is robust to a variety of realistic perturbations, including sensor loss and partial communication failure between individual leg controllers. Third, whereas centralized controllers are usually computationally expensive, our controller is relatively efficient, consisting of only 37 neurons that could easily be implemented in discrete analog components. Because of its simplicity, this circuit could be made to operate at high stepping frequencies. Indeed, our robotic implementation of this controller operates in real time even though it is simulated on a personal computer. Finally, we have attempted to show that this simulation may also have interesting implications for the biological system from which it was abstracted.

III. A Computer Model of the Cockroach Escape Response

Elucidating the neural basis of animal behavior is a technically and conceptually difficult endeavor. A computer simulation of a given neuroethological system can aid this endeavor in a number of important ways: (1) It synthesizes diverse experimental data into an interactively manipulable model in which the relationship between neural events and behavioral events can be directly observed and manipu-

Chapter XII Simulations of Cockroach Locomotion and Escape

lated; (2) it tests our functional understanding of the system; (3) it encourages synergistic interactions between experiment and modeling as manipulations of the simulation suggest additional experiments that in turn further refine the model; (4) it can support observations and manipulations that are technically difficult *in vivo*. In this section, we describe our work to date with a computer model aimed at elucidating the neural basis of the cockroach escape response.

A. The Cockroach Escape Response

It is becoming generally accepted that many behavioral and cognitive capabilities of the human brain must be understood as resulting from the cooperative activity of populations of nerve cells rather than the individual activity of any particular cell. For example, distributed representation of orientation by populations of directionally tuned neurons appears to be a common principle of many mammalian motor control systems (Georgopoulos *et al.*, 1988; Lee *et al.*, 1988). Although the general principles of distributed processing are evident in these mammalian systems, the details of their operation are not. Without deeper knowledge of the underlying neuronal circuitry and its inputs and outputs, it is difficult to answer such questions as how the population code is formed, how it is read out, and what precise role it plays in the operation of the nervous system as a whole. In this section, we describe work with an invertebrate system, the cockroach escape response, that offers the possibility of addressing these questions.

Any sudden puff of wind directed toward the American cockroach *(Periplaneta americana)*, such as from an attacking predator, evokes a rapid directional turn away from the wind source followed by a run (Ritzmann, 1984; Ritzmann, Chapter VI, this volume). The initial turn is generally completed in approximately 60 ms after the onset of the wind. During this time, the insect must integrate information from hundreds of sensors to direct a very specific set of leg movements involving dozens of muscles distributed among three distinct pairs of multisegmented legs. The leg movements underlying the turn have been characterized using a high-speed video camera (Nye and Ritzmann, 1992). In addition, the response exhibits various forms of plasticity, including adaptation to sensory lesions. This system has also been shown to exhibit context-dependent responses. For example, if the cockroach is in antennal contact with an obstacle, it may modify its escape movements accordingly (Ritzmann *et al.*, 1991).

The basic architecture of the neuronal circuitry responsible for the initial turn of the escape response is known (Daley and Camhi, 1988; Ritzmann and Pollack, 1988, 1990). Characteristics of the initiating wind puff are encoded by a population of several hundred broadly tuned wind-sensitive hairs located on the bottom of the insect's cerci (two antenna-like structures found at the rear of the animal). The sensory neurons that innervate these hairs project to a small population of four pairs of ventral giant interneurons (the vGIs). These giant interneurons excite a larger population of approximately 100 interneurons located in the thoracic ganglia associated with each pair of legs. These type A thoracic interneurons (the TI_As) integrate information from a variety of other sources as well, including leg proprioceptors, tactile inputs, auditory inputs, and light inputs (Murrain and Ritzmann, 1988; Ritzmann et al., 1991). Finally, the TI_As project to local interneurons and motor neurons responsible for the control of each leg.

Perhaps what is most interesting about this system is that, despite the complexity of the response it controls and despite the fact that its operation appears to be distributed across several populations of interneurons, the individual members of these populations are uniquely identifiable. For this reason, we believe that the cockroach escape response is an excellent model system for exploring the neuronal basis of distributed sensorimotor control at the level of identified nerve cells. As an integral part of that effort, we are constructing a computer model of the cockroach escape response.

B. Neural Network Model

Whereas a great deal is known about the overall response properties of many of the individual neurons in the escape circuit as well as their architecture of connectivity, little detailed biophysical data is currently available. For this reason, our initial models have employed simplified neural network models and learning techniques. This approach has proved to be effective for analyzing a variety of neuronal circuits (e.g., Lockery *et al.,* 1989; Anastasio and Robinson, 1989). Specifically, using backpropagation, we trained sigmoidal model neurons to reproduce the observed properties of identified nerve cells in the escape circuit.

To ensure that the resulting models were biologically relevant, we constrained backpropagation to produce solutions that were consistent with the known structural characteristics of the circuit. The most important constraints we have utilized to date are the existence or nonexistence of specific connections between identified cells and the signs of existing connections. It is important to emphasize that we employed backpropagation solely as a means of finding the appropriate connection weights given the known structure of the circuit, and no claim is being made about its biological validity.

Chapter XII Simulations of Cockroach Locomotion and Escape

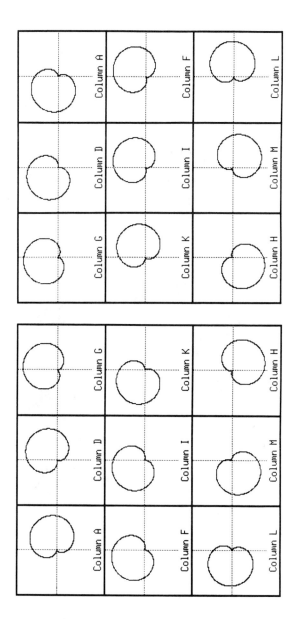

FIGURE 3. Model cercal wind fields. Each of the nine major hair columns on the left and right cerci is represented by a single model input.

Using this approach, we have reconstructed the observed wind fields of the eight ventral giant interneurons that serve as the first stage of interneuronal processing in the escape circuit. These wind fields, which represent the intensity of a cell's response to wind puffs from different directions, have been well characterized in the insect (Westin *et. al.*, 1977). The wind fields of individual cercal sensory neurons have also been mapped (Westin, 1979; Daley and Camhi, 1988). The response of each hair is broadly tuned about a single preferred direction, which we have modeled as a cardioid (Fig. 3). The cercal hairs are arranged in nine major columns on each cercus. All of the hairs in a single column share similar responses. Together, the responses of the hairs in all 18 columns provide overlapping coverage of most directions around the insect's body. The connectivity between each major cercal hair column and each ventral giant interneuron is known, as are the signs of these connections (Daley and Camhi, 1988). These constraints have also been incorporated into our model circuit (see first layer of Fig. 4).

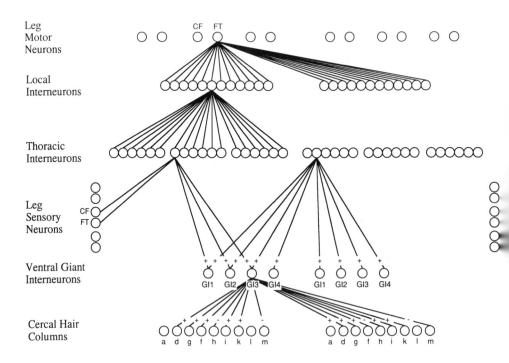

FIGURE 4. Neural network model of the cockroach escape circuitry. Only representative connection patterns between the layers of the model are shown. The signs of certain connections were constrained as shown.

Chapter XII   Simulations of Cockroach Locomotion and Escape

Using these data, each model vGI was trained to reproduce the corresponding wind field by constrained backpropagation.[1] The resulting responses of the left four model vGIs are shown in Fig. 5. The responses of the right model vGIs are mirror images of these. These model wind fields closely approximate those observed in the cockroach (Westin *et. al.,* 1977).

C. *Escape Turn Reconstructions*

Ultimately, we are interested in simulating the entire escape response. This requires some way to connect our neural models to behavior. In earlier work, Dowd and Comer demonstrated that simply integrating the responses of the vGIs on each side of the insect could account for the direction and amplitude of the escape turn (Dowd and Comer, 1988; Comer and Dowd, Chapter V, this volume). However, their model considered only angle of turn rather than actual leg movements. In contrast, we are interested in modeling the leg movements themselves, since it is only by properly controlling these movements that the insect's nervous system actually accomplishes a turn.

Toward that end, we have constructed a three-dimensional kinematic model of the insect's body that accurately represents the essential degrees of freedom of the legs during escape turns. For our purposes here, the essential joints are the coxal–femur (CF) and femur–tibia (FT) joints of each leg. The leg segment lengths and orientations, as well as the joint angles and axes of rotation, were derived from actual measurements (Nye and Ritzmann, unpublished data). The active leg movements during escape turns of a tethered insect, in which the animal is suspended by a rod above a greased plate, have been shown to be identical to those of a free-ranging animal (Camhi and Levy, 1988). Because an insect thus tethered is neither supporting its own weight nor generating appreciable forces with its legs, a kinematic body model can be defended as an adequate first approximation.

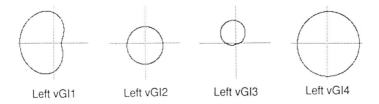

FIGURE 5. Wind fields of the four left model ventral giant interneurons.

---

[1] Strictly speaking, we are using only the delta rule here. The full power of backpropagation is not needed for this task since we are training only a single layer of weights.

The leg movements of the simulated body were controlled by a neural network model of the entire escape circuit (see Fig. 4). Where sufficient data were available, the structure of this network was constrained appropriately. The first layer of this circuit was described in the previous section and was not subject to further modification. There are six groups of six representative TI_As, one group for each leg. Within a group, representative members of each identified class of TI_A are modeled. Where known, the connectivity from the vGIs to each class of TI_As was enforced and all connections from vGIs to TI_As were constrained to be excitatory (Ritzmann and Pollack, 1988). Model TI_As also receive inputs from leg proprioceptors that encode the angle of each joint (Murrain and Ritzmann, 1988). The TI_A layer for each side of the body was fully connected to 12 local interneurons, which were in turn fully connected to motor neurons that encode the change in angle of each joint in the body model.

High-speed video films of the leg movements underlying actual escape turns in the tethered preparation for a variety of different wind angles and initial joint angles have been made (Nye and Ritzmann, 1992). The angles of each joint before wind onset and immediately after completion of the initial turn were used as training data for the model escape circuit. Only movements of the middle and hind legs were considered because the movements of the front legs were more complex. After training with constrained backpropagation, the model successfully reproduced the essential features of these data (Fig. 6). Wind from the rear always caused the rear legs of the model to thrust back, which would propel the body forward in a freely moving insect, and wind

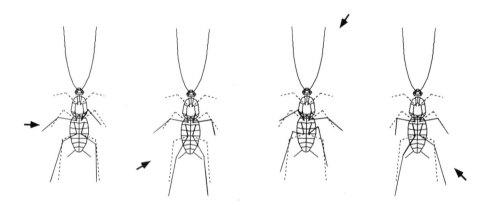

FIGURE 6. Ventral view of the leg movements produced by the model for wind puffs from different directions. Dashed lines indicated the leg positions before the simulated wind puff and solid lines indicated their position after the wind puff has been delivered. The arrows indicate the direction of the wind puff.

Chapter XII Simulations of Cockroach Locomotion and Escape

from the front caused the rear legs to move forward, pulling the body back. The middle legs always turned the body away from the direction of the wind.

D. Model Manipulations

The results we have described demonstrate that several neuronal and behavioral properties of this system can be reproduced using only simplified but biologically constrained neural network models. However, to serve as a useful tool for understanding the neuronal basis of the cockroach escape response, it is not enough for the model simply to reproduce what is already known about the normal operation of the system. In order to test and refine the model, we must also examine its responses to various lesions and compare them to the responses of the insect to analogous lesions. Here we report the results of two experiments of this sort.

Immediately following removal of the left cercus, cockroaches make a much higher proportion of incorrect turns (i.e., turns toward rather than away from the wind source) in response to wind from the left, while turns in response to wind from the right are largely unaffected (Vardi and Camhi, 1982a). These results suggest that, despite the redundant representation of wind direction by each cercus, the insect integrates information from *both* cerci in order to compute the appropriate direction of movement. As shown in Fig. 7, the response of the model to a left cercal ablation is

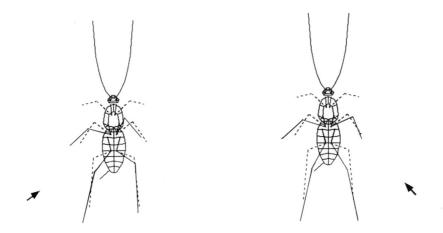

FIGURE 7. Ventral view of the model leg movements following left cercal ablation. (Left) Wind from the unlesioned side evokes leg movements consistent with a turn away from the wind direction. (Right) Wind from the side of the cercal ablation evokes leg movements consistent with a turn toward the direction of the wind.

consistent with these results. In response to wind from the unlesioned side, the model generates leg movements that would turn the body away from the source of wind. However, in response to wind from the lesioned side, the rear legs thrust back, the middle leg ipsilateral to the ablation pulls in, and the contralateral leg pushes out. These leg movements would turn the body toward the wind in a free-ranging insect.

It is interesting to note that, following a 30-day recovery period, the directionality of a lesioned insect's escape response is largely restored (Vardi and Camhi, 1982a). The mechanisms underlying this adaptation are not yet fully understood, but they appear to involve a reorganization of the vGI connections from the intact cercus (Vardi and Camhi, 1982b; Volman, 1989). After a cercal ablation, the wind fields of the vGIs on the ablated side are significantly reduced. Following the 30-day recovery period, however, these wind fields are largely restored. We have also examined these effects in the model. After cercal ablation, the model vGI wind fields show some similarities to those of similarly lesioned insects. In addition, using vGI retraining to simulate the adaptation process, we have found that the model can effect a similar recovery of vGI wind fields by adjusting the connections from the intact cercus.

A second experimental manipulation that has been performed on this system is selective lesion of individual ventral giant interneurons (Comer, 1985). The only result that we will describe here is the lesion of vGI1. In the animal, this results in a behavioral deficit similar to that observed with cercal ablation. Correct turns result for wind from the unlesioned side, but a much higher proportion of incorrect turns is observed in response to wind from the lesioned side. The response of the model to this lesion is also similar to its response to a cercal ablation (Fig. 7) and is thus consistent with these experimental results.

E. Discussion

As encouraging as the initial results reported in this section are, the current model is limited by the extreme simplicity of both its neural and peripheral components. Indeed, rather than being the final word, we view this model as only an initial step in a continuing cycle of refinement in which experimental and simulation work must continuously interact. We are currently including additional constraints in our model. For example, vGI models incorporating experimentally determined firing properties are being used to explore the mechanisms responsible for the vGI wind fields and their adaptation to cercal ablation (Beer *et al.*, 1991). In addition, we have begun to incorporate physical dynamics into our kinematic body model as a step toward making the periphery of our simulated cockroach more realistic.

Although the primary intent of this model is to elucidate the neural basis of the cockroach escape response, we believe that it may eventually have interesting engineering applications as well. Controlling multisegmented manipulators is not a trivial task. Yet the cockroach accomplishes such control very rapidly. This is true both for the escape turn itself and for the subsequent locomotion, both of which are likely to utilize many of the same local interneurons and motor neurons. In addition, the cockroach is able to integrate a diverse range of sensory information to make rapid context-dependent decisions. It is likely that working out the neural basis of these capabilities in the cockroach will serve to enrich our controller designs. If such biological principles do turn out to be of some engineering interest, the computer model can serve as an important first step in abstracting them for eventual application elsewhere.

Invertebrate systems offer the possibility of addressing important neurobiological questions at a much finer level than is generally possible in mammalian systems. In particular, the cockroach escape response is a complex, context-sensitive sensorimotor control system whose operation is distributed across several populations of interneurons but is nevertheless amenable to a detailed cellular analysis. Due to the overall complexity of such circuits and the wealth of data that can be extracted from them, modeling must play a crucial role in this endeavor. However, in order to be useful, models must make special efforts to remain consistent with known biological data and constantly be subjected to experimental test. Experimental work in turn must be responsive to model demands and predictions. This section has described our initial results with this cooperative approach to the cockroach escape response. We have also suggested that, in addition to its obvious biological interest, the development of this model may eventually have interesting engineering applications.

IV. Conclusion

Computational neuroethology has much to offer the emerging interdisciplinary collaboration between neuroethology and robotics. Building and manipulating simulations of neuroethological systems can provide new insights into the neural mechanisms of animal behavior, as well as suggest new approaches to the control of autonomous mobile robots.

In this chapter, we have presented two examples of such simulations. First, we described a novel neural network architecture for hexapod locomotion that was abstracted from work on the neural basis of insect walking. This circuit is fully distributed, efficient, and robust and has been used to control the locomotion of an actual hexapod robot. In addition, the locomotion simulation raised some issues of potential interest to neuroethologists.

Second, we described a model of the cockroach escape response that was capable of reproducing not only the normal leg movements observed during the cockroach escape turn but also those observed following two different lesions. Our experience with this simulation is currently serving as the basis for a second cycle of experimental work and model refinement in what we hope will ultimately become an ideal model system for studying the neural basis of distributed sensorimotor control and context-dependent responses. In addition, we argued that the principles underlying the operation of this system may be of significant engineering interest. Although these two models are currently limited in a number of ways, they both illustrate the potential for exciting interactions at the interface between neuroethology and robotics, and the role that computational neuroethology can play in fostering this interaction.

Acknowledgments

This work was supported in part by ONR grant N00014-90-J-1545 to R.D.B. Additional support was provided by the Howard Hughes Medical Institute and the Cleveland Advanced Manufacturing Program. H.J.C thanks the NSF for its support with grant BNS-8810757.

References

ACHACOSO, T. B., and YAMAMOTO, W. S. (1990). *Perspect. in Biol. Med.* **33,** 379–389.
ANASTASIO, T. J., and ROBINSON, D. A. (1989). *Neural Comp.* **1,** 230–241.
BEER, R. D. (1990). *Intelligence as Adaptive Behavior: An Experiment in Computational Neuroethology.* Academic Press, San Diego.
BEER, R. D., CHIEL, H. J., QUINN, R. D., ESPENSCHIED, K. S., and LARSSON, P. (1992). *Neural Comp.* **4,** 356–365.
BEER, R. D., CHIEL, H. J., and STERLING, L. S. (1989). In Touretzky, D. S. (ed.), *Advances in Neural Information Processing Systems 1* (pp. 577–585). Morgan Kaufmann.
BEER, R. D., KACMARCIK, G. J., CHAI, S., RITZMANN, R. E., and CHIEL, H. J. (1991). *Soc. Neurosci. Abstr.* **16,** 759.
BROOKS, R. A. (1989). *Neural Comp.* **1,** 253–262.
CAMHI, J. M., and LEVY, A. (1988). *J. Comp. Physiol.* **163,** 317–328.
CHIEL, H. J., and BEER, R. D. (1989). *Proc. First Int. J. Conf. on Neural Networks,* (pp. 407–414). IEEE.
CHIEL, H. J., BEER, R. D., QUINN, R. D., and ESPENSCHIED, K. S. (1992). *IEEE Trans. Robotics Autom.* **8,** 293–303.
CLIFF, D. (1991). In *From Animals to Animats* Meyer, J. A. and Wilson, S. W. (eds.), (pp. 29–39). MIT Press, Cambridge, MA.
COMER, C. M. (1985). *Brain Res.* **335,** 342–346.
DALEY, D. L., and CAMHI, J. M. (1988). *J. Neurophysiol.* **60,** 1350–1368.

DOWD, J. P., and COMER, C. M. (1988). *Biol. Cybern.* **60,** 37–48.
GEORGOPOULOS, A. P., KETTNER, R. E., and SCHWARTZ, A. B. (1988). *J. Neurosci.* **8,** 2928–2937.
GRAHAM, D. (1977). *Biol. Cybern.* **26,** 187–198.
GRAHAM, D. (1985). *Adv. Insect Physiol.* **18,** 31–140.
LEE, C., ROHRER, W. H., and SPARKS, D. L. (1988). *Nature* **332,** 357–360.
LOCKERY, S. R., WITTENBERG, G., and KRISTAN, W. B. (1989). *Nature* **340,** 468–471.
MURRAIN, M., and RITZMANN, R. E. (1988). *J. Neurobiol.* **19,** 552–570.
NYE, S. W., and RITZMANN, R. E. (1992). *J. Comp. Physiol.,* submitted.
PEARSON, K. G. (1976). *Sci. Am.* **235,** 72–86.
PEARSON, K. G. (1985). In Barnes, W. J. P., and Gladden, M. H. (eds.), *Feedback and Motor Control in Invertebrates and Vertebrates (pp.* 307–315). Croom Helm, London.
PEARSON, K. G., FOURTNER, C. R., and WONG, R. K. (1973). In Stein, R. B., Pearson, K. G., Smith, R. S., and Redford, J. B. (eds.), *Control of Posture and Locomotion* (pp. 491–514) Plenum, New York.
RITZMANN, R. E. (1984). In Eaton, R. C. (ed.), *Neural Mechanisms of Startle Behavior* (pp. 93–131). Plenum, New York.
RITZMANN, R. E., and POLLACK, A. J. (1988). *J. Neurobiol.* **19,** 589–611.
RITZMANN, R. E., and POLLACK, A. J. (1990). *J. Neurobiol.* **21,** 1219–1235.
RITZMANN, R. E., POLLACK, A. J., HUDSON, S., and HYVONEN, A. (1991). *Brain Res.* **563,** 175–183.
VARDI, N., and CAMHI, J. M. (1982a). *J. Comp. Physiol.* **146,** 291–298.
VARDI, N., and CAMHI, J. M. (1982b). *J. Comp. Physiol.* **146,** 299–309.
VOLMAN, S. F. (1989). *J. Neurobiol.* **20,** 762–783.
WESTIN, J. (1979). *J. Comp. Physiol.* **133,** 97–102.
WESTIN, J., LANGBERG, J. J., and CAMHI, J. M. (1977). *J. Comp. Physiol.* **121,** 307–324.
WILSON, D. M. (1966). *Annu. Rev. Entomol.* **11,** 103–122.
ZILL, S. (1985). In Barnes, W. J. P., and Gladden, M. H. (eds.), *Feedback and Motor Control in Invertebrates and Vertebrates (pp.* 187–208). Croom Helm, London.

Chapter XIII Lobster Walking as a Model for an Omnidirectional Robotic Ambulation Architecture

JOSEPH AYERS
Department of Biology and Marine Science Center
Northeastern University
East Point, Nahant, Massachusetts

JILL CRISMAN
Department of Electrical and Computer Engineering
Northeastern University
Boston, Massachusetts

I. Introduction

In this chapter we advocate the development of a biologically based robot architecture that draws on natural technology perfected by millions of years of evolution by natural selection. The physiological control systems of animals are among the most robust and adaptive systems known to humans. Because of evolution, physiological control systems for common functions are conserved between phyla. One of the most conserved functions is the central neuronal generation of locomotion. In every organism that has been subjected to critical experimentation, it has been demonstrated that the motor programs for species-specific behavioral acts are generated by central pattern generators resident in the central nervous system (Delcomyn, 1980). This central pattern generator model differs fundamentally from models based on reflex chains (Sherrington, 1910). For technical reasons, much of the early work on invertebrate motor systems was performed on decapod crustaceans (Hoyle, 1977; Kennedy and Davis, 1977). These early experiments led to a formal model of the central organization of segmental motor systems based on command systems, coordinating systems, and central pattern generators that we refer to as the CCCPG model (Kennedy and Davis, 1977; Stein, 1978; Grillner, 1983). The CCCPG model provides a formal framework for understanding how adaptive central pattern generators operate. The model addresses the functions to be controlled and transcends the underlying central mechanisms.

We will demonstrate that the command, coordinating, central pattern generator model can be extended to address the complexity inherent in the omnidirectional decapod walking systems. We draw on the available data on the macruran walking (Ayers and Davis, 1977a,b; Barnes, 1977; Clarac, 1985; Clarac and Barnes, 1985), postural (Bowerman and Larimer, 1974a,b), escape (Wine, 1984), swimmeret (Davis, 1973), and stomatogastric (Selverston and Moulins, 1987) systems to implement an ambulation controller based on the CCCPG model. We propose a layered architecture that can be readily implemented with a distributed microcontroller and bus architecture. The model currently operates on sequential machines to emulate the output of the lobster walking pattern generator within a graphical user interface. We are developing this implementation as a controller for an omnidirectional ambulatory robot.

The fundamental difference between this controller and existing biologically based robotic architectures (Beer, 1991; Brooks, 1986, 1989; Brooks and Flynn, 1989; Pennisi, 1991) is that it is based on the prevailing central model of crustacean locomotory systems (Hoyle, 1977; Kennedy and Davis, 1977). The CCCPG model is based on four major classes of components (Stein, 1978). The segmental central pattern generators can generate the patterns underlying locomotion in the absence of sensory feedback for each limb (Selverston and Moulins, 1987). The central pattern generators are in turn controlled by command systems that mediate the decision to execute the behavior as well as intersegmental modulation (Kupferman and Weiss, 1978; Kennedy and Davis, 1977). A third class of central intersegmental neurons, coordinating systems, specify the gaits or patterns of coordination between the different locomotory appendages (Stein, 1978; Paul and Mulloney, 1986). The sensory feedback component includes sensors and intersegmental sensory neurons that mediate adaptive feedback (Mill, 1976).

Robotics can gain knowledge from biology on the "natural engineering" perfected by billions of years of evolution, with regard to both the mechanics of the robot and the underlying neuronal control systems (Beer, 1991; Brooks, 1991). The CCCPG model is based on a limited set of central controlling elements that are extremely flexible in ultimate organization. Furthermore, crustaceans exhibit robust adaptive locomotion over rough terrain in the presence of considerable hydrodynamic perturbation. As a result, decapods have evolved robust proprioceptive and exteroceptive reflexes that mediate load compensation, righting, and optomotor reflexes as well as adaptations to hydrodynamic forces mediated by control surfaces in the chelipeds and abdomen (Kennedy and Davis, 1977). More important, lobsters exhibit a variety of skilled and adaptive behaviors including visually mediated investigation, tactically mediated searching, chemosensory taxes, complex manipulation, the complex movements underlying interindividual communication, attack, and defense (Sandeman and Atwood, 1982). These capabilities transcend those of the most complex robots.

Chapter XIII A Model for a Robotic Ambulation Architecture

Biologists can also benefit by a robotic implementation. The major challenge of motor systems physiologists is to formalize the relationship between the underlying mechanisms and the functions they control. Robots provide the highest-order models available for testing hypotheses of the organization of motor systems, for they incorporate segmental control systems, models of the effectors, sensors, and their interactions in response to a variety of predictable and unpredictable exteroceptive inputs and environmental perturbations. Robotic implementations can also specify the direction of further biological experiments to ensure more complete models.

This chapter is divided into two components. First we explore the biological CCCPG model and the experiments that resulted in this model. Then we describe the simulation of the CCCPG model and the results of the simulation.

II. Lobster Walking Behavior

Lobsters are interesting subjects for our ambulatory control model because they exhibit a wide range of behaviors and flexible walking alternatives. Although lobsters are capable of long-distance migrations, typical locomotory behavior is a subset of higher-order goal-oriented behavior, such as feeding, reproduction, investigation, and agonistic behavior. Locomotion involves the coordinated action of the chelipeds, the walking legs, the abdomen, and the swimmerets (see Fig. 1). Actual walking move-

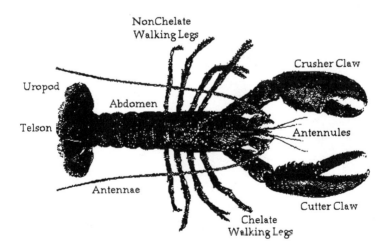

FIGURE 1. Gross anatomy of the American lobster. The cutter and crusher claws (chelipeds) are the first pair of periopods or walking legs. Walking legs 2 and 3 are dactylate and participate in searching and feeding movements. Walking legs 4 and 5 are nondactylate.

ments, however, are mediated by the second through fifth pairs of walking legs. The chelipeds and abdomen are utilized as control surfaces to control orientation relative to water currents, and the swimmerets function as thrusters during forward locomotion and during righting (Davis, 1968).

During locomotion, movements of only three major joints of the leg are required for walking in the four directions (Ayers and Davis, 1977a). The ranges of movement of these joints are summarized in Fig. 2. The most proximal joint, the thoracocoxal (ThC) joint, mediates protraction and retraction movements in the horizontal plane. The next joint, the coxobasal (CB) joint, mediates elevation and depression movements in the saggital plane. The fifth most distal joint, the merocarpopodite (MC) joint, mediates extension and flexion movements in the saggital plane. The other joints of the leg are involved in more complex behaviors such as grooming, righting, and manipulation (Evoy and Ayers, 1982).

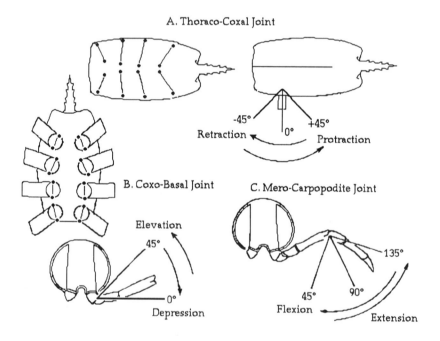

FIGURE 2. Principal degrees of freedom of the lobster leg during walking. (A) Thoracocoxal (ThC) joint. The left panel indicates the axis of rotation of walking legs 2–5. The right panel indicates the angular coordinates of ThC joint movement. (B) Coxobasal (CB) joint. The upper panel indicates the axis of rotation of walking legs 2–5. The lower panel indicates the angular coordinates of CB joint movement. (C) Angular coordinates of merocarpopodite (MC) joint movement. Adapted from Ayers and Davis (1977).

Chapter XIII A Model for a Robotic Ambulation Architecture

The coordination and timing of joint movements have been examined during both treadmill walking (Ayers and Davis, 1977a,b) and free walking (Ayers and Clarac, 1978). The walking step cycle is divided into a stance phase, in which the limb supports the body against gravity and provides translational propulsive force, and a swing phase, in which the limb elevates and returns to start the subsequent stance phase. The joint movements that underlie these two phases are illustrated in Fig. 3. Cyclic elevation and depression movements of the CB joint underlie respectively, the swing and stance phases of all four walking directions. The depression movement underlies the

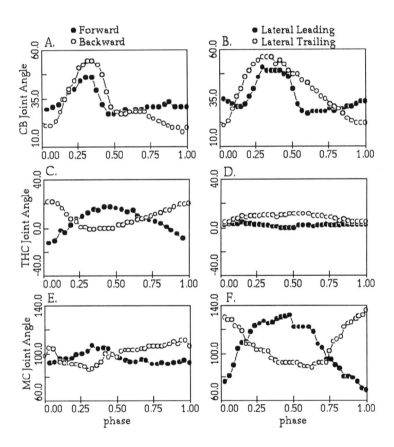

FIGURE 3. Joint angular movements during walking in the four directions. The left column indicates the movements during forward and backward walking and the right column represents the corresponding movements during lateral walking. The upper row indicates the CB joint movement, the middle row the ThC joint movements, and the lower row the MC joint motions.

antigravity component of stepping and can be divided into a descent phase and a support phase. The translational propulsive forces are generated by movements of the ThC and MC joints.

The movements of the ThC and MC joints exhibit at least three different patterns of coordination with CB joint, depending on the direction of walking (Fig. 3). During forward walking, the swing phase results from coordinated elevation and protraction. The stance phase results from coordinated depression and retraction. During backward walking, this pattern of coordination is reversed. During lateral walking on the leading side the swing phase results from simultaneous elevation and extension of the MC joint, while the stance phase results from coordinated depression and flexion. Again, these patterns of coordination are reversed during lateral walking on the trailing side.

The joints that are not actively involved in producing propulsive force exhibit reduced movements during walking in orthogonal directions. For example, during forward and backward walking the movements of the MC joint are reduced relative to those observed during lateral walking, and during lateral walking the movements of the ThC joint are reduced relative to those observed during forward and backward walking.

The timing of the two phases of stepping is temporally asymmetric (Ayers and Davis, 1977a,b). This relationship is most clearly demonstrated in the limb joints that generate propulsive force. For example, during ThC joint movements underlying forward and backward walking, the swing phase is always constant in duration. The duration of stance phase movements is linearly related to the period of stepping

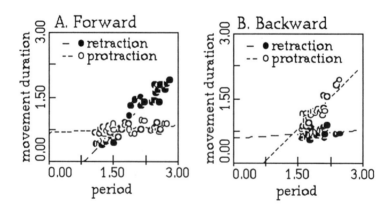

FIGURE 4. Temporal organization of the walking step cycle during forward and backward walking. The two graphs indicate the relative duration of the protraction and retraction movements of the thoracocoxal joint graphed as a function of step period. Note that the movements are constant in duration when they are swing phase movements and variable in duration when they are stance phase movements. Adapted from Ayers and Davis (1977a).

Chapter XIII A Model for a Robotic Ambulation Architecture

(Fig. 4). In other words, the duration of the stance phase determines the frequency of stepping, and the angular velocity of the propulsive force movement determines the velocity of walking.

III. A Neural Circuit Model for Omnidirectional Walking Pattern Generation

A. Segmental Circuitry

The motor patterns underlying walking have been examined at both the level of electromyograms (Ayers and Davis, 1977a,b; Ayers and Clarac, 1978) and the level of the underlying central programs (Chrachri and Clarac, 1989). The underlying temporal organization of the electromyographic patterns is that of a two-phase oscillation divided into a constant-duration elevator phase and a variable-duration depressor phase (Fig. 5). The underlying mechanisms generating this oscillation are unknown (Chrachri and Clarac, 1987, 1990; Sillar and Elson, 1986), largely because of lack of correlated studies using the specific command systems for walking in different directions (Bowerman and Larimer, 1974a,b). This temporal organization is analogous, however, to that proposed for the cockroach walking system as well as that found in endogenous pacemaker networks such as the pyloric system, where a synergy of endogenous pacemakers provides a common clock.

During locomotion as well as during postural behaviors, lobster leg motor units are activated in either of two synergies: the elevator or the depressor synergy (Evoy and Ayers, 1982). For example, during standing the depressors are activated with limb flexors, and during the defense posture all limb antagonists are coactivated to maintain limb rigidity. During escape swimming, the limbs are elevated and all limb joints extended (Wine and Krasne, 1974).

The simplest model that explains these findings is presented in Fig. 5C. The model is based on a common neuronal oscillator composed of the elevator and depressor synergies organized as an endogenous pacemaker network with reciprocal inhibition (Fig. 5B). The feature of an endogenous pacemaker model is that it separates the oscillator function (the clock of stepping) from the pattern generator function. This model allows a common clock for all directions of walking, which is fundamental to metastable coordination (Ayers and Davis, 1977a). Excitatory pathways from both the elevator and depressor synergies couple synergies of the other limb joints to either elevation or depression or both (Fig. 5C). This model of the fundamental oscillator is consistent with known central circuitry (Chrachri and Clarac, 1990). Very little is known about the actual pattern generating mechanisms underlying the neuronal oscillation, since most experiments have involved pharmacological stimulation and few have involved intracellular recording from net-

294   Part III   Computer Modeling

work components during selective stimulation of walking command systems (Bowerman and Larimer, 1974a,b).

In contrast to the reciprocal bursting that characterizes the elevator and depressor discharge of all walking directions, the discharge of muscles that control the ThC and MC joints exhibits at least three patterns (Ayers and Davis, 1977a; Ayers and Clarac, 1978). The electromyographic analysis indicates that walking leg motor synergies can be divided into swing or elevator phase synergies, stance or depressor phase synergies, and bifunctional synergies that can be active in the

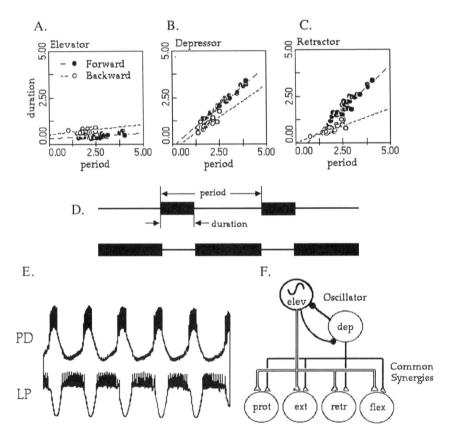

FIGURE 5. Temporal organization of motor pool bursts. In each graph the duration of the burst of electromyographic activity is shown as a function of the period of the oscillation during forward and backward walking. (A) CB elevator muscle. (B) CB depressor muscle. (C) ThC retractor muscle. (D) The plotted variables, period and duration. (E) Example of an endogenous pacemaker oscillator, the lobster pyloric system. PD, pyloric dilator motor neuron; LP, lateral pyloric motor neuron. (F) Network model of the elevator and depressor synergies.

swing phase or the stance phase or coactivated during both phases (Ayers and Davis, 1977a,b; Ayers and Clarac, 1978). When bifunctional muscles are activated during the depressor phase they exhibit variable-duration bursts, and when they are activated during the elevator phase they exhibit constant-duration bursts (Ayers and Davis, 1977a,b). Distributed velocity-sensitive reflexes provide peripheral reinforcing pathways to couple the elevator and depressor synergies with those controlling the other limb joints (Ayers and Davis, 1977a).

When limb joints serve primarily a postural component such as the behavior of the MC joint during forward and backward walking, the antagonistic muscles are coactivated to make the joint stiff and reduce angular excursion (Ayers and Clarac, 1978). This form of control ranges from muscles bursting twice during both the elevator and depressor phase to maintained tonic activity.

B. Walking Command Systems

The feature that distinguishes the crustacean limb pattern generator from most other pattern generation models is its ability to participate in a variety of behaviors, including posture, locomotion, feeding, defense, escape, and social behaviors (Evoy and Ayers, 1981). This flexibility demands that the segmental generators are more permissive than instructive (Kennedy and Davis, 1977; Pinsker and Ayers, 1983). The generation of walking movements in opposite directions defines mutually incompatible behavior, and the crustaceans can perform transitions between walking in different directions on a cycle-by-cycle basis (Ayers and Davis, 1977a,b; Ayers and Clarac, 1978). Thus the command systems for walking are fundamental to determination of the direction of walking and the intrasegmental coordination of the walking pattern generator. Bowerman and Larimer (1974a,b) identified a set of command neurons for both forward and backward walking. They also observed command-evoked movements characteristic of lateral walking (Larimer, personal communication). Locomotory command systems are among the least well understood of CCCPG model components. Their releasers, interactions, patterns of activation, and efferent effects have in most cases yet to be established. A synthetic model of their role in the walking system can be derived based on observations of both the walking and swimmeret systems (Davis and Kennedy, 1972). This model includes a parametric component, a pattern generating or coordinating component, and a recruiting component.

The fundamental role of walking command systems is as a parameter of the oscillation. This function is common to all directions of walking. As a parameter, the command turns the neuronal oscillator on and off and regulates its frequency (Fig. 6). Parameters may constitute the summed current of synapses or currents

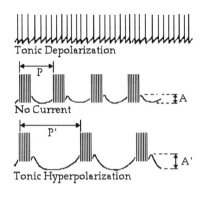

FIGURE 6. Parametric control of neuronal oscillators. This figure illustrates the role of parameters in the modulation of neuronal oscillators. The example is current injection into an endogenously bursting neuron. When hyperpolarizing, the parameter can terminate the oscillation. As the current changes from outward to inward, the oscillation is released and then increases in both amplitude and frequency. Excessive depolarization causes a state transition to a beating operation. Adapted from Pinsker and Ayers (1983).

due to neuromodulator gated channels with long time courses. In most cases in which this role has been examined, the command system can mediate a state transition of silent neurons to bursting or beating neurons (Pinsker and Ayers, 1983).

The second function of the walking command systems is modulation of the internal coordination of the pattern generator to generate the coordination patterns fundamental to walking in different directions. The most parsimonious model for coordination shifts underlying metastable coordination relies upon selective elimination of inappropriate connections (Ayers and Davis, 1977a,b). For example, during forward walking, the command and/or emergent oscillator properties must eliminate the protractor discharge that would occur during the depression phase and the retractor discharge that would occur during the elevation phase. The model proposed by Ayers and Davis (1977a,b) relied on presynaptic inhibition of the connections linking elevation and retraction and depression and protraction. Presynaptic inhibition of chordotonal afferents from the CB joint, which can perform this function as a reflex loop (Ayers and Davis, 1977b; Clarac et. al., 1978), has been confirmed (Cattaert et. al., 1990). Whether presynaptic inhibition of the excitatory connections linking the elevators and depressors with bifunctional motor neurons (Chrachri and Clarac, 1989) is also present is an important hypothesis to be tested.

Chapter XIII A Model for a Robotic Ambulation Architecture

The third function of the walking command systems is selective recruitment of propulsive phase synergies to regulate the angular velocity of the propulsive joint movements. Crustacean motor units are recruited in order of size and efferent effect (Davis, 1971). Chrachri and Clarac (1989) found that protractor and retractor motor unit pools are electrotonically coupled, which presumably underlies this recruitment order. During walking, limb motor unit discharge frequency is independent of the frequency of stepping (Ayers and Clarac, 1978). Motor unit recruitment occurs at the level of the bifunctional muscles that generate propulsive force and is independent of antigravity regulation (Fig. 7).

IV. The CCCPG-Based Ambulation Controller

We have implemented the CCCPG model for omnidirectional walking as a finite-state machine on a sequential processor. The ambulation controller program generates digital output for actuator control and real-time chart displays of motor programs imitating electromyograms. The program treats the CCCPG model components as objects that pass messages with regard to status changes. Changes in walking direction, speed, load, etc. are effected by menu selections or keystrokes. The program generates chart displays that are the basis of many of the figures of experimental results in this chapter.

At the single-limb control level, our ambulation controller relies on three major classes of components that control the elevator, depressor, protractor, retractor, extensor, and flexor synergies (Fig. 8). The oscillator component is a software clock that

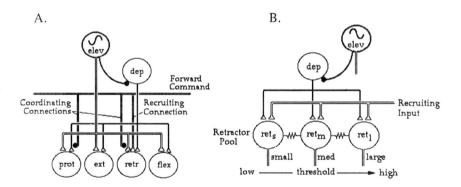

FIGURE 7. Recruiting and coordinating connections of walking command systems. (A)The hypothetical connections of the forward walking command systems operating on the elevator and depressor synergies. (B) The size principle of recruitment of propulsive force synergies on the basis of threshold in response to input from the recruiting input.

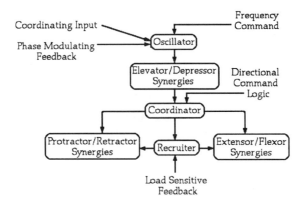

FIGURE 8. Message hierarchy of finite-state machine. The machine consists of three controlled objects, the elevator/depressor, protractor/retractor, and extensor/flexor synergies, which are controlled by the oscillator, coordinator, and recruiter. The oscillator receives messages from the frequency command, the coordinating and phase-modulating feedback layers. The coordinator receives messages from the oscillator and the directional command logic. The recruiter receives messages indirectly from the oscillator through the coordinator as well as from the load-sensitive feedback layer.

regulates the period of stepping as well as the duration of the swing or elevator phase fraction of the stepping cycle. The clock maintains step timing registers that contain the parameter associated with the desired stepping period. At the termination of each step cycle the oscillator loads the step timing registers with the clock tick values associated with the expected end of the elevator phase and the end of the step cycle based on the desired stepping period. During ongoing operation, the oscillator continually compares the processor clock ticks with the two registers and, when the target times are achieved, issues state transition messages to the pattern generator, the recruiter, and the neuronal oscillators of any ipsilateral or contralateral governed limbs.

The second major component of the ambulation controller is the pattern generator, which determines the pattern of discharge of bifunctional synergies. The pattern generator responds to the state transition message and desired period parameter from the oscillator, polls the walking command logic and determines through a truth table which synergies should be active, and sets or clears the booleans associated with different bifunctional synergies. The truth table implements the presynaptic inhibitory logic of our neuronal circuit model (see Figs. 7 and 10) and specifies the excitatory connections that would be disabled by the directional command.

The third major component of the ambulation controller is the recruiter, which determines which of the elements of the propulsive force synergies are active. The

Chapter XIII A Model for a Robotic Ambulation Architecture

recruiter responds to the desired period parameter as well as load-sensitive feedback and sets or clears the booleans associated with different elements of the active propulsive pool. In our current implementation, each unit within a synergy is represented as a boolean, although it could easily be represented as a scalar to provide more detailed amplitude control.

Fig. 9 is a graphical realization of the output of the ambulation controller demonstrating adaptation to speed during forward and backward walking. The output is realized as a strip-chart recording of the timing of activity in the different synergies in the same context as electromyograms or nerve recordings. In propulsive force synergies the width of the trace indicates the degree of recruitment within the pool. Notice that during increases in speed the period of stepping decreases while propulsive force synergies are selectively recruited.

In addition to controlling forward and backward walking, our ambulation controller implements omnidirectional locomotion including lateral walking on the leading and trailing sides (Ayers and Davis, 1977a,b). To complete this orthogonal symmetry, we have included directional and recruiting logic for backward, lateral leading, and lateral trailing walking (Fig. 10). In all cases, the directional logic

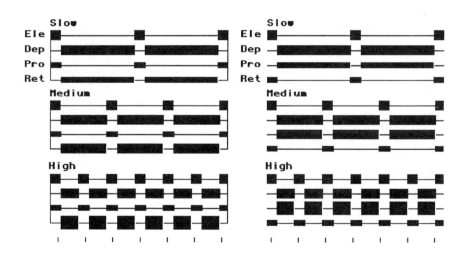

FIGURE 9. Simulated motor output patterns during adaptation to speed. Each trace represents the timing of activation of different limb synergies during, slow, medium, and fast walking. The height of the bar indicates the degree of recruitment that would be reflected in the angular velocity of joint movement of level of force output depending on whether the output was isometric or isotonic. Ele, elevator synergy; Dep, depressor synergy; Pro, protractor synergy; Ret, retractor synergy.

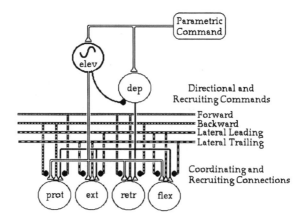

FIGURE 10. Complete limb control circuit for omnidirectional walking. This circuit adds the parametric modulation layer, directional, and recruiting logic for all four directions.

eliminates the inappropriate synergies by decoupling propulsive force synergies from the elevation/depression oscillation and adds speed-dependent recruitment to the propulsive force synergy. The added connections include those that decouple elevation from flexion and depression from extension during lateral leading walking and those that decouple elevation from extension and depression from flexion during lateral trailing walking.

The resulting realization of omnidirectional walking motor programs is indicated in Fig. 11. In this experiment, the simulation was switched to each of the four walking directions during medium-speed walking. Note that there is no change in the elevation depression oscillation, and when the protraction/retraction and extension/flexion synergies serve a postural function the antagonists at these joints are coactivated. The alterations in motor pattern in all cases result from selective elimination of inappropriate synergies and selective recruitment of propulsive force synergies.

An important feature of the lobster walking system is the ability of specimens to change their direction of walking on a cycle-by-cycle basis. We have termed this capability metastable coordination (Ayers and Davis, 1977a,b). Indeed, metastable coordination emerges as a property of the ambulation controller for omnidirectional walking (Fig. 12). In the two experiments illustrated the walking pattern is sequentially switched to each of the four directions during medium speed (Fig. 12A) and during high speed (Fig. 12B). Notice that there is an immediate alteration in the ThC and MC synergies with appropriate recruitment for the direction and period of walking.

Chapter XIII A Model for a Robotic Ambulation Architecture

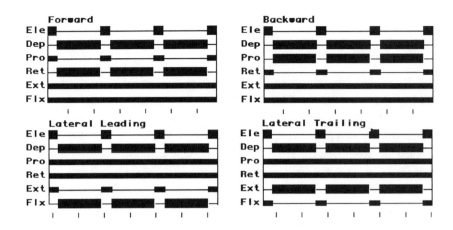

FIGURE 11. Motor programs for omnidirectional walking. Ext, extensor synergy; Flx, flexor synergy; all else as in Fig. 9.

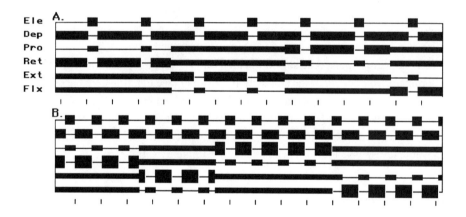

FIGURE 12. Metastable coordination of walking. In each panel the walking speed was held constant the the program switched sequentially to each of the four directions of walking: forward, lateral trailing, backward, lateral leading. (A) Medium-speed walking. (B) High-speed walking.

V. Proprioceptive Reflexes

Lobster walking is a highly adaptive behavior that is subject to both proprioceptive and exteroceptive modulation. Proprioceptive reflexes in the walking legs have been subjected to extensive analysis and can be divided into those mediated by chordotonal organs, the myochordotonal organ, the ThC joint receptor, and the cuticular stress detectors.

A. *Amplitude-Modulating Reflexes*

Chordotonal reflexes result from direct monosynaptic connections between sensory afferents and motor neurons (Clarac, 1990) and thus are capable of amplitude modulation of walking output. In terms of causality, they can be divided into resistance reflexes that oppose imposed movements, assistive or positive feedback reflexes that reinforce imposed movements, and distributed coordinating reflexes where imposed movements of one limb joint (e.g., CB joint) activate the synergies controlling other limb joints (e.g., ThC joint, Fig. 13). Distributed reflexes constitute peripheral pathways for the bifunctional synergies for walking in different directions (Ayers and Davis, 1977b). These peripheral loops are composed a fundamental movement (e.g., elevation or depression) that evokes reflex discharge and the chordotonally mediated feedback that excites the bifunctional synergies (e.g., coxal retractor, Fig. 13B). These reflexes are, in fact, tuned to the angular velocities for which they are appropriate (Ayers and Davis, 1978). For example, the angular velocity of the elevation movement of the CB joint is different during forward and backward walking. The retractor discharge that results from CB elevation occurs predominantly over the angular velocity of CB joint movement that characterizes the backward walking swing phase (Ayers and Davis, 1978).

We have implemented these coordinating reflexes at the level of the logic of the coordinator component of the ambulation controller (see Fig. 8) where the reflexes are considered a separate parallel excitatory connection between the elevator and depressor synergies and the bifunctional synergies. We are in the process of developing a graphical simulation of the actual limb movements and will implement the movement/response loop as such in the newer implementation.

B. *Phase-Modulating Reflexes*

In contrast to amplitude-modulating reflexes, reflexes resulting from input from the ThC joint receptor (Elson *et. al.*, 1992), the cuticular stress detector (Klärner and Barnes, 1986), and the funnel canal organs (Libersat *et. al.*, 1987) are capable of phase modulation of the walking rhythm to reset the timing of the step (Fig. 14). One of the goals of any omnidirectional walking controller is to implement omnidirectional phase-modulating control that is fundamental to proprioceptive reflexes as well as interleg coordination.

We implement phase modulation of the ambulation controller at the level of the oscillator (see Fig. 8). In response to perturbation, the oscillator loads the step tim-

Chapter XIII A Model for a Robotic Ambulation Architecture

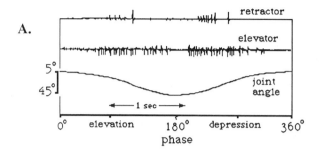

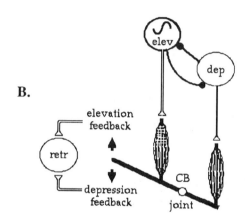

FIGURE 13. Amplitude-modulating reflexes from the CB joint. (A) Electromyographic recordings from the elevator muscle of the coxobasal joint and retractor of the thoracocoxal joint during passive sinusoidal movements of the coxobasal joint. (B) Causal nexus of coordinating reflexes. Movements of the CB joint resulting from elevator and depressor activity activate bifunctional muscles (here the retractor, retr) through chordotonal mediated movement feedback. (A) Adapted from Ayers and Davis (1977b).

ing registers with the clock tick values associated with the end of the perturbed elevator phase and step cycle. These reset timing values determine the time of messages sent to the coordinator and recruiter, as during unperturbed stepping.

C. Passive Traction

Experiments on treadmill-constrained locomotion have demonstrated that the sensory feedback provided by relative substrate motion in the open-loop situation is capable of both initiating locomotion and regulating its period on a cycle-by-cycle

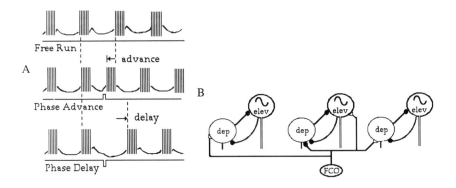

FIGURE 14. Perturbation of neuronal oscillators. (A) Phase-advancing and phase-delaying perturbations. The upper trace is a free run oscillation. The middle trace indicates a stimulus that results in a phase advance, and the lower trace indicates a stimulus that results in a phase delay. (B) Phase-modulating input provided by the funnel canal organ (FCO). This input excites the elevator synergy and inhibits the depressor synergy of the same leg while exciting the depressor synergies of adjacent legs. (A) Adapted from Pinsker and Ayers (1983). (B) Adapted from Libersat, et. al. (1987).

basis (Ayers and Davis, 1977a,b). The passive traction stimulus can be considered to be both a proprioceptive stimulus and an exteroceptive stimulus (Fig. 15). For example, during forward and backward walking, the ThC joint receptor is subject to direct modulation by movements resulting from the proprioceptive passive traction stimulus provided by the other legs. The ability of the passive traction stimulus to both initiate and change the direction of walking (Ayers and Davis, 1977a,b) implies that the stimulus supplies exteroceptive feedback to the walking command systems (Fig. 15).

Passive traction experiments reveal selective control of propulsive force muscles relative to antigravity muscles. As the speed of the treadmill is increased during forward walking, one observes that the burst duration of the coxal retractor decreases relative to that of the coxal depressor. This result is indicated graphically in Fig. 16. In other words, as the load on the retractor muscle is decreased due to the limb being retracted passively, the discharge of the retractor is terminated prematurely. This phenomenon is analogous to the positive feedback control of stroke muscles by campaniform sensillae in the cockroach walking system (Pearson, 1972). In our finite-state engine this reflex is incorporated as a segmental input to the stance phase recruiter (see Fig. 8). Increases in stepping speed associated with shorter-period step cycles cause the recruiter to activate sequentially larger motor units.

All these forms of passive traction modulation are implemented in our ambulation controller. Exteroceptive feedback is mediated at the level of the command systems,

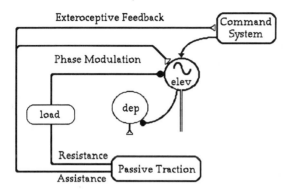

FIGURE 15. Three pathways that may mediate passive traction. Resistive feedback is capable of phase delaying the oscillation to slow the walking rhythm. Assistive feedback may phase advance the rhythm to reset individual step cycles as well as initiate and control the direction of walking of all walking legs. From Ayers and Davis (1977a).

which include the parametric levels (behavior on or off, frequency of oscillation, recruitment level, lateral biases) and the pattern generator level (direction of walking).

D. Load Compensation

An emergent property of the ambulation controller is omnidirectional load compensation. Decapods compensate for loads imposed during locomotion by increasing the period of stepping and recruiting additional propulsive force motor units in a similar fashion to insects (Fig. 16) (Grote, 1981; Pearson, 1972). At present, load is simulated as an on-or-off event We have implemented load compensation mechanisms at two levels. First, as described above, the load-sensitive feedback phase modulates the elevator clock to produce a phase delay in steps where the load is present. Second, the load stimulus amplitude modulates the propulsive phase motor units by altering the logic of the recruiter. During loaded steps, the recruiter engages the largest propulsive muscle motor units. An experiment in which a load stimulus is imposed during walking in all four directions is illustrated in Fig. 17. We are implementing the representation of transverse propulsive load as a scalar in future versions of our controller.

VI. Interlimb Coordinating Systems

One of the fundamental impediments to establishing the mechanisms of inter-limb coordination in decapods is that the complexity of the behavioral reper-

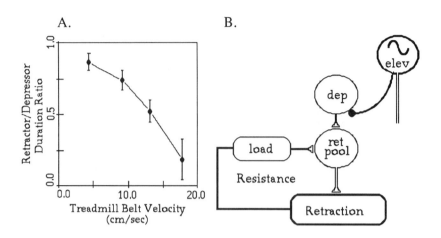

FIGURE 16. Selective control of propulsive versus antigravity synergies. (A) Selective regulation of the duration of propulsive force (retractor) versus antigravity force (depressor) synergies during forward walking. As the velocity of the treadmill is increased, decreasing the propulsive load, the retractor bursts are terminated prematurely. (B) Hypothetical circuit to explain this result. (A) Adapted from Ayers and Davis (1977a).

toire in which the limbs participate has yet to be established (Evoy and Ayers, 1982). The search for central programs in the thoracic ganglia has been limited to those evoked spontaneously or in response to pharmacologic stimulation. The only pattern of interlimb coordination that has been demonstrated in isolated central nervous system (CNS) preparations is an in-phase pattern that probably corresponds to the leg-waving behavior described by Pasztor and Clarac (1983). Walking decapods adopt an alternating tetrapod gait (Barnes, 1985; Chasserat and Clarac, 1980; Clarac, 1982; Cruse and Graham, 1985, 1986; Wilson, 1967). Isolated CNS preparations can exhibit a variety of intrasegmental motor patterns, none of which correspond closely to those exhibited during walking in their overall timing (Chrachri and Clarac, 1990). Isolated thoracic ganglion experiments have yet to employ the command systems utilized in behavioral experiments (Bowerman and Larimer, 1974a,b) to activate central motor rhythms (Sillar et. al., 1987). Thoracocoxal joint receptor stimulation can entrain the rhythms of adjacent limbs but only in the in-phase pattern (Sillar et. al., 1987). In other words, the "central motor program" for omnidirectional decapod walking has probably yet to be demonstrated due to the lack of employment of specific command pathways.

An alternative hypothesis is that interlimb coordination depends strongly on proprioceptive feedback (Clarac, 1981; Cruse and Muller, 1986). Clarac

Chapter XIII A Model for a Robotic Ambulation Architecture   307

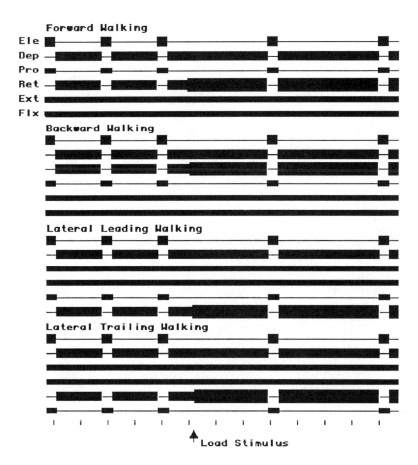

FIGURE 17. Omnidirectional load compensation. The four panels indicate the response of the ambulation controller to a load stimulus applied at the arrow during simulated walking in each of the four directions.

(1982) demonstrated that limb autonomy could change the pattern of ipsilateral limb coordination from an out-of-phase pattern to an in-phase pattern similar to that observed during leg waving (Pasztor and Clarac, 1983). Several sense organs have been implicated in this regulation, including the thoracocoxal joint receptors (Sillar et. al., 1987), the cuticular stress detector (Klärner and Barnes, 1986), and the funnel canal organs (Libersat et. al., 1987). Little is known about the direct interactions of these systems, but it is clear that their emergent interactions are to generate an ipsilateral phase leg of about 0.4 (Clarac and Barnes, 1985) and a similar contralateral phase lag. We have incorporated interlimb perturbation into our ambulation controller by coupling a set of eight neuronal oscillators by ipsilateral and contralateral phase-modulating connections

We have modeled interlimb control with a perturbation model (Ayers and Selverston, 1979; Pearson and Iles, 1972) based on control of the elevator synergy (Fig. 18). According to this model, each elevator clock sends central timing messages to contralateral pattern generators, and proprioceptive feedback from the funnel canal organs indirectly inhibits adjacent elevator clocks through excitation of the depressor synergies that reset their timing relative to the governing clock. Our implementation allows us to regulate coordination on a phase constant basis or a latency constant basis and to vary the phase lag between adjacent and contralateral control centers.

On the basis of experimentation with this model, we have found that a stable alternating tetrapod gait obtains as long as the contralateral and ipsilateral phase lags are equivalent in the range of 0.3 to 0.5 over a broad range of stepping frequencies (Fig. 19). The stable alternating tetrapod gait obtains whether the phase coupling is directed from posterior to anterior segments or from anterior to posterior segments.

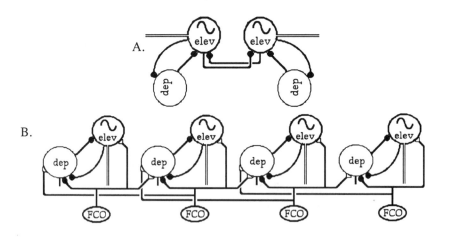

FIGURE 18. Interlimb coordinating systems. (A) Reciprocal inhibition between the elevator control synergies of contralateral limbs. (B) Peripheral control of ipsilateral limb elevator synergies by input from the funnel canal organ sensory system (FCO). The FCO afferents of one limb excite the depressors of adjacent limbs, thereby indirectly inhibiting the adjacent elevator synergies.

Chapter XIII A Model for a Robotic Ambulation Architecture

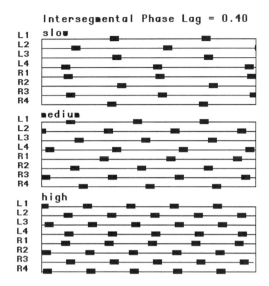

FIGURE 19. Maintenance of the alternating tetrapod gait during adaptation to speed. Each of the horizontal traces indicates the timing of elevator synergy activity in the left (L1–4) and right (R1–4) legs. In the three panels the stepping frequency was gradually increased while the ipsilateral and contralateral phase lags were held constant at 0.4.

VII. Exteroceptive Reflexes

The higher-order controllability of the ambulation controller may be best illustrated by extension to the command level of control, namely exteroceptive optomotor control of locomotory orientation in the pitch plane. Crustaceans exhibit three forms of yaw-correcting optomotor reflexes. The first form occurs in response to pure angular rotation stimuli and results in forward locomotion on one side and backward locomotion on the other (Loeb, 1918). The second form results from purely translational stimuli (where each eye sees the mirror image of the other) and results in forward or backward walking (Davis and Ayers, 1972). The third form results from angular rotation superimposed on a translational motion. All three forms of exteroceptive reflex can be realized by a simple taxic circuit based on the control of forward and backward command systems by unidirectional visual motion-sensitive afferents. The decapod optic nerve contains higher-order visual interneurons that discharge in response to unidirectional motion in the lateral visual field (Wiersma and Yamaguchi, 1967). The units consist of those that respond to medially directed motion (rear to front or from lateral fields to anterior midline) or laterally directed motion (front to rear or from anterior midline to lateral fields). These motion detectors apparently activate the descending command systems to mediate yaw-correcting locomotory optomotor reflexes (Davis and Ayers, 1972; Kennedy and Davis, 1977)

The complete circuit of the yaw control layer is indicated in Fig. 20. During front-to-rear translational motion, the lateral motion detectors would be activated symmetrically, activating the forward command systems on the contralateral sides as well as the stepping frequency controllers. This stimulus would therefore acti-

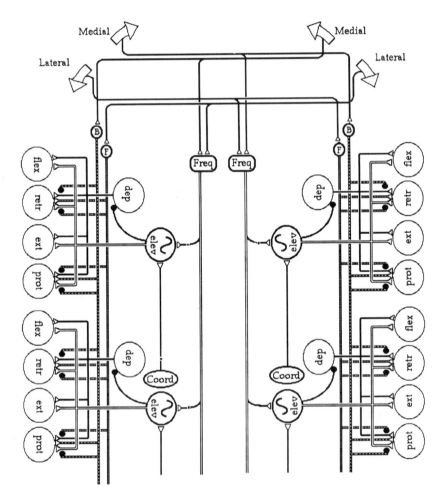

FIGURE 20. Control system for optomotor yaw control. Medial and lateral visual motion detectors activate both parametric and directional logic commmands for forward and backward walking.

vate forward walking if the motion was from front to rear and backward walking if the translational motion was from rear to front (Davis and Ayers, 1972).

Pure angular rotational stimulation would activate the medial motion detector on one side and the lateral motion detector on the opposite side. This would cause the specimen to walk backward on one side and forward on the opposite side, thereby rotating in place and minimizing the rotational stimulus in a bidirectionally symmetric fashion (Kennedy and Davis, 1977).

A third emergent reflex resulting from this simple network is course correction during forward or backward locomotion. During forward locomotion, the two eyes see symmetrical rear-to-front or front-to-rear translational motion. During rotational perturbations (perhaps caused by water currents and surges, etc.), an angular rotation is superimposed on the translational motion. In other words, one eye sees translational plus rotational motion while the other eye sees translational minus rotational motion. A turning perturbation to the right during forward walking would cause the left eye to see more rapid visual motion than the right eye. Due to the contralateral projections of the lateral motion detectors to the forward command of the opposite side in the yaw compensation network (Fig. 20), the velocity command to the two sides would be biased to cause the right side to walk faster than the left side and bring the system back on course. Given the inherent symmetry, this feature would also operate during backward walking.

VIII. Future Directions

The ambulation controller we have described could be readily implemented in an autonomous robot. The architecture is ideally suited to be implemented by distributed processors, each of which corresponds to the controller for a single limb. The bus logic would correspond to the intersegmental interneurons of the lobster CNS, such as the walking direction logic and frequency controllers. The frequency controller could be as simple as 2 or 3 bits, and an 8-bit address and data bus would be quite adequate for passing scalars to registers of the distributed processors. A more complex sequential processor might correspond to the "brain" and set the logic of the bus lines controlling the individual limb processors.

Perhaps the most daunting challenge in the implementation of such a robot is that of the actuators and power supplies. Crustacean muscle is different from most linear actuators in that it is capable of continuously switching from isometric to isotonic control and back (Atwood, 1976). Successful emulation will probably involve hybrid mechanical systems involving linear actuators, pneumatic, and/or hydraulic interfaces. A combination of low-pressure hydraulic and pneumatic controllers would be ideally suited to operation of such a robot under water, where a nearly neutrally buoyant robot would have limited force generation requirements. Power supplies would probably be constrained to battery technology with inherent limitations in the power output of the robot. Ambulatory locomotion is one of the most efficient forms of locomotion, especially in marine organisms (Clarac, 1982).

We are in the process of developing an eight-legged ambulatory robot based on the architecture described here. One of the challenges of such an implementation is that of emulating the sensors inherent in crustacean systems. Crustaceans have a variety of proprioceptors, including contact, position, and force transducers, which can be readily emulated by standard mechanical transducers (Mill, 1976). Vestibular sensors can be implemented as simple inclinometers. In marine environments, the most advantageous medium for communications is sonar. We are developing communication systems for biotelemetry that are suitable for communications and control of an autonomous robot (Massa et. al., 1992). Sonar systems are ideal for beacon tracking and docking procedures, such as replacing power supplies in situ. We are also exploring the use of sonar arrays for an underwater acoustic imaging system (Massa et. al., 1992).

A final area of development that is essential for an underwater walking machine is roll and pitch control. Control in these planes involves two interrelated problems. The first has to do with compensatory reflexes for orientational perturbations due to substrate irregularity. The second has to do with perturbations resulting from water currents. Solutions to both of these problems are inherent in the behavior of the lobster, which has evolved to deal with these problems on a daily basis. Decapods have an elaborate repertoire of compensatory reflexes that have been well documented (Neil, 1985). Their implementation in our proposed architecture is easily realizable (Fig. 20). Furthermore, the chelipeds and abdomen can be used extensively as hydrodynamic control surfaces (Bowerman and Larimer, 1974a,b). Integration of current sensors into rheotaxic control reflexes will realize this important aspect of orientational control.

References

ATWOOD, H. L. (1976). "Organization and Synaptic Physiology of Crustacean Neuromuscular Systems," *Prog. Neurobiol.* **7,** 291–391.
AYERS, J., and CLARAC, F. (1978). "Neuromuscular Strategies Underlying Different Behavioral Acts in a Multifunctional Crustacean Leg Joint," *J. Comp. Physiol.* **128,** 81–94.
AYERS, J., and DAVIS, W. J. (1977a). "Neuronal Control of Locomotion in the Lobster, *Homarus americanus*. I. Motor Programs for Forward and Backward Walking," *J. Comp. Physiol.* **115,** 1–27.

Chapter XIII A Model for a Robotic Ambulation Architecture   313

AYERS, J., and DAVIS, W. J. (1977b). "Nervous Control of Locomotion in the Lobster, *Homarus americanus*. II. Types of Walking Leg Reflexes," *J. Comp. Physiol.* **115,** 29–46.

AYERS, J., and DAVIS, W. J. (1978). "Neuronal Control of Locomotion in the Lobster, *Homarus americanus*. III. Dynamic Organization of Walking Leg Reflexes," *J. Comp. Physiol.* **123,** 289–298.

AYERS, J., and SELVERSTON, A. I. (1979). "Monosynaptic Entrainment of an Endogenous Pacemaker Network: A Cellular Mechanism for von Holst's Magnet Effect," *J. Comp. Physiol.* **129,** 5–17.

BARNES, W. J. P. (1975). "Leg Coordination during Walking in the Crab, *Uca pugnax*," *J. Comp. Physiol.* **96,** 237–256.

BARNES, W. J. P. (1977). "Proprioceptive Influences on Motor Output during Walking in the Crayfish," *J. Physiol. (Paris)* **73,** 543–564.

BEER, R. D. (1991). *Intelligence as Adaptive Behavior: An Experiment in Computational Neuroethology.* Academic Press, New York.

BOWERMAN, R. F., and LARIMER, J. L. (1974a). "Command Fibres in the Circumoesophageal Connectives of Crayfish. I. Tonic Fibres," *J. Exp. Biol.* **60,** 95–117.

BOWERMAN, R. F., and LARIMER, J. L. (1974). "Command Fibres in the Circumoesophageal Connectives of Crayfish. II. Phasic Fibres," *J. Exp. Biol.* **60,** 119–134.

BROOKS, R. A. (1986). "A Robust Layered Control System for a Mobile Robot," *IEEE J. Robot. Autom.* **RA2**(1), 14–23.

BROOKS, R. A. (1989). "A Robot That Walks; Emergent Behaviors from a Carefully Evolved Network," *Neural Comput.* **1,** 253–262.

BROOKS, R. A. (1991). "New Approaches to Robotics," *Science* **253,** 1227–1232.

BROOKS, R. A., and FLYNN, A. M. (1989). "Fast, Cheap and Out of Control: A Robot Invasion of the Solar System," *J. Br. Interplanet. Soc.* **42,** 478–485.

CATTAERT, D., ELMANIRA, E., MARCHAND, A., and CLARAC, F. (1990). "Central Control of the Sensory Afferent Terminal from a Leg Chordotonal Organ in Crayfish in Vitro Preparation," *Neurosci. Lett.* **108,** 81–87.

CHASSERAT, C., and CLARAC, F. (1980). "Interlimb Coordinating Factors during Driven Walking in Crustacea. A Comparative Study of Absolute and Relative Coordination," *J. Comp. Physiol.* **139,** 293–306.

CHASSERAT, C., and CLARAC, F. (1986). "Basic Processes of Locomotor Coordination in the Rock Lobster. II. Simulation of Leg Coupling," *Biol. Cybern.* **55,** 171–185.

CHRACHRI, A., and CLARAC, F. (1987). "Induction of Rhythmic Activity in Motoneurons of Crayfish Thoracic Ganglia by Cholinergic Agonists," *Neurosci. Lett.* **77,** 49–54.

CHRACHRI, A., and CLARAC, F. (1989). "Synaptic Connections between Motor Neurons and Interneurons in the Fourth Thoracic Ganglion of the Crayfish, *Procambarus clarkii*," *J. Neurophysiol.* **62**(6), 1237–1250.

CHRACHRI, A., and CLARAC, F. (1990). "Fictive Locomotion in the Fourth Thoracic Ganglion of the Crayfish, *Procambarus clarkii*," *J. Neurosci.* **10**(3), 707–719.

CLARAC, F. (1978). "Locomotory Programs in Basal Leg Muscles after Limb Autotomy in the Crustacea," *Brain Res.* **145,** 401–405.

CLARAC, F. (1981). "Postural Reflexes Coordinating Walking Legs in a Rock Lobster," *J. Exp. Biol.* **90,** 333–337.

CLARAC, F. (1982). "Decapod Crustacean Leg Coordination during Walking," in Herreid, C. F., and Fourtner, C. R. (eds.), *Locomotion and Energetics in Arthropods* (pp. 31–71). Plenum, New York.

CLARAC, F. (1985). "Stepping Reflexes and the Sensory Control of Walking in Crustacea," in Barnes, W. J. P., and Gladden, M. H. (eds.), *Feedback and Motor Control in Invertebrates and Vertebrates* (pp. 379–400). Croom Helm, London.

CLARAC, F. (1990). "Proprioception from Chordotonal Organs in Crustacean Limbs," in Weise, K. (ed.), *Frontiers in Crustacean Neurobiology* (pp. 262–270). Birkhäuser Verlag, Basel.

CLARAC, F., and BARNES, W. J. P. (1985). "Peripheral Influences on the Coordination of the Legs during Walking in Decapod Crustaceans," in Bush, B. M. H., and Clarac, F. (eds.), *Coordination of Motor Behavior.* (pp. 249–269). Cambridge University Press, Cambridge.

CLARAC, F., VEDEL, J. P., and BUSH, B. M. H. (1978). "Intersegmental Reflex Coordination by a Single Joint Receptor Organ (CB) in Rock Lobster Walking Legs," *J. Exp Biol.* **73,** 29–46.

CRUSE, H., and GRAHAM, D. (1985). "Models for the Analysis of Walking in Arthropods," in Bush, B. M. H., and Clarac, F. (eds.), *Coordination of Motor Behavior.* (pp. 283–302). Cambridge University Press, Cambridge.

CRUSE, H., and MÜLLER, W. (1986). "Two Couplings Which Determine the Coordination of Ipsilateral Legs in the Walking Crayfish," *J. Exp. Biol.* **121,** 349–369.

DAVIS, W. J. (1968). "Lobster Righting Responses and their Neural Control," *Proc. Roy. Soc. B.* **70,** 435–436.

DAVIS, W. J. (1971). "Functional Significance of Motor Neuron Size and Soma Position in Swimmeret System of the Lobster," *J. Neurophysiol.* **34,** 274–288.

DAVIS, W. J. (1973). "Neuronal Organization and Ontogeny in the Lobster Swimmeret System, in Stein, R. B., Pearson, K. G., Smith, R. S., and Redford, J. B. (eds.), *Control of Posture and Locomotion: Advances in Behavioral Biology,* Vol. 7. (pp. 437–456). Plenum, New York.

DAVIS, W. J., and AYERS, J. (1972). "Locomotion: Control by Positive-Feedback Optokinetic Responses," *Science* **177,** 183–185.

DAVIS, W. J., and KENNEDY, D. (1972). "Command Interneurons Controlling Swimmeret Movements in the Lobster. 1. Types of Effects on Motorneurons." *J. Neurophysiology* **35,** 1–12.

DELCOMYN, F. (1980). "Neural Basis of Rhythmic Behavior in Animals," *Science* **210,** 492–498.

ELSON, R. C., SILLAR, K. T., and BUSH, B. M. H. (1992). "Identified Proprioceptive Afferents and Motor Rhythm Entrainment in the Crayfish Walking System," *J. Neurophysiol.*, in press.

EVOY, W., and AYERS, J. (1982). "Locomotion and Control of Limb Movements," in Sandeman, D. C., and Atwood, H. (eds.), *The Biology of Crustacea,* Vol. 4. *Neural Integration and Behavior* (pp. 62–106). Academic Press, New York.

GRILLNER, S. (1985). "Neurological Bases of Rythmic Motor Acts in Vertebrates," *Science* **228,** 143–149.

GROTE, J. R. (1981). "The Effect of Load on Locomotion in Crayfish," *J. Exp. Biol.* **92,** 277–288.

HOYLE, G. (ed). (1977). *Identified Neurons and Behavior of Arthropods.* Plenum, New York.

KENNEDY, D., and DAVIS, W. J. (1977). "Organization of Invertebrate Motor Systems," in Geiger, R., Kandel, E., and Brookhart, J. M. (eds.), *Handbook of Physiology.* (pp. 1023–1087). American Physiology Society. Bethesda, MD.

KLÄRNER, D., and BARNES, W. J. P. (1986). "The Cuticular Stress Detector (CSD2) of the Crayfish. II. Activity during Walking and Influences on Leg Coordination," *J. Exp. Biol.* **122,** 161–175.

KUPFERMAN, I., and WEISS, K. R. (1978). "The Command Neuron Concept," *Behav. Brain Sci.* **1,** 3–39.

LIBERSAT, F., CLARAC, F., and ZILL, S. (1987). "Force Sensitive Mechanoreceptors of the Dactyl of the Crab: Single Unit Responses during Walking and Evaluation of Function," *J. Neurophysiol.* **57,** 1601–1607.

LOEB, J. (1918). "Forced Movements, Tropisms, and Animal Conduct," *Monographs on Experimental Biology.* J. B. Lippincott Company.

MASSA, D., AYERS, J., and CRISMAN, J. (1992). "Acoustic Communication, Navigation and Sensing Systems for a Biologically-Based Controller for a Shallow Water Walking Machine," *Oceans* **92,** in press.

MILL, P. J. (ed.). (1976). *Structure and Function of Proprioceptors in the Invertebrates.* Chapman and Hall, London.

NEIL, D. (1985). "Multisensory Interactions in the Crustacean Equilibrium System," in Barnes, W. J. P., and Gladden, M. H. (eds.), *Feedback and Motor Control in Invertebrates and Vertebrates.* Croom Helm, Ltd., 379–400.

PASZTOR, V. M., and CLARAC, F. (1983). "An Analysis of Waving Behaviour: An Alternative Motor Programme for the Thoracic Appendages of Decapod Crustacea," *J. Exp. Biol.* **102,** 59–77.

PAUL, D., and MULLONEY, B. (1986). "Intersegmental Coordination of Swimmeret Rhythms in Isolated Nerve Cords of Crayfish," *J. Comp. Physiol.* **158,** 215–224.

PEARSON, K. G. (1972). "Central Programming and Reflex Control of Walking in the Cockroach," *J. Exp. Biol.* **56,** 173–193.

PEARSON, K. G., and ILES, J. F. (1973). "Nervous Mechanisms Underlying Intersegmental Coordination of Leg Movements during Walking in the Cockroach," *J. Exp. Biol.* **58,** 725–744.

PENNISI, E. (1991). "Robots Go Buggy," *Science News* **140,** 361–363.

PINSKER, H. M., and AYERS, J. 1983. *The Clinical Neurosciences, Chapter 9, Neuronal Oscillators.* (W. D. Willis, ed.). Churchill Livingston, New York, 203–266.

SANDEMAN, D. C., and ATWOOD, H. L. (1982). *The Biology of Crustacea,* Vol. 4. *Neural Integration and Behavior.* Academic Press, New York.

SELVERSTON, A. I., and MOULINS, M. (1987). *The Crustacean Stomatogastric System.* Springer-Verlag, Berlin.

SILLAR, K. T., and ELSON, R. C. (1986). "Slow Active Potentials in Walking-Leg Motor Neurones Triggered by Non-spiking Proprioceptive Afferents in the Crayfish," *J. Exp. Biol.* **126,** 445–452.

SILLAR, K. T., CLARAC, F., and BUSH, B. M. H. (1987). "Intersegmental Coordination of Central Neural Oscillators for Rhythmic Movements of the Walking Legs in Crayfish, *Pacifastacus lenuisculus*," *J. Exp. Biol.* **131,** 245–264.

STEIN, P. S. G. (1978). "Motor Systems, with Specific Reference to the Control of Locomotion," *Annu. Rev. Neurosci.* **1,** 61–81.

WIERSMA, C. A. G., and YAMAGUCHI, T. (1967). "A Neuronal Component of the Optic Nerve of the Crayfish as Studied by Single Unit Analysis," *J. Comp. Neurol.* **128,** 333–358.

WILSON, D. M. (1967). "Stepping Patterns in Tarantula Spiders," *J. Exp. Biol.* **47,** 133–151.

WINE, J. J. (1984). "The Structural Basis of an Innate Behavioral Pattern," *J. Exp. Biol.* **112,** 283–319.

PART IV

ROBOTICS

Chapter XIV Legged Robots

Marc H. Raibert and Jessica K. Hodgins[1]
The Leg Laboratory
Artificial Intelligence Laboratory
Massachusetts Institute of Technology
Cambridge, Massachusetts

Currently at the IBM T. J. Watson Research Center, Hawthorne, NY.

I. Introduction

One reason for the study of legged robots is the hope of building vehicles that can travel on terrain that is too rough for conventional wheeled and tracked vehicles. Such terrain includes a large fraction of the earth's surface, including rocky, hilly, and mountainous regions, wooded areas, jungles, swamps, and river banks. It also includes much of the terrain that humans have created, such as the interiors of our factories, ships, nuclear reactors, and mines. The mobility of people and animals on rough terrain encourages us to explore what might be achieved by legged vehicles. Such vehicles could find application in agriculture, space, defense, and public service. Although practical legged vehicles that travel on rough terrain with significant mobility do not yet exist, a great deal of progress is being made.

A second reason to study legged robots is the prospect of using them to understand more about the legged behavior of people and animals. Most animals perform walking and running tasks with surprising agility, speed, efficiency, and grace. To understand walking and running in animals, one must learn about their nervous systems, their mechanical systems, and their control algorithms, as well as the principles of legged locomotion itself. Even though the study of legged machines does not reveal directly the mechanisms used by living systems, it can provide powerful and concrete models that give insight into the sorts of problems that all legged systems must solve. Working legged robots also provide a completeness check on our models. Successful robot implementations require that we identify and attend to every important aspect of the locomotion problem. This is not the case when studying animals, which can continue to perform whether or not the important mechanisms have been identified. We believe that advances in legged machines, in their mechanics, control, sensing, and planning, will provide insight into the mysteries of human and animal locomotion. Progress in understanding locomotion in animals will, in turn, make it possible to build robots that are more agile, dextrous, and fundamentally more interesting than those we see today.

A third reason for interest in legged robots is the rich dynamical behavior that legged systems exhibit. Consider, for example, the rhythmic oscillations found in walking and running, the ballistic phases that occur during gymnastic maneuvers, the energetics of locomotion at moderate and high speeds, the control of active balance, the intermittent opportunity for control in running, and the planning problems involved in locomotion on rough terrain. Each of these phenomena involves interesting dynamical behavior whose study can bring new ideas and directions to robotics. The importance of dynamical phenomena in legged locomotion is related to the large ratio of payload mass to limb strength found in typical legged systems. This ratio is generally larger for legged systems than for other sorts of robots. We hope that the study of legged locomotion will lead to new paradigms for building dynamic robots and ultimately to a deeper understanding of robots as dynamical systems.

This chapter gives a brief survey of work on legged robots, with focus on four areas: statically stable walkers, dynamic walkers, dynamic runners, and locomotion on rough terrain.

II. Statically Stable Walkers

The scientific study of legged locomotion began just over a century ago when Leland Stanford, then Governor of California, commissioned Eadweard Muybridge to find out whether or not a trotting horse left the ground with all four feet at the same time (see Table 1). Stanford had wagered that it never did. After Muybridge proved him wrong with a set of stop-motion photographs that appeared in *Scientific American* in 1878, Muybridge went on to document the walking and running behavior of over 40 mammals, including humans (Muybridge 1901/1955, 1899/1957). Even now, his photographic data are of considerable value and beauty and survive as a landmark in locomotion research.

The study of machines that walk also had its origin in Muybridge's time. An early walking model appeared in about 1870 (Lucas, 1894). It used a linkage to

Chapter XIV Legged Robots

Table I. Milestones in the Development of Legged Robots

1836	Weber and Weber	Measurements in corpses show that natural frequency of leg as compound pendulum is similar to cadence of live walker.
1850	Chebyshev	Designs linkage used in early walking mechanism (Lucas, 1894).
1872	Muybridge	Develops stop-motion photography to document running animals.
1893	Rygg	Patents human-powered mechanical horse.
1942	Wallace	Patents hopping tank with reaction wheels that provide stability.
1961	Space General	Eight-legged linkage machine walks in outdoor terrain (Morrison, 1968).
1963	Cannon, Higdon and Schaefer	Control system balances single, double, and limber inverted pendulums.
1968	Frank and McGhee	Simple digital logic controls walking of Phony Pony.
1968	Mosher	GE quadruped truck climbs rainroad ties under control of human driver.
1969	Bucyrus-Erie Co.	Big Muskie, a 15,000-ton walking dragline, is used for strip mining. It moves in soft terrain at a speed of 900 ft/hr (Sitek, 1976).
1977	McGhee	Digital computer coordinates leg motions of hexapod robot walks with wave gait using digital computer to coordinate leg motion.
1977	Gurfinkel	Hybrid computer controls hexapod walker in USSR.
1977	McMahon and Greene	Human runners set new speed records on *tuned track* at Harvard. Its compliance is adjusted to mechanics of human leg.
1980	Hirose and Umetani	Quadruped machine climbs stairs and climbs over obstacles using simple sensors and reflex-like control. The leg mechanism simplifies control.
1980	Kato	Hydraulic biped walks with quasidynamic gait.
1980	Matsuoka	Mechanism balances in the plane while hopping on one leg.
1981	Miura and Shimoyama	Biped balances actively while walking in three dimensional space.
1983	Sutherland	Self-contained hexapod carries human rider. Computer, hydraulics, and human share computing task.
1983	Odetics	Self-contained hexapod lifts and moves back end of pickup truck (Russell, 1983).
1983	Raibert	One-legged machine balances as it hops in place, travels at a specified rate, tolerates distrubances, and jumps over small obstacles.
1984	Furusho	Planar five-link biped starts walking from a standing position, and travels at 0.8 m/s.
1987	Waldron and McGhee	Three-ton self-contained hexapod carrying human driver travels at 5 mph, travels in irregular terrain, and pulls a load.
1988	Hodgins and Koechling	Planar biped climbs short stairway, jumps over obstacles, and runs with top speed of 13.1 mph.
1989	Raibert	Quadruped runs with trotting, pacing, and bounding gaits, and changes between gaits.
1990	McGeer	Planar biped with knees walks passively down sloping surface.

move the body along a straight horizontal path while the feet moved up and down to exchange support during stepping. The linkage was based on a design by the Russian mathematician Chebyshev. During the years that followed, many workers have viewed the task of building walking machines as the task of designing linkages that would generate stepping patterns when driven by a source of power. Many designs were proposed (e.g., Rygg, 1893; Nilson, 1926; Snell, 1947; Shigley, 1957; Morrison, 1968) (see Fig. 1). The performance of such machines is limited by their fixed patterns of motion, since they could not adjust to variations in the terrain.

A second approach to providing control for legged locomotion was to harness a human. Ralph Mosher used this approach in building a four-legged walking truck at General Electric in the mid-1960s (Liston and Mosher, 1968). The project was part of a decade-long campaign to build advanced teleoperators capable of providing better dexterity through high-fidelity force feedback. The walking machine Mosher built stood 11 feet tall, weighed 3000 pounds, and was powered hydraulically. It is shown in Fig. 2. A human driver controlled the motion. Each of the driver's limbs was connected to a handle or pedal that controlled one of the truck's four legs. Whenever the driver caused a truck leg to push against an obstacle, force feedback would cause the handle or pedal to push back on the human, letting the driver feel the obstacle as though it were his or her own arm or leg doing the pushing.

After about 20 hours of training, Mosher was able to handle the machine with surprising agility. Films of the machine operating under his control show it ambling along at about 5 miles per hour (mph), climbing a stack of railroad ties, pushing a foundered jeep out of the mud, and maneuvering a large drum onto some hooks. Despite its dependence on a well-trained human for control, the GE Walking Truck was a milestone in legged technology.

A third approach to controlling legged locomotion became feasible in the 1970s: use of a digital computer. Robert McGhee's group at the Ohio State University was the first to do so (McGhee, 1983). In 1977 they built an insectlike hexapod that would eventually walk with a number of gaits, turn, walk sideways, and negotiate simple obstacles. The computer's primary task was to solve kinematic equations in order to coordinate the 18 electric motors driving the legs. The coordination ensured that the machine's center of mass stayed over the polygon of support provided by the supporting feet while allowing the legs to sequence through a gait (Fig. 3). The machine traveled quite slowly, covering several yards per minute. The hexapod provided McGhee with an experimental means of following up on his earlier theoretical findings on the combinatorics and selection of gait (McGhee, 1968; McGhee and Jain, 1972; Koozekanani and McGhee, 1973).

At about the same time, Gurfinkel and his co-workers in the USSR built a machine with characteristics and performance quite similar to McGhee's (Okhotsimski *et al.*,

Chapter XIV Legged Robots

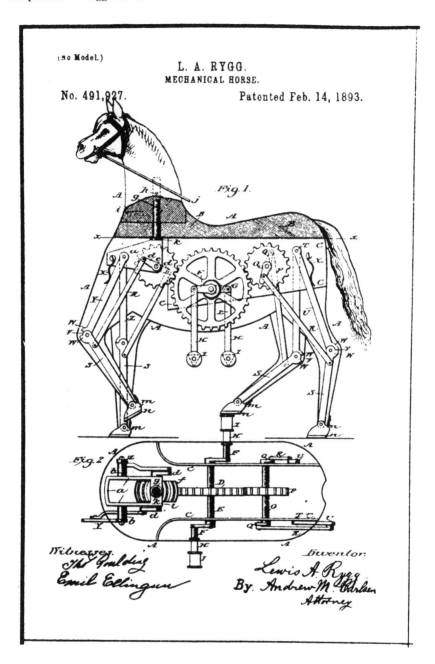

FIGURE 1. Mechanical horse patented by Lewis A. Rygg in 1893. The stirrups double as pedals so the rider can power the stepping motions. The reins move the head and forelegs from side to side for steering. Apparently the machine was never built.

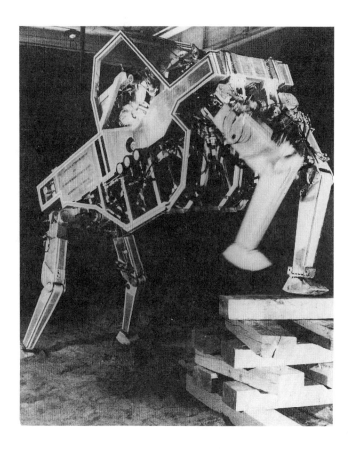

FIGURE 2. Walking truck developed by Ralph Mosher at General Electric in about 1968. The human driver controlled the machine with four handles and pedals that were connected to the four legs hydraulically. Photograph courtesy of General Electric Research and Development Center.

1977; Gurfinkel *et al.*, 1981; Devjanin *et al.*, 1983). It used a hybrid computer for control, with analog computation aiding in kinematic calculations.

The group at Ohio State subsequently built a much larger hexapod, called the Adaptive Suspension Vehicle (Fig. 4). It was designed for self-contained operation on natural terrain (Waldron *et al.*, 1984; Pugh *et al.*, 1990). It carries a gasoline engine for power, several computers and a human operator for control, and a laser range sensor for terrain preview. This machine walked at about 5 mph, negotiated simple obstacles on rough terrain, and pulled heavy loads.

Hirose realized that the three basic approaches to controlling legged locomotion—mechanical linkage, human teleoperation, and computer control—are not mutu-

Chapter XIV   Legged Robots   325

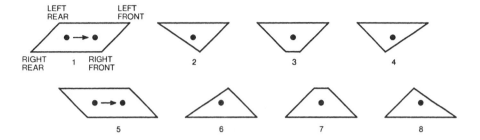

FIGURE 3. Statically stable gait. The diagram shows the sequence of support patterns provided by the feet of a quadruped walking with a crawling gait. The body and legs move to keep the projection of the center of mass within the polygon defined by the feet. A supporting foot is located at each vertex. The dot indicates the projection of the center of mass. Data adapted from McGhee and Frank (1968).

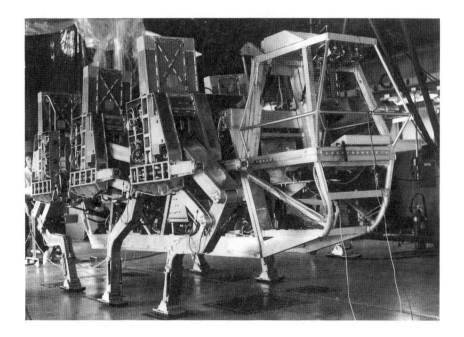

FIGURE 4. The hexapod walking machine developed at Ohio State University. It stands about 10 feet tall, 15 feet long, and weighs 3 tons. A 90-horsepower motorcycle engine provides power to 18 variable displacement hydraulic pumps that drive the joints. The legs use pantographs linkages to improve energy efficiency. The operator normally provides steering and speed commands while computers control the stepping motions of the legs.

ally exclusive. His experience with clever and unusual mechanisms—he had built seven versions of a mechanical snake—led to a leg with a special mechanical structure that simplified the control of locomotion and could improve efficiency (Hirose and Umetani, 1980; Hirose *et al.*, 1984). The leg was a three-dimensional pantograph that translated the motion of each actuator into a pure Cartesian translation of the foot. With the ability to generate x, y, and z translations of each foot by merely choosing an actuator, the control computer was freed from the task of performing kinematic solutions. The mechanical linkage helped to perform the calculations needed for locomotion. The linkage was efficient because the actuators performed only positive work in moving the body forward and no work was done against gravity when moving on the level. Hirose used the pantograph leg in a small quadruped, about 1 yard long.

These three walking machines, McGhee's, Gurfinkel's, and Hirose's, represent a class called *static crawlers*. Each differs in the details of construction and in the computing technology used for control, but they share a common approach to balance and stability. They all keep enough feet on the ground to guarantee a broad base of support at all times, and the body and legs move to keep the center of mass over this support base. The forward velocity is kept low enough so that kinetic energy can be ignored in the stability calculation. Other machines in this class have also been studied (e.g., see Russell, 1983; Sutherland and Ullner, 1984; Ooka *et al.*, 1985; Bares and Whittaker, 1990).

III. Dynamic Walkers

We now consider the study of machines that balance actively as they walk. These systems operate in a regime where the velocities and kinetic energies of the masses are important to the behavior in addition to the geometric configuration.

The first machines that balanced actively were automatically controlled inverted pendulums. A person can balance a broom on his or her finger with relative ease. Why not use automatic control to build a broom that can balance itself? Claude Shannon was probably the first to do so. In 1951 he used the parts from an erector set to build a machine that balanced an inverted pendulum atop a small powered truck. The truck drove back and forth in response to the tipping movements of the pendulum, as sensed by a pair of switches at its base. In order to move from one place to another, the truck first had to drive away from the destination to unbalance the pendulum, then proceeded toward the destination. In order to balance again at the destination, the truck moved past the destination until the pendulum was again upright with no forward velocity, then moved back to the destination.

Chapter XIV Legged Robots

At Shannon's urging, Robert Cannon and his students at Stanford University set about demonstrating controllers that balanced two pendulums at once. In one case the pendulums were mounted side by side on the cart, and in the other they were mounted one on top of the other (Fig. 5). They also demonstrated balance for a flexible inverted pendulum (Schaefer, 1965; Schaefer and Cannon, 1966). Cannon's group was interested in the single-input, multiple-output problem and in the limitations of achievable balance: how could they use the single force that drove the cart's motion to control the angles of two pendulums as well as the position of the cart? How far from balance could the system deviate before it was impossible to return to equilibrium, given such parameters of the mechanical system as the cart motor's strength and the pendulum lengths? These studies of balance for inverted pendulums were important precursors to later work on locomotion. The inverted pendulum model for walking would become a primary tool for studying balance in legged systems (e.g., Hemami and Weimer, 1974; Vukobratovic and Stepaneko, 1972; Vukobratovic, 1973; Hemami and Golliday, 1977; Kato *et al.*, 1983; Miura and Shimoyama, 1984).

Despite the recognized importance of active balance in legged locomotion, the first dynamic legged systems did not appear until about 1980. Kato and his co-workers built a biped that walked with a *quasi-dynamic* gait (Ogo *et al.*, 1980; Kato *et al.*, 1983). The machine had 10 hydraulically powered degrees of freedom

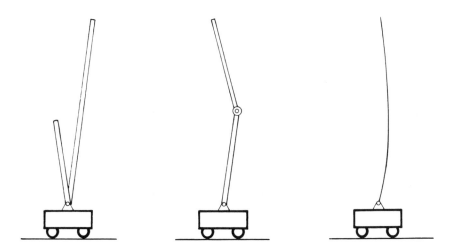

FIGURE 5. Cannon and his students built machines that balanced inverted pendulums on a moving cart. They balanced two pendulums side by side, one pendulum on top of another, and a long limber inverted pendulum. Only one input, the force driving the cart horizontally, was available for control. Data adapted from Schaefer and Cannon (1966).

and two large feet. This machine was usually a static crawler, moving along a preplanned trajectory to keep the center of mass over the base of support provided by the large supporting foot. Once during each step, however, the machine temporarily destabilized itself to tip forward so that support would be transferred quickly from one foot to the other. Before the transfer took place on each step, the *catching* foot was positioned to return the machine to equilibrium passively. No active response was required. The inverted pendulum model was used to plan the tipping and catching motions. This machine walked with a quasi-dynamic gait, taking about a dozen 0.5-m steps per minute.

Kato's approach was an interesting way to achieve dynamic behavior. The system was not dynamic in the sense of reacting at run-time to the progress of the motion. Instead, an off-line analysis of the dynamics of the system specified how to position the catching foot statically to get run-time dynamic behavior. Knowledge of dynamics of the system was preprocessed to form a simple run-time strategy.

Miura and Shimoyama (1984) built a walking machine that may have been the first to balance itself actively. Their *stilt biped* was patterned after a human walking on stilts. Each foot provided only a point of support, and the machine had three actuators: one for each leg that moved the leg sideways and a third that separated the legs fore and aft. Because the legs did not change length, the hips were used to pick up the feet. This gave the machine a pronounced shuffling gait reminiscent of Charlie Chaplin's stiff-kneed walk.

Control for the stilt biped relied, once again, on the inverted pendulum model of its behavior. Each time a foot was placed on the floor, its position was chosen according to the tipping behavior expected from an inverted pendulum. Actually, the problem was broken down as though there were two planar pendulums, one in the pitching plane and one in the rolling plane. The choice of foot position along each axis took the current and desired state of the system into account. The control system used tabulated descriptions of planned leg motions together with linear feedback to perform the necessary calculations. Unlike Kato's machine, which came to static equilibrium before and after each dynamic transfer, the stilt biped tipped all the time.

Furusho and Masubuchi (1987a, 1987b) used an inverted pendulum model in conjunction with a hierarchical control scheme. A low-level servo moved each joint toward a desired position. An inverted pendulum model was derived from the dominant modes of the closed-loop system. The setpoints for the closed-loop system were selected to produce a walking motion. Their five-link planar biped started from standing, walked at 0.8 m/s, and then returned to a standing position. Furusho and his colleagues later built machines with seven and nine links that walked in three dimensions and used ankle torques to control forward speed (Furusho and Sano, 1990).

Chapter XIV   Legged Robots   329

Later work in Kato's lab, especially that of Takanishi (Takanishi *et al.,* 1985, 1989, 1990b), resulted in a series of actively balanced walking bipeds. They modeled the electric motors as point masses and the links as massless. Because electric motors are so much heavier than the links of the robots, this simplified model yielded useful results. One of these machines had a large upper trunk that moved back and forth and side to side to stabilize the walking motion and to balance when pushed (Fig. 6).

McGeer (1989, 1990) took an unusual and elegant approach to dynamic walking, called *passive dynamic walking.* He was motivated by the stable behavior of toy animals that walk downhill by waddling from side to side and by Mochon and

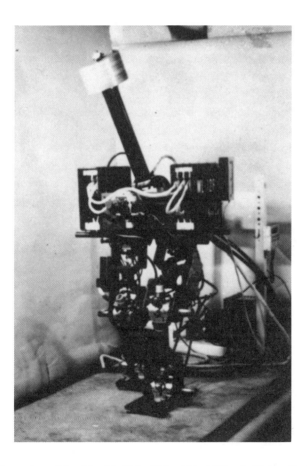

FIGURE 6. Dynamic walking biped built by Takanishi at Waseda University. The control system stabilizes the walking motion by moving the upper trunk back and forth.

McMahon's (1980, 1981) work on ballistic walking, which suggested that humans swing their legs passively during the swing phase of walking. McGeer designed machines that travel downhill using no sensors, no actuators, and no computers. They rely on their mechanical geometry—link lengths, link masses, joint damping, walking surface slope, foot shape—to do the computing. One can think of McGeer's machines as mechanical computers whose algorithms are embedded in their form. These machines are remarkable because they obtained stable dynamic walking with such a parsimonious design. Future legged machines will probably combine the passive characteristics of McGeer's design with the active control found in most other robots.

In order to predict and influence the behavior of a dynamic system, one must consider the energy stored in the system: in the elevation of each mass, in the deflection of each compliant element, and in the velocity of each mass. The exchange of energy among its various forms is also important in dynamic legged systems. For example, in dynamic walking the kinetic and gravitational potential energies oscillate out of phase throughout the cycle. During running the body's potential energy of elevation changes into kinetic energy during falling, then into strain energy when parts of the leg deform elastically during rebound with the ground, then into kinetic energy again as the body accelerates upward, and finally back into potential energy of elevation. This sort of dynamic exchange is central to legged locomotion.

Dynamics play a role in giving legged systems the ability to balance actively. A statically balanced system avoids tipping and the ensuing horizontal accelerations by keeping its center of mass over the polygon of support formed by the feet. In contrast, a dynamic legged system is always tipping. The control system avoids tipping too far by manipulating body and leg motions to ensure that each tipping interval is brief and that each tipping motion in one direction is compensated by a tipping motion in the opposite direction. An effective base of support is thus maintained over time.

The ability of an actively balanced system to depart from static equilibrium relaxes the rules governing how legs can be used for support. For example, if a legged system can tolerate tipping, then it can position its feet away from the center of mass in order to use footholds that are widely separated or erratically placed. On the other hand, by keeping the feet near the centerline, the system can travel where there is only a narrow path of good support. Animals routinely exploit active balance to travel on difficult terrain.

IV. Dynamic Runners

If a legged system can tolerate intermittent support, then it can move all its legs to new footholds at one time, to jump onto or over obstacles, and to use short periods of ballis-

Chapter XIV Legged Robots

tic flight for increased speed. Such behavior is called running because it includes a ballistic flight phase. In animals, running is usually associated with travel at high speed. Matsuoka (1979) was the first to build a machine that ran. His goal was to model repetitive hopping in humans. He formulated a model with a body and one massless leg, and he simplified the problem by assuming that the duration of the support phase was short compared with the ballistic flight phase. This extreme form of running, in which nearly the entire cycle is spent in flight, minimizes the influence of tipping during support. This model permitted Matsuoka to derive a time-optimal state feedback controller that provided stability for hopping in place and for low-speed translations.

To test his method for control, Matsuoka built a planar one-legged hopping machine. The machine operated at low gravity by rolling on ball bearings on a table that was inclined 10° from the horizontal in an effective gravity field of 0.17 g. An electric solenoid provided a rapid thrust at the foot. The machine hopped in place at about 1 hop per second and balanced itself as it traveled back and forth on the table.

The authors of this chapter and their colleagues built a series of running machines, including several one-legged hoppers, planar and three-dimensional bipeds, and a quadruped. Our goal was to explore basic issues in legged locomotion, particularly balance and the use of ballistic flight phases. We started by studying one-legged machines that ran by hopping (Raibert *et al.*, 1984; Raibert, 1986). See Figs. 7 and 8. The study of machines with one leg permitted us to concentrate on active balance and dynamics, while avoiding the difficult task of coordinating many legs. The one-legged hopping machines were designed to have springy legs in an effort to emulate the compliance of biological legs.

The control algorithms for the hopping machines concentrated on generating a resonant hopping motion using the compliance of the legs, on controlling the machine's forward speed as it hopped, and on keeping the body in an upright posture. In a typical experiment, a robot would travel around a circular track[2] in the laboratory or run on a treadmill. Control programs regulated the behavior of each running robot. The control programs acquired data from the machine's sensors, accepted commands from a human operator, performed calculations according to control algorithms, issued commands to the machine's actuators, and recorded data for later analysis.

Here is a brief account of the control algorithms used in these robots. A control system for running must perform three primary functions:

- *Hopping control:* cause the legs to step, exchanging support.
- *Speed control:* provide balance to regulate the running speed.
- *Posture control:* maintain the body in an upright posture.

---

[2] Some of the machines were constrained mechanically to move with just three degrees of freedom: fore–aft, up-and-down, pitch rotation. We mapped these motions to the surface of a sphere, which gave the overall machine the ability to travel along a circular path.

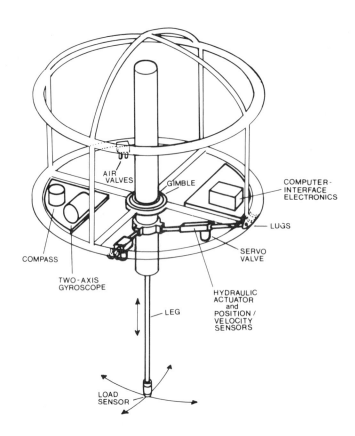

FIGURE 7. Diagram of 3D one-legged machine used for experiments. It has two primary parts: a body and a leg. The body is made of an aluminum frame, on which are mounted hip actuators, valves, gyroscopes, and computer interface electronics. The leg is a pneumatic cylinder with a padded foot at one end and a linear potentiometer at the other end. Two two-way pneumatic valves control the flow of compressed air to and from the lower end of the leg actuator. A pressure regulator and check valve control the pressure in the upper end of the leg actuator. The leg is springy because air trapped in the leg actuator compresses when the leg shortens. The leg is connected to the body by a gimbal-type hip with two degrees of freedom. A pair of low-friction hydraulic actuators powered by pressure control servo valves acts between the leg and body to determine the hip angles. Sensors measure the length of the leg, the length and velocity of each hydraulic actuator, contact between the foot and the floor, pressures in the leg air cylinder, and the pitch, roll, and yaw angles of the body. Analog measurements are digitized on the machine and transmitted to the control computer over a parallel bus. An umbilical connects the machine to hydraulic, pneumatic, and electrical power supplies and to the control computer, all of which are located nearby in the laboratory.

Chapter XIV Legged Robots   333

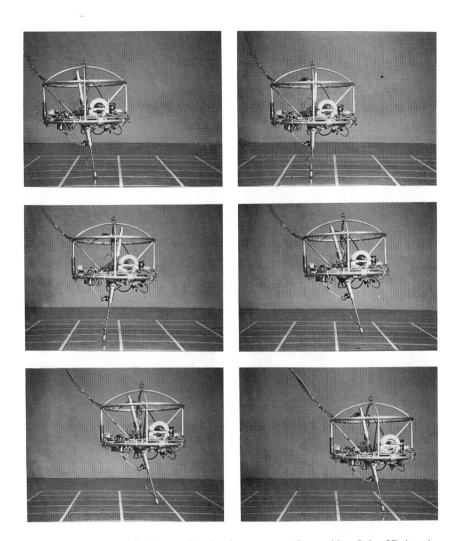

FIGURE 8. Sequence of photographs showing one complete stride of the 3D hopping machine running from left to right. Grid on floor indicates 0.5 m intervals. Running speed is about 1.75 m/sec, with stride length 0.63 m, and stride period 0.380 sec. Adjacent frames separated by 76 msec.

A. *Hopping Control*

An idea that has developed in biomechanics over the last 15 years is that animals use elastic structures in their limbs to improve the energetic efficiency of their locomotion and to match impedance when feet strike the ground. Tendons and ligaments in the legs and feet stretch during each collision with the ground, converting some the system's kinetic energy into elastic strain energy. The stored energy is returned during the next step, when the elastic structures rebound. A significant fraction of the total running energy, perhaps 20 to 40%, recirculates from one step to the next without needing resupply from the muscles. Kangaroos use their substantial Achilles tendons to perform this energy recovery function, whereas Alexander argues that humans store energy in their Achilles tendons and the ligaments that support the arch of the foot (Alexander, 1988). Compliant legs and feet also reduce peak loads that occur in running when the feet strike the ground at the end of each flight phase (McMahon, 1984; Alexander, 1988).

The legged robots we built use compliance in the legs to produce the vertical oscillations needed in running. The control algorithms allow the mass of the body to rebound on the springy leg during ground collisions and to be drawn back to earth by gravity during the flight phase. Control of the leg spring rest length is used to inject or remove energy from the system in order to initiate hopping oscillations, modulate them, or stop them. For vertical hopping with a massless leg, the altitude of a particular hop is predicted by the sum of the potential strain energy in the leg spring, the potential energy (PE) of elevation of the system mass, and the kinetic energy (KE) due to motion of the body:

$$h = (PE_{strain} + PE_{elevation} + KE) / Mg$$

where h is the expected altitude of the hop, M the system mass, and g the acceleration due to gravity. The control system can inject or remove energy to influence this outcome. This hopping control mechanism takes advantage of the dynamic interaction between the mechanical system and the control to generate the motion. No trajectory is specified.

B. Speed Control

Legged systems are like inverted pendulums: they tip and accelerate whenever the point of support is displaced from the projection of the center of mass (Hemami *et al.*, 1973). If the average point of support is kept under the average location of the center of mass, the system may tip for short periods without tipping over entirely. One way to achieve such a balancing relationship between the feet and the center of mass is to move the body in a symmetric fashion over the supporting feet during each support period. When the control system places the foot to obtain a symmetric sweeping pattern, the forward speed will be the same at liftoff as it was at touchdown. We call this position of the foot the *neutral point.* When the control system displaces the foot from the neutral point, the body accelerates, with the magnitude

Chapter XIV Legged Robots

and direction of acceleration related to the magnitude and direction of the displacement, as shown in Fig. 9. The control system displaces the foot from the neutral point by a distance proportional to the difference between the actual speed and the desired speed. The control system computes the desired foot position as

$$x_{fh,d} = \frac{\dot{x} T_s}{2} + k_{\dot{x}} (\dot{x} - \dot{x}_d)$$

where $x_{fh,d}$ is the forward displacement of the foot from the projection of the center of gravity, $\dot{x}$ is the forward speed, $\dot{x}_d$ is the desired forward speed, T_s is the predicted duration of the next support period, and $k_{\dot{x}}$ is a gain. The first term of Eq. (2) is an estimate of the neutral point and the second term is a correction for any error in forward speed or for a desired acceleration. The duration of the next support period is predicted to be the same as the measured duration of the previous support period. After the control system finds $x_{fh,d}$ a kinematic transformation determines the joint angles that will position the foot as specified.

C. Posture Control

Depending on the number of legs, the gait, and whether there is a tail, the trunk may pitch and roll during running. The long-term attitude of the trunk must be stabilized if the system is to remain upright. The control system we implemented regulates the orientation of the trunk by applying torques to the body during the support phase. In the biped and quadruped models, the hip actuators are used to

FIGURE 9. When the foot is positioned at the neutral point, the body travels along a symmetric path that leaves the system unaccelerated in the forward direction. Displacement of the foot from the neutral point accelerates the body by skewing the symmetry of the body's trajectory. When the foot is placed closer to the hip than the neutral point, the body accelerates forward during stance and the forward speed at liftoff is higher than the forward speed at touchdown (left). When the foot is placed further from the hip than the neutral point, the body decelerates during stance and the forward speed at liftoff is slower than the forward speed at touchdown (right). Horizontal lines under each figure indicate the distance the body travels during stance, and the curved lines indicate the path of the body.

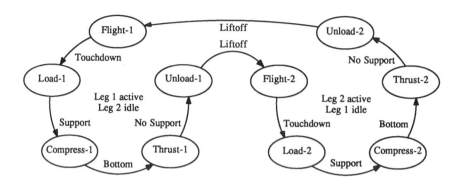

State	Actions
Flight	
Active leg leaves ground	Interchange active, idle legs
	Lengthen active leg for landing
	Position active leg for landing
	Shorten idle leg
Loading	
Active leg touches ground	Zero active hip torque
	Keep idle leg short
Compression	
Active leg spring shortens	Servo pitch with active hip
	Keep idle leg short
Thrust	
Active leg spring lengthens	Extend active leg
	Servo pitch with active hip
	Keep idle leg short
Unloading	
Active leg spring approaches full length	Shorten active leg
	Zero hip torques active leg
	Keep idle leg short

FIGURE 10. Finite-state machine that coordinates running. The state shown in the left column is entered when the sensory event, listed just below the state name, occurs. Actions are listed on the right. The controller advances through the states in sequence. The diagram is for a two-legged gait.

Chapter XIV Legged Robots

FIGURE 11. Photograph of quadruped running machine used for experiments. The body is an aluminum frame, on which are mounted legs, hip actuators, gyroscopes, and computer interface electronics. Each hip has two low friction hydraulic actuators that position the leg fore and aft, and sideways. Sensors measure the position, velocity, and force of the hydraulic hip actuators, hydraulic leg length, overall leg length, leg spring length, contact between the feet and the floor, and the pitch, roll, and yaw orientations of the body. An umbilical cable connects the machine to hydraulic, pneumatic, and electrical power supplies, and to a control computer, all of which are located nearby in the laboratory. An arrangement of spring in series with position source in each leg was motivated by simple muscle models. Physical parameters of the quadruped machine are given in (Raibert 1990).

apply the torques required for attitude control. In the kangaroo model, the knee is used to perform this function. Vertical loading on the feet keeps the leg from slipping when the torque is applied. The posture control torques are generated by a linear servo:

$$\tau = -k_p(\phi - \phi_d) - k_v(\dot{\phi})$$

where τ is the leg torque, ϕ is the angle of the body, ϕ_d is the desired angle of the body, $\dot{\phi}$ is the angular rate of the body, and k_p and k_v are gains.

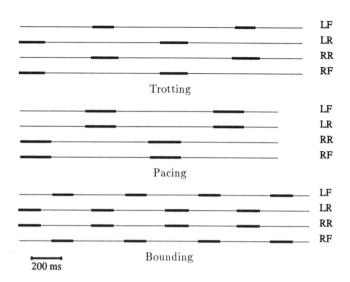

FIGURE 12. Pair gaits. Gait diagrams showing the pattern of leg use for the trot, pace, and bound, as executed by the quadruped robot in experiments. Each of these gaits use the legs in pairs. The bars indicate periods of ground contact, as measured by load switches in the feet. (Trotting data Q87.335.3, pacing data Q87.142.4, bounding data Q87.167.1.)

The control systems for running use separate algorithms for stabilizing hopping, forward speed, and posture of the trunk. Each of these parts of the control acts independently, behaving as though it influences just one component of the behavior. Interactions due to imperfect decoupling are treated as disturbances. The decoupling of the control system simplifies the control implementation significantly.

In addition to the control algorithms described so far, each implementation uses a finite state machine to track the ongoing behavior of the model, to synchronize the control actions to the running behavior, and to do some bookkeeping. Fig. 10 shows a state machine for a running biped.

The control algorithms for one-legged hopping machines were adapted for control of planar and three-dimensional biped runners, a quadruped machine that trots, paces, and bounds, machines with articulated legs, and simulated kangaroos and ostriches. Although no one machine or model performed all these tasks, taken together they traveled at specified speeds, changed gait during running, ran fast (13 mph), jumped, maintained balance when disturbed, climbed simplified stairways, and performed rudimentary gymnastic maneuvers (see Figs. 11–13) (Hodgins *et al.*, 1985; Raibert *et al.*, 1986; Koechling and Raibert, 1988; Hodgins and Raibert, 1990b; Raibert, 1990).

Chapter XIV Legged Robots

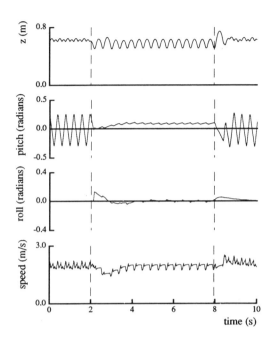

FIGURE 13. Quadruped model during two gait transitions, one from bounding to trotting ($t = 2$ s) and one from trotting to bounding ($t = 7.95$ s). (Data File QS91.314.0.)

D. Kangaroo Simulation

In recent work we have been interested in the possibility of using knowledge of robot control algorithms to study animal locomotion. Kangaroo hopping is the obvious candidate for study, because one-legged hopping is so much like kangaroo hopping. We implemented a simplified kangaroo simulation and controlled its behavior with an adaptation of the one-legged hopping machine control algorithms. The kangaroo model is shown in Fig. 14. The model is based on measurements made by Alexander and Vernon (1975) on a real kangaroo.

At the highest level, the control is quite similar to the hopping machine control, using the three parts described earlier (see Table II and Fig. 15). At a lower level, the leg control level, a constraint was introduced to account for three joints in the leg. The constraint requires the control to keep the knee on the line connecting the toe to the system center of mass, to minimize knee torque during bouncing. Differences between the control for one-legged robot hopping and simulated kangaroo hopping include provisions for control of the tail, arms, and spine and the presence of extra flight states to accommodate folding of the leg during swing.

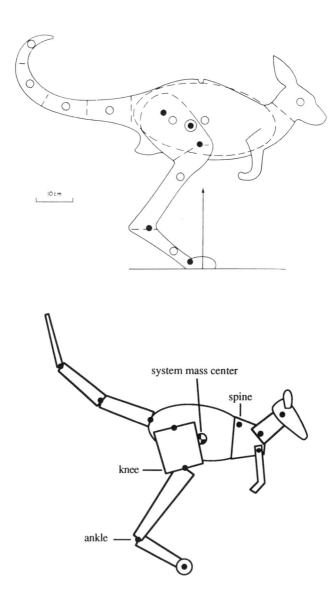

FIGURE 14. (Top) Drawing of real kangaroo used as the basis of the kangaroo model. It was a juvenile red kangaroo weighing 6.6 kg. Reprinted by permission from "The Mechanics of Hopping by Kangaroos," by R. McN. Alexander and A. Vernon, in *J. Zool. London* (1975), The Zoological Society of London, **177,** 265–303. (Bottom) Diagram of kangaroo model. Except for the trunk, each link was modeled as the frustum of a cone, with pivots at the base and tip. The two legs of the real kangaroo were combined into one model leg. All dimensions were chosen to match the real kangaroo in link length and link mass, assuming the kangaroo was the density of water. Mass centers

Chapter XIV Legged Robots

TABLE II. Comparison of the Behavior of a Real Kangaroo and a Simulated Kangaroo.

Quantity	Animal	Simulation
Kangaroo hop		
Peak vertical acceleration (g)	5	4.6
Pitch magnitude (deg)	10	12
Magnitude of tail wag relative to trunk (deg)	30	31
Stride length (m)	2.2	1.5
Stride duration (s)	0.35	0.35
Flight duration (s)	0.25	0.24
Stance duration (s)	0.10	0.11
Desired running speed (m/s)	6.2	5.0

The kangaroo data are from Alexander and Vernon (1975). (Data: KN91.3.20.)

1. Kangaroo Jump. The kangaroo jump is not very different from the kangaroo's normal hop. It is different in that the altitude of the hop is greater and the leg, spine, and arm are used to increase clearance during the jump. The similarity of the two motions permitted us to reuse many of the elements already defined for hopping, some with modified parameters.

We designed the simulated jump by observing the motion of a real kangaroo jumping over a fence, as depicted in the videotape "Animal Olympians" (Boswell, 1981). The characteristics we found most notable and tried to copy were:

- Prepare for jump: the legs compress more than normal. The back arches, arms swing back, and tail swings down more than normal.
- Thrust for jump: The leg joints each extend to make the leg very nearly straight at liftoff. The arms are thrown forward and up, and the back is hyperextended.
- Ascent of jump: The knee bends, swinging the leg back to provide clearance. The ankle flexes to about 90° (presumably because of multijoint muscles). The back is maintained in the hyperextended state, with the arms forward.
- Descent of jump: Leg swings forward, spine is flexed hunching the back, arms swing back.
- Continue hopping.

The state machine and actions used for jumping are given in Table III. Fig. 16 shows data from two jumps with different jump heights but with fixed desired

FIGURE 14. (cont.) and moments of inertia were calculated from the geometric model. The model is planar. Contact between the foot and ground was modeled as a pair of spring–damper elements, one acting vertically and one tangent to the ground. A torque source acts at each joint.

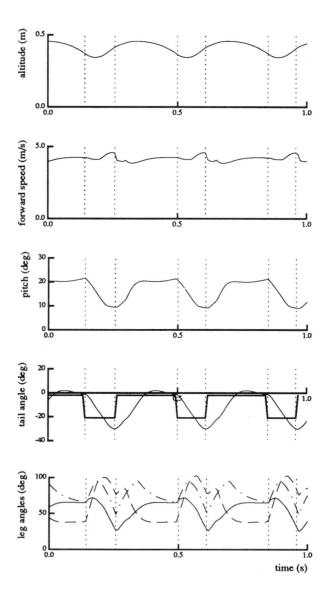

FIGURE 15. Data recorded from planar kangaroo model during three steps of running with a desired speed of 5 m/s. Motion of the tail base joint is driven by a pulse waveform filtered by the spring/damper actuator model (fourth plot from top). The shoulder and spine joints are driven in a similar fashion. The vertical dashed lines bracket the stance phase. The leg joint angles are defined in Fig. 14. Key to joints in leg angle plot: (solid) hip, (dashed) knee, (dot-dashed) ankle. Data from Raibert and Hodgins (1991).

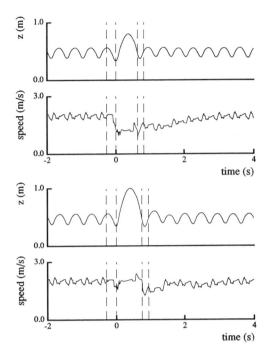

FIGURE 16. Data from planar kangaroo model during two jumps. z is the altitude of the trunk center of mass. The top two curves are for a shallow jump (z_{max} = 0.8 m) and the bottom two curves are a higher jump (z_{max} = 1.0 m), both with an approach speed of 2 m/s. The vertical dashed lines indicate the approach step, the jump step, and the recovery step. (Data Files K91.198.38 and K91.198.9.)

approach speeds. We also manipulated approach speed while holding the jump height constant (data not shown).

Most of the control elements used in the jump are the same as those used for normal hopping, or they are parameterized versions of the same controls (see Table III). Three parameters were ankle thrust, foot placement bias, and pitch bias. These parameters had fixed values during normal running but were adjusted according to the nature of the jump. The action "fold leg back for clearance" did not occur in normal hopping.

The control system for jumping relies on the ability of the normal hopping algorithms to reject disturbances. When the model lands after a jump in a configuration that is not ideal, the error can be rejected by the controls that keep the body level, control the running speed, and regulate the altitude of hopping. This ability of the control system to eliminate variability in the behavior of the sys-

Table III. Finite-State Machine Used for Kangaroo Jump, Simplified

State	Action Units
Flight-descent	
Trunk descending	Extend leg for landing
	Position foot for landing (jump)
	Rotate tail up, arms back, and hunch spine
Compression	
Foot touches	Ankle folks in compression
	Position knee on thrust line
	Servo pitch using knee joint
	Rotate tail down, arms forward, and arch spine
Thrust	
Foot under hip	Extra thrust with ankle and hip
	Position knee on thrust line
	Servo pitch using knee joint
	Keep tail down, arms forward, and spine arched
Flight-ascent	
Foot leaves ground	Fold leg back for clearance
	Keep tail down, arms forward, and spine arched
	Position active leg for landing
	Shorten idle leg
Flight-descent	
Trunk descending s	Rotate hip to swing leg passively
	Keep tail down, arms forward, and spine arched
Flight-approach	
Wait 0.1 s	Extend leg for landing
	Position foot for landing (recovery)
	Rotate tail up, arms back, and hunches spine
Compression	
Foot touches ground	Negative thrust with ankle and hip
	Position knee on thrust line
	Servo pitch using knee joint
	Rotate tail down, arms forward, and arch spine
Continue normal hopping	

The action units are similar to those used for normal hopping, with a few modifications.

tem after a maneuver was important for the execution of this and other maneuvers.

In order to get folding and swinging motions of the leg during the jump that are fluid, we added several states in the flight phase of the state machine. Each of the extra states specifies a set of leg joint positions to be held for a specific time. This was a bad way to implement coordinated fluid motion. Coordination should have been provided with an algorithmic constraint on the joint motion, by a passive coordination of the joints, or by using a control that tracked a set of trajectories.

Chapter XIV Legged Robots   345

The kangaroo jump as implemented does not adjust the stride to place the foot in the proper position with respect to an obstacle. Such positioning, a crucial component of real animal jumping, maximizes use of available clearance.

V. Locomotion on Rough Terrain

To many, the practical value of legged machines is tied to the promise of mobility on rough terrain. Static and dynamic locomotion systems can both play useful roles in this context. Because they generally move slowly, statically stable systems can acquire data from terrain sensors before they move, interpret the sensor information, and find sequences of feasible footholds. Static systems may also tolerate the slow acquisition of terrain data typical of currently available sensory systems. Dynamically stable systems can use their kinetic energy to bridge large gaps that may separate one foothold from the next or to jump over obstacles. Static and dynamic approaches to rough terrain are discussed in this section.

There are several ways that terrain becomes rough and therefore difficult to negotiate. The terrain may

- not be level
- provide limited traction (be slippery)
- include areas of poor or nonexistent support (holes)
- have vertical variations that:
  cause uneven altitude of footholds
  include large obstacles between footholds (poles)
- offer only intricate footholds (e.g., rungs of a ladder)

For a statically stable system to travel on rough terrain the control must decide which locations on the terrain provide suitable footholds and then place the feet on the chosen footholds. A suitable foothold is one that provides adequate support, allows the system to keep its balance, and permits progress toward the goal. It is assumed that the task of placing a foot on a desired foothold, once chosen, is easy for a statically stable system.

Okhotsimski and Platonov (1973, 1975) devised such an algorithm for a simulated six-legged walker. Information from a range finder provided terrain information to identify feasible footholds. Their algorithm took the physical structure of the walking machine into account so as to choose footholds that minimized the maximum force exerted by any leg and to keep each foot's reaction force as nearly vertical as possible. Their simulations traveled on three-dimensional poles and holes terrain using the sequence of support polygons found by the foothold selection algorithms.

Several workers have developed algorithms for statically stable rough-terrain locomotion. These algorithms first choose a desired path along which the body is to move; this is followed by heuristic selection of reachable footholds along the path. We will call this approach the *body motion-then-footholds* paradigm. McGhee and Iswandhi (1979) used such a paradigm for six-legged walking. After specifying the body motion, their algorithm lifted the legs with the least kinematic travel available in the direction of travel, while putting legs into support with the largest available travel. This heuristic extended each support state to increase the probability that it would overlap with the next support state. Avoidance of deadlock was emphasized over stability by maximizing the number of legs in the air.

Hirose applied the body motion-then-footholds paradigm to quadruped rough-terrain locomotion (Hirose, 1984). He developed a hierarchical algorithm with one level providing gait control and another level providing basic motion regulation. The motion regulation performed such functions as controlling the pitch and height of the body and preventing collisions between the legs. In Hirose's computer simulations, a quadruped walked across terrain with holes, crossed a river, and made local modifications to the motion trace to avoid a large hole.

Okhotsimski and his co-workers built a six-legged walking machine that could climb up onto a small ledge (Devjanin *et al.*, 1983). The machine climbed by raising its body and then placing each foot up on the ledge. Care was taken to keep the body level during climbing. They also implemented algorithms to control the forces in the legs during locomotion on soft soil (Gorinevsky and Shneider, 1990).

The work on rough-terrain locomotion just described assumes that global knowledge of the terrain is available, such as would be provided by computer vision or scanning range data. Such information is called *terrain preview* information, and the assumption of its availability and interpretability is common. It is possible, however, for rough-terrain locomotion to be accomplished without global information. Sensors in each foot can inform the control computer whether the foot is pressing on anything and, if so, how hard. In this case, the legs can act like sensing probes.

Two machines introduced earlier in this chapter relied on this approach. Hirose and Umetani's (1980) quadruped used contact sensors on the feet to probe the terrain directly in the path of the robot. If a foot switch signaled contact as the foot advanced forward, a reflex-like algorithm caused the foot to be pulled back, lifted, then advanced forward again. Another reflex-like algorithm caused support legs to push downward if a load cell in the foot indicated that it was not bearing an adequate vertical load. A third reflex caused the relative altitude of the feet to be adjusted so the body remained level, as indicated by an oil-damped pendulum. Hirose's quadruped used these reflexes to climb up and down steps without a model of the terrain and without human intervention.

Chapter XIV Legged Robots

The Ohio State University Adaptive Suspension Vehicle could also travel on rough terrain without global terrain information (Pugh *et al.,* 1990). It used force sensors in the leg actuators and a force distribution control algorithm to accommodate variations in the terrain, without foothold selection or planning. They demonstrated their machine's ability to travel on gentle slopes, through a muddy cornfield, and over railroad ties, all without a visual sensor or human foothold selection.

So far we have considered rough terrain in the context of static slow-moving systems. Dynamic legged systems should be able to traverse more difficult terrain than static systems of comparable size and reach. In principle, dynamic legged systems should be able to use balance to travel where available footholds provide only a narrow base of support, to use kinetic energy to travel where the available footholds form an erratic pattern of support, and to use ballistic flight phases to leap over regions of terrain that offer no good support at all. Generally, a dynamic legged system can use its kinetic energy as a bridge from one foothold to the next.

These potential advantages of dynamic legged systems are obtained at the cost of requiring more complicated control for placing the feet. Foot placement is straightforward in statically stable systems, once a reachable foothold has been chosen. In dynamic legged systems, however, the act of positioning the feet with respect to available footholds interacts with the need for stability. Each placement of a foot on the ground causes the body to accelerate and influences the forward speed and direction of travel. The algorithm responsible for placing the feet must manipulate the dynamic parameters of the system to simultaneously balance the machine and keep it moving as desired.

Takanishi, Kato and their colleagues (1990a) built several robots that walk dynamically on simple rough terrain. Their most recent machine walks up and down 0.1-m stairs and on $\pm 10°$ slopes. The control algorithms are based on a generalization of zero moment point control, as introduced by Vukobratovic and Stepaneko (1972). When the terrain is not smooth the zero moment point moves along a surface connecting the point of contact of the feet rather than along the surface of the ground. Given knowledge of the terrain, trajectories for the feet, and the zero moment point, the control system precomputes trunk and waist motions that execute the planned trajectories and cause the machine to walk.

Hodgins studied algorithms for running on rough terrain (Hodgins and Raibert, 1990a). She implemented three methods for adjusting the robot's stride to place the feet on desired footholds, while at the same time maintaining balance. Each method adjusted a different parameter of the running cycle: the forward running speed, the running height, and the duration of ground contact. All three control methods were successful in manipulating step length in laboratory experiments on a planar biped running machine, but the method that adjusted forward speed provided the widest range of step lengths with accurate control of step length. Hodgins used these control methods to demonstrate a robot placing its feet on a target, leaping over an obstacle, and running up and down a short flight of stairs (Fig. 17).

VI. Discussion

In addition to myriad low-level details and implementation decisions that are important in legged locomotion, a number of broader issues motivate the design of legged machines and their control. One issue is that of task-level control. Control systems for legged locomotion are challenging to design because they must provide a bridge between the relatively high-level commands used to specify desired behavior and the low-level signals needed to drive the joint motions. Unfortunately, high-level specifications of desired behavior do not generally include specific information about which joints should move, when they should

FIGURE 17. Planar biped running up and down a flight of three stairs. The control system adjusts the length of each step so that the feet land approximately in the center of each stair. Sensors in the planarizing boom connecting the biped machine to the floor tell the control computer where the machine is with respect to the stairs. The machine is shown running from right to left at about 0.5 m/s. Light sources indicate the paths of the feet. Each step is 0.18 m high and 0.30 m deep. Photograph from Hodgins and Raibert (1990a).

Chapter XIV Legged Robots 349

move, or how they should move. For example, the desired behavior "Run forward at 2 m/s using a trotting gait" has no explicit information about how the hip joint on leg 2 should move at various times throughout the locomotion cycle. The coordinate systems of joint space and task space are generally quite different, as are the time constants of behavior in each space.

We are developing a framework for building control systems for locomotion that decompose behavior into intermediate-level control elements called "action units" (Raibert *et al.*, 1992). Action units are elements of control that have a clear functional goal and a specific joint-level implementation. Ideally, action units are sufficiently high level that the control system designer can use them to accomplish a desired behavior, such as maintain balance, change speed, increase ground clearance. At the same time, action units are intended to be sufficiently low level that each can be translated into specific control action at specific joints. The joint controller moves a joint to a particular angle, causes the joint to track a trajectory, exerts a specified torque, or lets the joint coast. This mixture of high- and low-level attributes makes action units useful as bridging elements. Examples of action units used for steady running are:

- Deliver vertical thrust
- Keep knee on thrust line
- Keep body level
- Let leg coast
- Retract leg
- Position foot with respect to hip

Each of these action units has a functional objective and a specific means of execution. For example, the objective of "deliver vertical thrust" is to propel the body upward. Thrust can be accomplished by servoing the joints of the leg to a specified value. The objective of "knee on thrust line" is to constrain the leg motion while minimizing knee torque caused by thrust. The action is to calculate appropriate kinematics and servo the hip to the appropriate angle. The use of action units can organize the task of building effective control systems for locomotion.

Breaking down a behavior into constituent action units is not necessarily a difficult task, but doing so in a way that results in successful control requires one to have detailed knowledge of the behavior, a theory of how it can be performed, and physical intuition. At the moment, this process is largely a "black art," with the only guarantee of successful control being an empirical test. An experienced practitioner with an understanding of mechanics, physical intuition, and detailed knowledge of a behavior can frequently design a set of action units and use them to build a control system for a behavior. In addition to steady running, we have used the action unit approach to control a planar one-legged kangaroo jumping, a biped robot somersault, and a pair of quadruped robot gait transitions (Raibert *et al.*, 1992).

Although action units may simplify the control system design process, they also raise many questions. How easily can we combine large numbers of action units in a single control system? Do their functions and actions superpose? Can action units be devised that are high level enough to help solve general control problems, or will the control system designer continue to bear responsibility for generating a control strategy?

A second broad issue concerns the relative roles of the computational system and the mechanical system in determining behavior. Many researchers in neural motor control think of the nervous system as a source of commands that are issued to the body as "direct orders." We believe that the mechanical system has a mind of its own, governed by the physical structure and the laws of physics. Rather than issuing commands, the nervous system can only make "suggestions" which are reconciled with the physics of the system and the task. The use of a spring–mass/gravity–mass oscillation as a basis for the hopping mechanism in our robots addresses this issue, as does McGeer's downhill walker. The use of symmetry in achieving balance is another example because it depends on letting the body coast on the leg during the stance phase, without control. It is ironic that while workers in neural motor control tend to minimize the importance of the mechanical characteristics of an animal's body, few workers in biomechanics seem very interested in the role of the nervous system. We think that the nervous system and the mechanical system should be designed to work together, sharing responsibility for the behavior that emerges.

Acknowledgements

This research was supported by a contract from the Defense Advanced Research Projects Agency and grants from the System Development Foundation, the C. S. Draper Laboratory, and IBM.

References

ALEXANDER, R. MCN. (1988). *Elastic Mechanisms in Animal Movement.* Cambridge University Press, New York.

ALEXANDER, R. MCN., and VERNON, A. (1975). "The Mechanics of Hopping by Kangaroos (Macropodidas)," *J. Zool. (Lond.)* **177,** 265–303.

BARES, J. E., and WHITTAKER, W. L. (1990). "Walking Robot with a Circulating Gait," *IEEE International Workshop on Intelligent Robots and Systems,* pp. 809–816.

Chapter XIV   Legged Robots

BOSWELL, J. (1981). "Animal Olympians, Videotape," *NOVA, WGBH Educational Foundation,* Distributed by Vestron Video, Stamford, CT.

DEVJANIN, E. A., GURFINKEL, V. S., GURFINKEL, E. V., KARTASHEV, V. A., LENSKY, A. V., SHNEIDER, A. YU., and SHTILMAN, L. G. (1983). "The Six-Legged Walking Robot Capable of Terrain Adaptation," *Mechanisms and Machine Theory* **18,** 257–260.

FURUSHO, J., and MASUBUCHI, M. (1987a). "A Theoretically Motivated Reduced Order Model for the Control of Dynamic Biped Locomotion," *J. Dynamic Syst. Control* **109,** 155–163.

FURUSHO, J., and MASUBUCHI, M. (1987b). "Control of a Dynamical Biped Locomotion System for Steady Walking," in Miura, A., and Shimoyama, I (eds.), *Study on Mechanisms and Control of Bipeds* (pp.116–127). University of Tokyo, Tokyo.

FURUSHO, J., and SANO, A. (1990). "Sensor-Based Control of a Nine-Link Biped," *Int. J. Robot. Res.,* **9**(2), 83–98.

GORINEVSKY, D. M., and SHNEIDER, A. Y. (1990). "Force Control in Locomotion of Legged Vehicles over Rigid and Soft Surfaces," *Int. J. Robot. Res.* **9**(2), 4–23.

GURFINKEL, V. S., GURFINKEL, E. V., SHNEIDER, A. YU., DEVJANIN, E. A., LENSKY, A. V., and SHITILMAN, L. G. (1981). "Walking Robot with Supervisory Control," *Mechanism and Machine Theory* **16,** 31–36.

HEMAMI, H., and GOLLIDAY, C. L., JR. (1977). "The Inverted Pendulum and Biped Stability," *Math. Biosci.* **34,** 95–110.

HEMAMI, H., and WEIMER, F. C. (1974). "Further Considerations of the Inverted Pendulum," *Proc. Fourth Iranian Conference on Electrical Engineering,* Pahlavi University, Shiraz, Iran, pp. 697–708.

HEMAMI, H., WEIMER, F. C., and KOOZEKANANI, S. H. (1973). "Some Aspects of the Inverted Pendulum Problem for Modeling of Locomotion Systems," *IEEE Trans. Autom. Control* **AC-18,** 658–661.

HIROSE, S. (1984). "A Study of Design and Control of a Quadruped Walking Vehicle," *Int. J. Robot. Res.* **3,** 113–133.

HIROSE, S., and UMETANI, Y. (1980). "The Basic Motion Regulation System for a Quadruped Walking Vehicle," *ASME Conference on Mechanisms,* Beverly Hills, CA, Paper No. 80-DET-34.

HIROSE, S., NOSE, M., KIKUCHI, H., and UMETANI, Y. (1984). "Adaptive Gait Control of a Quadruped Walking Vehicle," *Int. J. Robot. Res.* **1,** 253–277.

HODGINS, J., and RAIBERT, M. H.. (1990a). "Adjusting Step Length for Rough Terrain Locomotion," *IEEE Trans. Robot. Automat.* **7**(3), 289–298.

HODGINS, J., and RAIBERT, M. H. (1990b). "Biped Gymnastics," *Int. J. Robot. Res.* **9**(2),115–132.

HODGINS, J., KOECHLING, J., and RAIBERT, M. H. (1986). "Running Experiments with a Planar Biped," in *Third International Symposium on Robotics Research* (G. Giralt and M. Ghallab, eds.). MIT Press, Cambridge, MA, 349–355.

KATO, T., TAKANISHI, A., JISHIKAWA, H., and KATO, I. (1983). "The Realization of the Quasi-Dynamic Walking by the Biped Walking Machine," in Morecki, A., Bianchi, G., Kedzior, K. (eds.). *Fourth Symposium on Theory and Practice of Robots and Manipulators.* (pp. 341–351). Polish Scientific Publishers, Warsaw.

KOECHLING, J., and RAIBERT, M. H. (1988). "How Fast Can a Legged Robot Run?" in Youcef-Toumi, K., and Kazerooni, H. (eds.), *Symposium in Robotics,* DSC-Vol. 11. American Society of Mechanical Engineers, 241–249.

KOOZEKANANI, S. H., and MCGHEE, R. B. (1973). "Occupancy Problems with Pairwise Exclusion Constraints—An Aspect of Gait Enumeration," *J. Cybern.* **2**, 14–26.

LISTON, R. A., and MOSHER, R. S. (1968). "A Versatile Walking Truck," *Proc. Transportation Engineering Conference,* Institution of Civil Engineers, London.

LUCAS, E. (1894). "Huitieme recreation—la machine a marcher," *Recreations Mathematiques* **4**, 198–204.

MATSUOKA, K. (1980). "A Mechanical Model of Repetitive Hopping Movements," *Biomechanisms* **5**, 251–258.

MCGEER, T. (1989). "Powered Flight, Child's Play, Silly Wheels, and Walking Machines," *IEEE Conference on Robotics and Automation,* Scottsdale, AZ.

MCGEER, T. (1990). "Passive Dynamic Walking," *Int. J. Robot. Res.* **9**(2), 62–82.

MCGHEE, R. B.. (1968). "Some Finite State Aspects of Legged Locomotion," *Math. Biosci.* **2**, 67–84.

MCGHEE, R. B. (1983). "Vehicular Legged Locomotion," in Saridis, G. N. (ed.), *Advances in Automation and Robotics.* JAI Press.

MCGHEE, R. B., and ISWANDHI, G. I.. (1979). "Adaptive Locomotion of a Multilegged Robot over Rough Terrain," *IEEE Trans. Syst. Man Cybern.* **SMC–9**, 176–182.

MCGHEE, R. B., and JAIN, A. K. (1972). "Some Properties of Regularly Realizable Gait Matrices," *Math. Biosci.* **13**, 179–193.

MCMAHON, T. A. (1984). *Muscles, Reflexes, and Locomotion.* Princeton University Press, Princeton, NJ.

MIURA, H., and SHIMOYAMA, I. (1984). "Dynamic Walk of a Biped," *Int. J. Robot. Res.* **3**, 60–74.

MOCHON, S., AND MCMAHON, T. A. (1980). "Ballistic Walking," *J. Biomech.* **13**, 49–57.

MOCHON, S. AND MCMAHON, T. A. (1981). "Ballistic Walking: An Improved Model," *Math. Biosci.* **52**, 241–260.

MORRISON, R. A. (1968). "Iron Mule Train," *Proc. Off-Road Mobility Research Symposium,* pp. 381–400. International Society for Terrain Vehicle Systems, Washington, DC.

MUYBRIDGE, E. (1955). *The Human Figure in Motion.* Dover, New York. First edition, 1901, Chapman & Hall, London.

MUYBRIDGE, E. (1957). *Animals in Motion.* Dover, New York. First edition, 1899, Chapman & Hall, London.

NILSON, F. A. (1926). "Supporting and Propelling Mechanism for Motor Vehicles," Patent Number 1,574,679.

OGO, K., GANSE, A., and KATO, I. (1980). "Dynamic Walking of Biped Walking Machine Aiming at Completion of Steady Walking," in Morecki, A., Bianchi, G., and Kedzior, K. (eds.), *Third Symposium on Theory and Practice of Robots and Manipulators.* Elsevier Scientific Publishing, Amsterdam.

OKHOTSIMSKI, D. E., and PLATONOV, A. K. (1973). "Control Algorithm of the Walker Climbing over Obstacles," *International Joint Conference on Artificial Intelligence,* Stanford, CA.

OKHOTSIMSKI, D. E., and PLATONOV, A. K. (1975). "Perceptive Robot Moving in 3D World," *Proc. IV International Joint Conference on Artificial Intelligence,* Tbilisi, USSR, September 3–8.

OKHOTSIMSKI, D. E., GURFINKEL, V. S., DEVJANIN, E. A., and PLATONOV, A. K. (1977). "Integrated Walking Robot Development," *Conference on Cybernetic Models of the Human Neuromuscular System.* Engineering Foundation.

Chapter XIV Legged Robots

OOKA, A., OGI, K., WADA, Y., KIDA, Y., TAKEMOTO, A., OKAMOTO, K., and YOSHIDA, K. (1985). "Intelligent Robot System II," in Hanafusa, H., and Inoue, H., (eds.), *Second International Symposium on Robotics Research* (pp. 341–347). MIT Press, Cambridge, MA.

PUGH, D. R., RIBBLE, E. A., VOHNOUT, V. J., BIHARI, T. E., WALLISER, T. M., PATTERSON, M. R., and WALDRON, K. J. (1990). "A Technical Description of the Adaptive Suspension Vehicle," *Int. J. Robot. Res.* **9** (2), 24–42.

RAIBERT, M. H. (1986a). *Legged Robots That Balance.* MIT Press, Cambridge, MA.

RAIBERT, M. H. (1986b). "Symmetry in Running," *Science* **231**, 1292–1294.

RAIBERT, M. H. (1990). "Trotting, Pacing, and Bounding by a Quadruped Robot," *J. Biomech.* **23** (Suppl.1), 79–98.

RAIBERT, M. H., and HODGINS, J. (1991). "Animation of Dynamic Legged Locomotion," *SIGGRAPH '91,* Las Vegas, 25:4, 349–358.

RAIBERT, M. H., and SUTHERLAND, I. E. (1983). "Machines That Walk," *Sci. Am.* **248**, 44–53.

RAIBERT, M. H., BROWN, H. B., JR., and CHEPPONIS, M. (1984). "Experiments in Balance with a 3D One-Legged Hopping Machine," *Int. J. Robot. Res.* **3**, 75–92.

RAIBERT, M. H., CHEPPONIS, M., and BROWN, H. B. JR. (1986). "Running on Four Legs as Though They Were One," *IEEE J. Robot. Autom.* **2** (2), 70–82.

RAIBERT, M. H., HODGINS, J., PLAYTER, R. R., and RINGROSE, R. P. (1992). "Animation of Maneuvers: Jumps, Somersaults, and Gait Transitions," *Imagina,* Monte Carlo.

RUSSELL, M., JR. (1983). "Odex I: The First Functionoid," *Robot. Age* **5**, 12–18.

RYGG, L. A. (1893). "Mechanical Horse," Patent Number 491,927.

SCHAEFER, J. F. (1965). "On the Bounded Control of Some Unstable Mechanical Systems.," Ph.D Thesis, Stanford University, Stanford, CA.

SCHAEFER, J. F., and CANNON, R. H., JR. (1966). *On the Control of Unstable Mechanical Systems.* (6c.1–6c.13). International Federation of Automatic Control. London.

SHIGLEY, R. (1957). "The Mechanics of Walking Vehicles," Land Locomotion Laboratory, Report 7, Detroit, MI.

SNELL, E. (1947). "Reciprocating Load Carrier," Patent Number 2,430,537.

SUTHERLAND, I. E., and ULLNER, M. K. (1984). "Footprints in the Asphalt," *Int. J. Robot. Res.* **3**, 29–36.

TAKANISHI, A., NAITO, G., ISHIDA, M., and KATO, I. (1985). "Realization of Plane Walking by a Biped Walking Robot WL-10R," in Morecki, A., Bianchi, G., Kedzior, K., (eds.), *Fifth Symposium on Theory and Practice of Robots and Manipulators* (pp. 383–394). MIT Press, Cambridge, MA.

TAKANISHI, A., TOCHIZAWA, M., KARAKI, H., and KATO, I. (1989). "Dynamic Biped Walking Stabilized with Optimal Trunk and Waist Motion," *IEEE/RSJ International Workshop on Intelligent Robots and Systems '89* pp. 187–192.

TAKANISHI, A., LIM, H, TSUDA, M., and KATO, I. (1990a). "Realization of Dynamic Biped Walking Stabilized by Trunk Motion on a Sagittally Uneven Surface," *IEEE International Workshop on Intelligent Robots and Systems,* pp. 323–330.

TAKANISHI, A., TAKEYA, T., KARAKI, H., and KATO, I. (1990b). "A Control Method for Dynamic Biped Walking Under Unknown External Force," *IEEE International Workshop on Intelligent Robots and Systems,* pp. 795–801.

VUKOBRATOVIC, M. (1973). "Dynamics and Control of Anthropomorphic Active Mechanisms," in Morecki, A., Bianchi, G., Kedzior, K. (eds.), *Theory and Practice of*

Robots and Manipulator Systems, Proceedings of RoManSy '73 (pp. 313–332). Elsevier Scientific Publishing, Amsterdam.

VUKOBRATOVIC, M., and STEPANEKO, Y. (1972). "On the Stability of Anthropomorphic Systems," *Math. Biosci.* **14,** 1–38.

WALDRON, K. J., VOHNOUT, V. J., PERY, A., and MCGHEE, R. B. (1984). "Configuration Design of the Adaptive Suspension Vehicle,". *Int. J. Robot. Res.* **3,** 37–48.

WEBER, WEBER (1836). *Mechanik der menschlichen Gehwerkzeuge,* Dieterich'sche Buchhandlung, Goettingen, Germany.

Chapter XV   A Robot that Walks; Emergent
                          Behaviors from a Carefully
                          Evolved Network

RODNEY A. BROOKS
MIT Artificial Intelligence Laboratory
Cambridge, Massachusetts

I. Introduction

In earlier work (Brooks, 1986; Brooks and Connell, 1986), we have demonstrated complex control systems for mobile robots built from completely distributed networks of augmented finite state machines. In this chapter we demonstrate that these techniques can be used to incrementally build complex systems integrating relatively large numbers of sensory inputs and large numbers of actuator outputs. Each step in the construction is purely incremental, but nevertheless along the way viable control systems are left at each step, before the next little piece of network is added. In addition, we demonstrate how complex behaviors, such as walking, can emerge from a network of rather simple reflexes with little central control. This contradicts vague hypotheses made to the contrary during the study of insect walking (for example, Bässler, 1983, page 112).

II. The Subsumption Architecture

The subsumption architecture (Brooks, 1986) provides an incremental method for building robot control systems linking perception to action. A properly designed network of finite state machines, augmented with internal timers, provides a robot with a certain level of performance and a repertoire of behaviors. The architecture provides mechanisms to augment such networks in a purely incremental way to improve the robot's performance on tasks and to increase the range of tasks it can perform. At an architectural level, the robot's control system is expressed as a series of layers, each specifying a behavior pattern for the robot and each implemented as a network of message passing augmented finite state machines. The network can be thought of as an explicit wiring diagram connecting outputs of some machines to inputs of others with wires that can transmit messages. In the implementation of the architecture on the walking robot the messages are limited to 8 bits.

Each augmented finite state machine (AFSM) (Fig. 1) has a set of registers and a set of timers, or alarm clocks, connected to a conventional finite state machine that can control a combinatorial network fed by the registers. Registers can be written by attaching input wires to them and sending messages from other machines. The messages written into them replace any existing contents. The arrival of a message, or

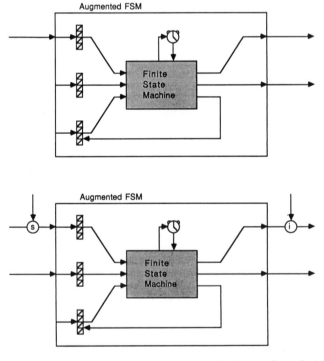

FIGURE 1. An augmented finite state machine consists of registers, alarm clocks, a combinatorial network, and a regular finite state machine. Input messages are delivered to registers, and messages can be generated on output wires. AFSMs are wired together in networks of message-passing wires. As new wires are added to a network, they can be connected to existing registers, they can inhibit outputs, and they can suppress inputs.

Chapter XV A Robot That Walks

the expiration of a timer, can trigger a change of state in the interior finite state machine. Finite state machine states can either wait on some event, conditionally dispatch to one of two other states based on some combinatorial predicate on the registers, or compute a combinatorial function of the registers, directing the result either back to one of the registers or to an output of the augmented finite state machine. Some AFSMs connect directly to robot hardware. Sensors deposit their values to certain registers, and certain outputs direct commands to actuators.

A series of layers of such machines can be augmented by adding new machines and connecting them into the existing network in the ways shown in Fig. 1. New inputs can be connected to existing registers, which might previously have contained a constant. New machines can inhibit existing outputs or suppress existing inputs by being attached as side-taps to existing wires (Fig. 1, circled "i"). When a message arrives on an inhibitory side-tap no messages can travel along the existing wire for some short time period. To maintain inhibition there must be a continuous flow of messages along the new wire. [In previous versions of the subsumption architecture (Brooks, 1986) explicit, long, time periods had to be specified for inhibition or suppression with single shot messages. Recent work has suggested this better approach (Connell, 1988).] When a message arrives on a suppressing side-tap (Fig. 1, circled "s"), again no messages are allowed to flow from the original source for some small time period, but now the suppressing message is gated through and masquerades as having come from the original source. Again, a continuous supply of suppressing messages is required to maintain control of a side-tapped wire. One last mechanism for merging two wires is called defaulting (indicated in wiring diagrams by a circled "d"). This is just like the suppression case, except that the original wire, rather than the new side-tapping wire, is able to wrest control of messages sent to the destination.

All clocks in a subsumption system have approximately the same tick period (0.04 seconds on the walking robot), but neither they nor messages are synchronous. The fastest possible rate of sending messages along a wire is one per clock tick. The time periods used for both inhibition and suppression are two clock ticks. Thus, a side-tapping wire with messages being sent at the maximum rate can maintain control of its host wire.

III. The Networks and Emergent Behaviors

The six-legged robot is shown in Fig. 2. We refer to the motors on each leg as an α motor (for *advance*), which swings the leg back and forth, and a β motor (for *balance*), which lifts the leg up and down.

FIGURE 2. The six-legged robot is about 35 cm long, has a leg span of 25 cm, and weighs approximately 1 kg. Each leg is rigid and is attached at a shoulder joint with two degrees of rotational freedom, driven by two orthogonally mounted model airplane position-controllable servo motors. An error signal has been tapped from the internal servo circuitry to provide crude force measurement (5 bits, including sign) on each axis, when the leg is not in motion around that axis. Other sensors are two front whiskers, two four-bit inclinometers (pitch and roll), and six forward-looking passive pyroelectric infrared sensors. The sensors have approximately 6° angular resolution and are arranged over a 45° span. There are four onboard 8-bit microprocessors linked by a 62.5-Kbaud token ring. The total memory usage of the robot is about 1 Kbyte of RAM and 10 Kbytes of EPROM. Three silver–zinc batteries fit between the legs to make the robot totally self-contained.

Figure 3 shows a network of 57 augmented finite state machines that was built incrementally and can be run incrementally by selectively deactivating later AFSMs. The AFSMs without bands on top are repeated six times, once for each leg. The AFSMs with solid bands are unique and comprise the only central control in making the robot walk, steer, and follow targets. The AFSMs with striped bands are duplicated twice each and are specific to particular legs.

The complete network can be built incrementally by adding AFSMs to an existing network producing a number of viable robot control systems itemized below.

Chapter XV A Robot That Walks

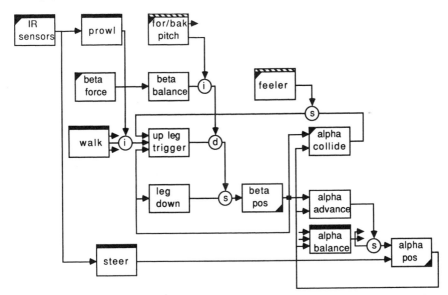

FIGURE 3. The final network consists of 57 augmented finite state machines. The AFSMs without bands on top are repeated six times, once for each leg. The AFSMs with solid bands are unique and comprise the only central control in making the robot walk, steer, and follow targets. The AFSMs with striped bands are duplicated twice each and are specific to particular legs. The AFSMs with a filled triangle in their bottom right corner control actuators. Those with a filled triangle in their upper left corner receive inputs from sensors.

All additions are strictly additive with no need to change any existing structure. Figure 4 shows a partially constructed version of the network.

1. **Standup.** The simplest level of competence for the robot is achieved with just two AFSMs per leg, *alpha pos* and *beta pos*. These two machines use a register to hold a set position for the α and β motors respectively and ensure that the motors are sent those positions. The initial values for the registers are such that on power up the robot assumes a stance position. The AFSMs also provide an output that reports the most recent commanded position for their motor.

2. **Simple walk.** A number of simple increments to this network result in one that lets the robot walk. First, a *leg down* machine for each leg is added that notices whenever the leg is not in the down position and writes the appropriate *beta pos* register in order to set the leg down. Then, a single *alpha balance* machine is added that monitors the α position, or forward swing of all six legs, treating straight out as zero, forward as positive, and backward as negative. It sums these six values and sends out a single identical message to all six *alpha pos* machines, which, depending on the sign of the sum, is either null or an increment or decrement to the current α position of each leg. The

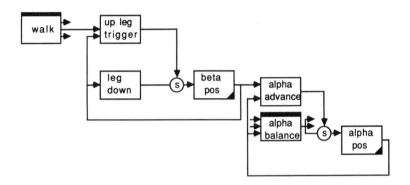

FIGURE 4. A strict subset of the full network enables the robot to walk without any feedback. It pitches and rolls significantly as it walks over rough terrain. This version of the network contains 32 AFSMs. Thirty of these comprise six identical copies, one for each leg, of a network of five AFSMs that are purely local in their interactions with a leg. The last two machines provide all the global coordination necessary to make the machine walk; one tries to drive the sum of leg swing angles (α angles) to zero, and the other sequences lifting of individual legs.

alpha balance machine samples the leg positions at a relatively high rate. Thus if one leg happens to move forward for some reason, all legs will receive a series of messages to move backward slightly.

Next, the *alpha advance* AFSM is added for each leg. Whenever it notices that the leg is raised (by monitoring the output of the *beta pos* machine) it forces the leg forward by suppressing the signal coming from the global *alpha balance* machine. Thus, if a leg is raised for some reason it reflexively swings forward, and all other legs swing backward slightly to compensate (notice that the forward-swinging leg does not even receive the backward message due to the suppression of that signal). Now a fifth AFSM, *up-leg trigger*, is added for each leg, which can issue a command to lift a leg by suppressing the commands from the *leg down* machine. It has one register that monitors the current β position of the leg. When it is down, and a trigger message is received in a second register, it ensures that the contents of an initially constant third register are sent to the *beta pos* machine to lift the leg.

With this combination of local leg-specific machines and a single machine trying to globally coordinate the sum of the α position of all legs, the robot can very nearly walk. If an *up-leg trigger* machine receives a trigger message it lifts its associated leg, which triggers a reflex to swing it forward, and then the appropriate *leg-down* machine will pull the leg down. At the same time all the other legs still on the ground (those not busy moving forward) will swing backward, moving the robot forward.

Chapter XV A Robot That Walks

The final piece of the puzzle is to add a single AFSM that sequences walking by sending trigger messages in some appropriate pattern to each of the six *up-leg trigger* machines. We have used two versions of this machine, both of which complete a gait cycle once every 2.4 seconds. One machine produces the well-known alternating tripod (Wilson, 1966), by sending simultaneous lift triggers to triples of legs every 1.2 seconds. The other produces the standard back-to-front ripple gait by sending a trigger message to a different leg every 0.4 seconds. Other gaits are possible by simple substitution of this machine. The machine walks with this network, but is insensitive to the terrain over which it is walking and tends to roll and pitch excessively as it walks over obstacles. The complete network for this simple type of walking is shown in Fig. 4.

3. **Force balancing.** A simpleminded way to compensate for rough terrain is to monitor the force on each leg as it is placed on the ground and back off if it rises beyond some threshold. The rationale is that if a leg is being placed down on an obstacle it will have to roll (or pitch) the body of the robot for the leg β angle to reach its preset value, increasing the load on the motor. For each leg a *beta force* machine is added which monitors the β motor forces, discarding high readings coming from servo errors during free space swinging, and a *beta balance* machine which sends out lift-up messages whenever the force is too high. It includes a small deadband where it sends out zero move messages, which trickle down through a defaulting switch on the *up-leg trigger* and eventually suppress the *leg down* reflex. This is a form of active compliance that has a number of known problems on walking machines (Klein *et al.,* 1983). On a standard obstacle course (a single 5 cm-high obstacle on a plane) this new machine reduced the standard deviation, over a 12-second period, of the readings from onboard 4-bit pitch-and-roll inclinometers (with approximately a 35° range), from 3.592 and 0.624 respectively to 2.325 and 0.451 respectively.

4. **Leg lifting.** There is a trade-off between how high each leg is lifted and overall walking speed. But low leg lifts limit the height of obstacles that can be easily scaled. An eighth AFSM for each leg compensates for this by measuring the force on the forward swing (α) motor as it swings forward and writing the height register in the *up-leg trigger* at a higher value setting up for a higher lift of the leg on the next step cycle of that leg. The *up-leg trigger* resets this value after the next step.

5. **Whiskers.** In order to anticipate obstacles better, rather than waiting until the front legs are rammed against them, each of two whiskers is monitored by a *feeler* machine and the lift of the left and right front legs is appropriately upped for the next step cycle.

6. **Pitch stabilization.** The simple force-balancing strategy above is by no means perfect. In particular, in high-pitch situations the rear or front legs (depending on the direction of pitch) are heavily loaded and so tend to be lifted slightly, causing the robot to sag and increase the pitch even more. Therefore one *forward pitch* and one *backward pitch* AFSM are added to monitor high-pitch conditions on the pitch inclinometer and to inhibit the local *beta balance* machine output in the appropriate circumstances. The pitch standard deviation over the 12-second test is reduced to 1.921 with this improvement, while the roll standard deviation stays around the same at 0.458.
7. **Prowling.** Two additional AFSMs can be added so that the robot bothers to walk only when there is something moving nearby. The *IR sensors* machine monitors an array of six forward-looking pyroelectric infrared sensors and sends an activity message to the *prowl* machine when it detects motion. The *prowl* machine usually inhibits the leg-lifting trigger messages from the *walk* machine except for a little while after infrared activity is noticed. Thus the robot sits still until a person, say, walks by, and then it moves forward a little.
8. **Steered prowling.** The single *steer* AFSM takes note of the predominant direction, if any, of the infrared activity and writes into a register in each *alpha pos* machine for legs on that side of the robot, specifying the rear swinging stop position of the leg. This is reset on every stepping cycle of the leg, so the *steer* machine must constantly refresh it in order to reduce the leg's backswing and force the robot to turn in the direction of the activity. With this single additional machine the robot is able to follow moving objects such as a slow-walking person.

IV. Conclusion

This exercise in synthetic neuroethology has successfully demonstrated a number of things, at least in the robot domain. All these demonstrations depend on the manner in which the networks were built incrementally from augmented finite state machines.

Robust walking behaviors can be produced by a distributed system with very limited central coordination. In particular, much of the sensory–motor integration that goes on can happen within local asynchronous units. This has relevance, in the form of an existence proof, to the debate on the central versus peripheral control of motion (Bizzi, 1980) and in particular in the domain of insect walking (Bässler, 1983).

Higher-level behaviors (such as following people) can be integrated into a system that controls lower level behaviors, such as leg lifting and force balancing, in a completely seamless way. There is no need to postulate qualitatively different sorts of structures for different levels of behaviors and no need to postulate unique forms of network interconnect to integrate higher-level behaviors.

Coherent macro behaviors can arise from many independent micro behaviors. For instance, the robot following people works even though most of the effort is being done by independent circuits driving legs, and these circuits are getting only very indirect pieces of information from the higher levels, and none of this communication refers at all to the task in hand (or foot).

There is no need to postulate a central repository for sensor fusion to feed into. Conflict resolution tends to happen more at the motor command level, rather than the sensor or perception level.

Acknowledgments.

Grinnell More did most of the mechanical design and fabrication of the robot. Colin Angle did much of the processor design and most of the electrical fabrication of the robot. Mike Ciholas, Jon Connell, Anita Flynn, Chris Foley, and Peter Ning provided valuable design and fabrication advice and help.

This report describes research done at the Artificial Intelligence Laboratory of the Massachusetts Institute of Technology. Support for the research is provided in part by the University Research Initiative under Office of Naval Research contract N00014-86-K-0685 and in part by the Advanced Research Projects Agency under Office of Naval Research contract N00014-85-K-0124.

References

BASSLER, U. (1983). *Neural Basis of Elementary Behavior in Stick Insects.* Springer-Verlag, New York.

BIZZI, E. (1980). "Central and Peripheral Mechanisms in Motor Control," in Stelmach, G. E., and Requin, J. (eds.), *Tutorials in Motor Behavior.* North-Holland, Amsterdam.

BROOKS, R. A. (1986). "A Robust Layered Control System for a Mobile Robot," *IEEE J. Robot. Autom.* **RA-2,** 14–23.

BROOKS, R. A., AND CONNELL, J. H. (1986). "Asynchronous Distributed Control System for a Mobile Robot," *Proc. SPIE,* Cambridge, MA, pp. 77–84.

CONNELL, J. H. (1988). "A Behavior-Based Arm Controller," MIT AI Memo. 1025.

KLEIN, C. A., OLSON, K. W., and PUGH, D. R. (1983). "Use of Force and Attitude Sensors for Locomotion of a Legged Vehicle," *Int. J. Robot. Res.* **2**(2), 3–17.

WILSON, D. M. (1966). "Insect Walking," *Annu. Rev. Entomol.* **II,** 1966; reprinted in *The Organization of Action: A New Synthesis,* C. R. Gallistel, Lawrence Erlbaum, 1980, pp. 115–142.

Chapter XVI

Control of a Hexapod Robot Using a Biologically Inspired Neural Network

ROGER D. QUINN AND KENNETH S. ESPENSCHIED

Department of Mechanical and Aerospace Engineering
Case Western Reserve University
Cleveland, Ohio

I. Introduction

In rough terrain, legged walking machines promise much greater mobility than their wheeled counterparts. Hence, researchers have been investigating this mode of transportation for decades. Raibert and Hodgins (this volume) review the history of walking machines.

Clearly, a legged vehicle has many more degrees of freedom that need to be actively controlled than typical wheeled vehicles. Each actively controlled joint requires at least one actuator and sensor for control in addition to the sensors required for navigation and other high-level tasks. As a result, the major problems encountered in walking-vehicle development are system control and sensor reliability. The control system must process a high flow rate of sensory data, which may be conflicting, and coordinate the motions of multiple legs, each with multiple joints, while maintaining system stability.

Centralized control, in which all decisions are based on all sensory information and all performance requirements, requires that computational speed be extremely high relative to walking speed. Distributed control approaches, in which some control decisions are based on localized information, may speed the overall system. A distributed, hierarchical approach to system control permits distribution of lower-level control functions, freeing the central processor for higher-level control decisions. Brooks (1989 and Chapter V, this volume) has successfully used this type of approach to control the locomotion of hexapods.

Neural networks offer the possibility of the most highly distributed control. Each neuron can be viewed as a processor working in parallel with other neurons. In fact, neural networks have been successfully applied to complex pattern transformation and control problems (Sejnowski and Rosenberg, 1987; Narendra and Parthasarathy, 1990).

However, most research on artificial neural nets has emphasized homogeneous architectures in which all neurons are of the same design. As a consequence, neural connection weights must be adjusted by "learning" algorithms for most tasks.

In contrast, biological nervous systems have many different kinds of neurons, and this heterogeneity is important for their function. An example is provided by insects, which have nervous systems that are orders of magnitude more complex than the most advanced artificial neural nets. As they walk, insects solve the problem of coordination of their six multisegmented legs in real time in the presence of variations in terrain. In addition, because of the heterogeneity and architecture of their nervous systems, insects display remarkable coordination when they first reach their adult stage; they do not require learning to perform basic functions. Also, insect nervous systems are extremely robust; after suffering some types of severe damage, insects can immediately function surprisingly well (Delcomyn, 1985; Graham, 1985).

Previously, a heterogeneous artificial neural network had been developed to control the kinematic problem of locomotion of a simple computer model of a hexapod in the presence of the static stability constraint (Beer *et al.,* 1989; Beer, 1990; Beer and Chiel, Chapter XII, this volume). The basic controller architecture was inspired by neurobiology; the artificial neural net is heterogeneous and learning was not necessary for locomotion. The kinematic locomotion controller consists of a network of 6 neurons for each leg and 1 central command neuron, for a total of 37 neurons. Beer and Chiel's Fig. 1 (page 270, this volume) shows a schematic of the network. There are three motor neurons per leg: the backward-swing neuron and the forward-swing neuron cause the leg to swing backward and forward, respectively, and the foot neuron causes the foot to be raised. The output currents of the swing motor neurons determine the speed of locomotion. Changing a single input caused the hexapod to change its gait and corresponding walking speed. In Beer's simulation, the controller generated a continuum of statically stable, insect-like gaits.

It was thought that the gaits occurred in Beer's simulation environment as a result of the interaction between the neural controller and the simple kinematic model. This led to the question: Could this neural network control the locomotion

Chapter XVI Control of a Hexapod Robot

of an actual hexapod robot in the presence of real-world effects such as coupled kinematic motions, inertia, friction, and time delays?

To answer this question, we built a small mechanical hexapod robot and developed an implementation of the neural network to control its locomotion. In developing this implementation, new results were revealed concerning the neural network which had previously not been discovered.

II. Design of the Mechanical Hexapod

Figure 1 is a photograph of the hexapod robot. It is approximately 50 cm long by 30 cm wide and its mass is about 1 kg. For simplicity, and to match the simulated model for which the controller was developed as much as possible, the mechanical hexapod was designed with two degrees of freedom per leg. The kinematics of each of its legs may best be described in terms of polar coordinates (r, θ) in the vertical plane. Each leg swings (θ) front to back along the body in the vertical plane and extends and retracts (r) radially along the leg's axis. This kinematic configura-

FIGURE 1. Photograph of the hexapod robot.

tion permits the robot hexapod the freedom of motion to walk efficiently forward and backward. As with insects, the joint motions must be coupled and move simultaneously for translation of the body.

Lightweight, 2-W, dc motors power the leg joints through integral transmissions; the ratios were 35.6:1 and 12.9:1 for the angular and radial motions, respectively. The radial leg motion is driven with the use of a final rack-and-pinion transmission; the pinion diameter is 9.5 mm. The angular motion θ of the leg is driven with an additional spur gear reduction of 4:1. The overall transmission ratios for each joint were chosen to permit walking at a 1-Hz leg rate while carrying a battery pack that would permit a half-hour of operation. However, in the present implementation, power is transmitted from a remote source through an umbilical cord.

A single-turn potentiometer senses the orientation of the leg swing angle θ and a 10-turn potentiometer senses the angular position of the pinion that drives the radial extension r.

III. Implementation of the Locomotion Controller

We simulated the neural network on a personal computer for ease of implementation and development. The locomotion controller from Beer's simulation was translated from LISP to the C programming language due to its superiority for real-time control. A user interface was developed that permitted the instantaneous stepping pattern and leg angle ranges (or neuron outputs) to be displayed on the computer monitor in real time. Data acquisition and control boards permit communication between the robot circuits and the computer through an umbilical cord.

Two major differences between the robot and Beer's simulated hexapod needed to be addressed in the controller implementation: the leg kinematics differed, and Beer's simulation neglected kinetics.

The leg kinematics were different: the simulated leg swung in the horizontal plane, but the robot leg swings in the vertical plane. Also, in Beer's simulation the foot moved up or down independently of leg swing, and there was a passive radial extension so that the foot position remained fixed to the ground during contact. In addition, the simulated foot had only two vertical positions, which depended on the foot neuron output. The robot kinematics require only two motors to produce the desired motion, whereas the simulated leg kinematics would have required three motors or more complicated mechanisms. Note that a swing degree of freedom was preserved in the robot leg kinematics so that the angle sensors on the robot could provide input to the angle sensor neurons on the simulated hexapod.

Chapter XVI Control of a Hexapod Robot

Inertia and friction were neglected in the simulated environment. Ideal position control was assumed: the legs arrived at their prescribed position at the next simulation time step. The leg swing velocities were taken to be proportional to swing motor neuron outputs. In our implementation, we used the swing outputs as velocity commands and integrated them to determine the desired angular position of the leg.

A functional transformation was necessary to accommodate the differences between the simulated hexapod and the robot so that the network controller could be left intact and function as originally designed. We refer to the part of the robot control system that accomplishes this transformation as the "virtual leg."

Figure 2 is a schematic describing the leg control system, where the virtual leg is denoted by a box formed by a dashed line. The x and z foot positions and velocities are relative to the body. The desired forward velocity of the body is proportional to the sum of the outputs from the backward swing neurons of the supporting legs. The desired velocity $\dot{x}$ (relative to the body) of a foot in stance (contact with the ground) is opposite the desired velocity of the body relative to the ground. During the forward swing phase, the desired foot velocity is taken to be proportional to the forward swing neuron output minus the backward swing neuron output. In each case, the velocity of the foot relative to the body is integrated to determine the desired instantaneous x position of the foot.

The desired z position of the foot relative to the body is related to the output from the foot neuron. The foot travels along lines parallel to the body (see Fig. 3). When the foot neuron is below threshold, the foot is up and follows a prescribed "up line" that is fixed from step to step. When the foot neuron passes threshold, the leg extends to an initially prescribed length and the foot then follows the "down line" parallel to the body starting at that point. As long as the foot neuron is above threshold, the foot remains on that down line. Note that, although the initially prescribed length of the leg is fixed, the length and height of the down line may vary from step to step because its starting and end points vary, whereas the up line is fixed in height.

The task space x and z position of the foot is converted in software to the desired configuration space coordinates, swing angle θ_d and radial length r_d, through the kinematic transformation: $x = r \sin \theta$ and $z = r \cos \theta$. Through the use of digital-to-analog interfacing, the desired configuration of each leg is communicated to analog reflex position controllers.

Sensorimotor reflex control was developed and implemented using analog circuitry. This reflex controller uses feedback proportional to the difference between the desired trajectory and the sensed position. The result is muscle-like stiffness and equilibrium trajectory following. The idea of a moving equilibrium for biological muscles was reviewed by Bizzi and Mussa-Ivaldi (1990) and has been successfully implemented in simulation for multibody mechanical systems (Quinn and Lin, 1989).

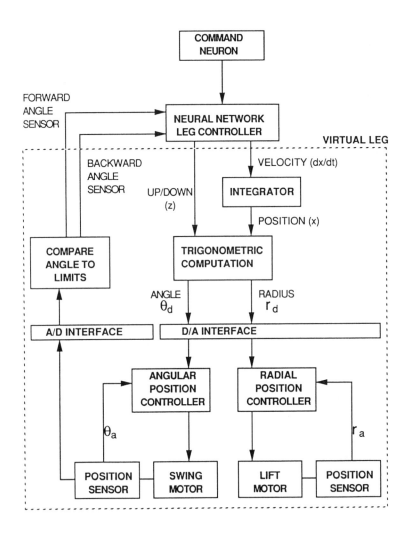

FIGURE 2. Leg control system.

The reflex controllers compare the desired joint positions r_d and θ_d of each leg with the actual positions r_a and θ_a sensed by the potentiometers and supply current to the motors based on the difference. The feedback control law is linear but with a saturation level (5 V) and a short deadband (about 2% of the signal that causes saturation). The joint reflex position controllers are implemented in the form of analog circuits with potentiometers for adjustment of the slope of the linear portion of the feedback law. The joint positions prescribed by the neural controller vary over

Chapter XVI Control of a Hexapod Robot

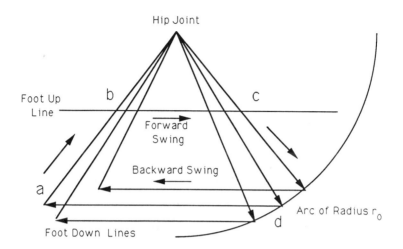

FIGURE 3. Foot trajectory.

time, so the feedback controllers are functioning as tracking controllers as well as vibration controllers.

Note that there are two feedback control loops in our implementation: one for analog position control and another for angle sensor feedback to the neural network (Fig. 2). The feedback in both loops is provided by the position-sensing potentiometers. An analog reflex position control loop for each joint is completely external to the computer. The outputs from the six angle sensors (potentiometers) are converted to digital data and fed back to the simulated controller and serve as inputs to the forward and backward angle sensor neurons.

Each of the 12 motors has a separate integrated circuit sensorimotor control board, and these boards are housed in a single servo unit. The power stage for each motor is integrated with its position control circuit. An umbilical cord at the rear of the hexapod connects the robot with the servo unit. Other cables connect the data acquisition and control boards with the servo unit.

IV. Experimental Results

The simulated neural net controller implemented in the C language generates stepping patterns very similar to the gaits previously observed in the LISP implementation (Beer, 1990; Beer and Chiel, Chapter XV, this volume). Typical stepping patterns generated by the neural network are shown in Fig. 4a. The network gener-

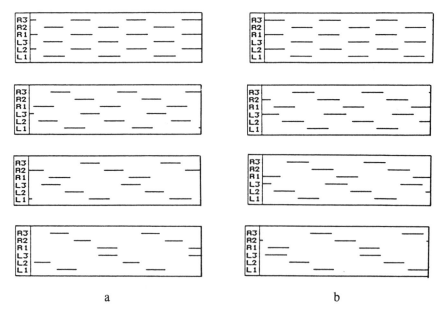

FIGURE 4. (a) Typical gaits with robot out of control loop. (b) Typical gaits with robot in control loop.

ates a continuum of gaits, from which we have chosen four that closely match natural insect gaits reported by Wilson (1966).

With the robot in the control loop, the control system causes the hexapod to follow the network-generated motions and walk in a manner similar to an insect. Typical robot stepping patterns are shown in Fig. 4b.

The particular gait of the robot at any time and the associated walking speed depend on the value of a single external input to the command neuron. This command input is issued by the experimenter through the computer keyboard and stimulates the command neuron that supplies current to the neural net as depicted in Beer and Chiel's Fig. 1 (page 270). The command neuron output C varies over the range $0 \leq C \leq 1$. Fig. 5 is a plot of steady-state walking vehicle speed versus C. Note that for command levels of about 0.2 to 0.3 the walking speed is dual valued. This occurs because more than one gait is possible and the prescribed walking speed depends on the number of feet that are in stance.

The duty factor is defined as the time that a foot is in stance divided by the total stride time, where the stride is the complete gait cycle of a leg (McKerrow, 1991). Fig. 6 is a plot of the duty factor (an average of all legs) versus command value. The tripod gait ($C = 1$) has the lowest duty factor.

Chapter XVI Control of a Hexapod Robot

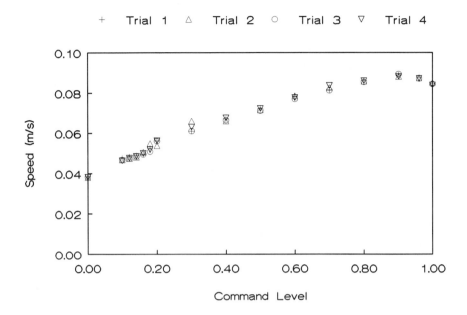

FIGURE 5. Walking speed versus command level.

The tripod gait is the fastest statically stable walking pattern for insects. Referring to Fig. 5, the maximum speed of the robot occurs just before the tripod gait at a command level of about 0.9. This can be explained as follows: The network-generated speed is proportional to the sum of the outputs from the backward swing neurons of the stancing legs. Increasing the command value tends to increase the output from the backward swing neurons, which tends to accelerate the body. An increase in the command value also changes the gait such that it has a lower duty factor, which tends to decrease the network-generated speed. Hence, the speed is a nonlinear function of command value and the tripod gait, having the lowest duty factor, is slower than the gait just preceding it. This is also true for the prescribed gaits of the network controller with the robot not in the control loop.

When the command neuron's output quickly changes from one extreme to another, the gait and speed change smoothly. Fig. 7a shows the gait transition as the command neuron output rises from a steady-state value of 0.2 to 1 and the stepping pattern progresses from a slow metachronal wave gait to the tripod gait. At the end of the plot (the length of the plot corresponds to 8 seconds), the hexapod has nearly settled into the tripod gait. Fig. 7b shows a gait transition as the command neuron output decreases from a steady-state value of 1 to 0.21. At the beginning of the plot, the set point (steady-state value) of the command neuron was changed abruptly

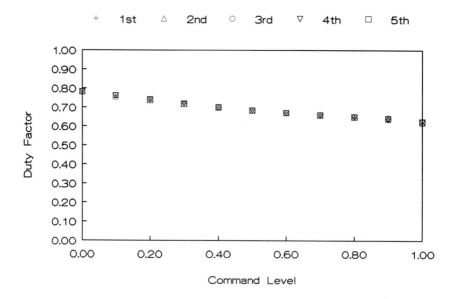

FIGURE 6. Duty factor versus command level.

from 1.0 to 0.2. The time constant of the command neuron is 1.67 seconds. At the end of the plot, the command neuron output is near steady state (0.2) and the stepping pattern is near the corresponding steady state metachronal wave gait.

Lesion and perturbation studies have shown that the robustness of the network controller is maintained when the robot is in the control loop; the controller is robust to the severing of any single central or sensory neural connection. The controller's robustness properties and implications for future work are described in detail by Chiel *et al.* (1992).

V. The Effects of Time Delays on the Control System

The implementation of the neural network into a system controller for the robot led to some new and important findings about the effects of time delays on the network. These results were discovered during the process of debugging the system. We found that it was necessary to tune parameters in the virtual leg portion of the control system in order to achieve robot performance closely matching simulation performance. Note that the neural network itself was not changed.

Our initial choices of virtual leg parameters resulted in a continuum of statically stable gaits. However, these gaits were not identical to those generated by the net-

Chapter XVI   Control of a Hexapod Robot

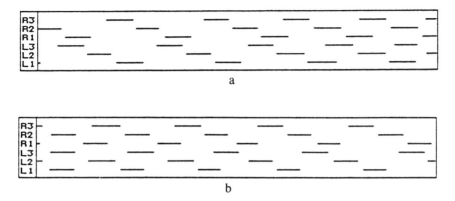

FIGURE 7. (a) Transition from slow to fast gait. (b) Transition from fast to slow gait.

work controller when the robot was not in the loop. Metachronal waves formed slowly, then spread apart quickly, and swing times varied noticeably from step to step. The robot also favored one side of its body, swinging a leg on one side of its body immediately following the swing of the corresponding leg on the opposite side (see Fig. 9a and b). Ideally, for stability and smoothness, the two sides should be 180° out of phase. In simulation this was indeed the case. Investigation revealed that this unfavorable operation of the robot could be explained by two related problems: First, there was an additional time delay that did not exist in Beer's simulation, and second, the delay was slightly asymmetric.

A time delay results from the position error between the actual angular position of the leg and the time-varying equilibrium position prescribed by the neural controller. Because we are using proportional feedback position control, the actual leg position lags the equilibrium position. Hence, input to the forward and backward angle sensor neurons is delayed.

For the purpose of measuring system time delays, the robot was operated open-loop with its feet off the ground. In this case, open-loop operation means that there is no feedback from the robot to the network controller. However, the analog position controllers continue to attempt to track the prescribed joint trajectories output from the network controller.

Fig. 8a shows plots of the left, front leg swing angle versus time (the other leg plots are similar) and the gait pattern. When the system is running open-loop, the network operates with the assumption that the prescribed angles are reached at the next computational loop. A positive slope denotes the forward swing phase and a negative slope denotes the backward swing phase. The dotted lines show the controller angle ranges (not mechanical limits). During forward (or backward) swing, when the plotted line reaches the top (or bottom) dotted line, current is injected into the forward (or back-

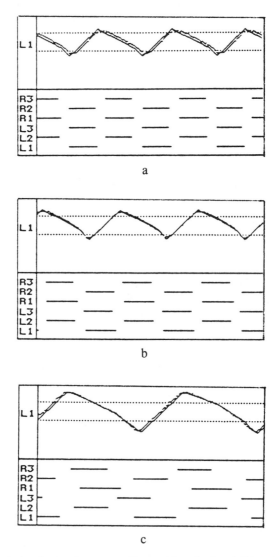

FIGURE 8. (a) Actual and network–generated leg positions for front left leg. Command output = 1.0. (b) Network-generated positions delayed by four program loops. Command output = 1.0. (c) Network-generated positions delayed by seven program loops. Command output = 0.48.

ward) angle sensor neuron. In Fig. 8a there are two plots for each leg. The forward plot in time is the prescribed angle and the lagging plot is the measured leg angle. The delay is clearly visible at this real-time scale and is slightly larger during backward swing.

To confirm our belief that time delays may cause irregularities in the robot locomotion, artificial delays were inserted into the network simulation and the robot

Chapter XVI Control of a Hexapod Robot

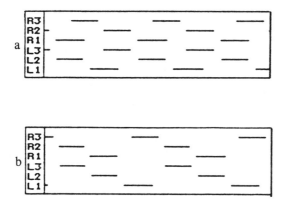

FIGURE 9. (a) Network-generated gait with right rear leg delayed by one program loop. Command output = 0.74. (b) Network-generated gait with right rear leg delayed by one program loop. Command output = 0.3.

was controlled open-loop. Fig. 8b is similar to Fig. 8a, where the robot is running open-loop but the simulated feedback to the controller has been artificially delayed by four computational loops (66.8 ms). In other words, the prescribed angles are not reached in simulation until four computational loops have passed. This amount of delay caused the virtual (simulated) leg angles to match closely the actual leg angles on the plot during the forward swing. However, the actual backward swing angle visibly lags the simulated swing angle. This amount of delay did not significantly degrade the operation of the controller; stable metachronal wave gaits were formed.

Fig. 8c is similar to Fig. 8b, but the delay has been increased to seven computational loops or 117 ms. Note that the virtual and actual leg angles are aligned during the backward swing, whereas the actual leg angles are now ahead of the virtual angles during forward swing. This amount of delay caused the network simulation to malfunction in the same manner as the robot running closed-loop with the original choices of system parameters: metachronal waves formed slowly, then spread apart quickly, and the robot favored one side of its body. Based on Fig. 8b and c and the above results, the actual delay was probably about seven computational loops (117 ms), which is about equal to the time constant of the pacemaker (111 ms).

The time delays also differed slightly from leg to leg. This asymmetry was caused by variations in hardware, particularly stiffness differences in position control modules.

Inserting a much smaller delay that was asymmetric resulted in the hexapod favoring one side of its body, but metachronal waves remained intact. Figs. 9a and b show gait patterns with the command set at 0.74 and 0.3, respectively, where the robot is again operating open-loop and an artificial delay of one computational loop was added to the virtual angle feedback from the right rear leg only. The robot favors the right side of its body; each left leg swings forward immediately following the forward swing of the corresponding right leg on the opposite side of the body.

The combination of large symmetric delays (seven loops) and small asymmetric delays (one loop for the right rear leg) caused the network controller to generate gaits that were irregular and very similar to the gaits of the robot in the initial implementation.

The time delay problems were addressed in three ways. For a given load, the position error can be decreased by increasing the stiffness of the position control circuit. However, there is a maximum permissible stiffness. Beyond this limit, the position control system becomes unstable and the legs develop a tremor. The joint stiffnesses in all the position control modules were raised to reduce the time delays. Care was taken to measure and balance the stiffnesses to reduce asymmetries. The final stiffnesses of the angular and radial position control modules were about 8 V/radian and 4 V/inch.

Because of the stability limit of the joint stiffnesses, a second method of reducing the symmetric delay was required. Simply slowing the neural network simulation produces the same time delay reduction benefit as increasing leg stiffness. Because position error depends on the rate of change of the prescribed position, reducing this rate helps to reduce the relative lag in feedback to the neural network. The program speed was reduced to 60% of its original value. The drawback to this approach is a corresponding slowing of the walking speed of the robot.

Although our effort to reduce the asymmetry problem by balancing stiffnesses helped significantly, there remained a slight tendency for the robot to swing its left legs forward soon after the corresponding right legs finished swinging forward. A measurement of the time delays revealed that the right rear leg experienced a slightly larger delay than the other legs. We had adjusted the stiffnesses to the best of our ability; the remaining asymmetries may have been due to hardware differences that are not adjustable (such as component tolerances) or to variations in sensor calibration. The angle ranges, which are set in the software, are the most accessible means of making small adjustments to render the system symmetric. The angle range of the right rear leg was reduced by about 5%. After this final adjustment, the robot walked with the desired insect-like gaits as shown in Fig. 4b.

It is not surprising that the neural net controller malfunctions when the feedback is delayed symmetrically for about one pacemaker time constant, because this delay should be considered relatively large. However, even in the case of this large symmetric delay, the controller performance is not degraded catastrophically. Furthermore, the 180° across-the-body phasing is relatively sensitive to small *asymmetric* delays, but this is also not a catastrophic failure.

Chapter XVI Control of a Hexapod Robot

VI. Summary

A biologically inspired, heterogeneous, artificial neural network had been designed previously. When this network was used to control the locomotion of a simple kinematic computer model of a hexapod walking vehicle, a continuum of statically stable insect-like gaits resulted.

There was concern that the dynamics of a mechanical hexapod and other real-world effects might pose problems for the controller. Instead of creating a more accurate dynamic model and continuing with computer simulations, we chose to build a mechanical hexapod to assess the utility of the controller. The neural network controller was translated from the LISP computing environment to the C programming language, but was otherwise unchanged. Analog feedback sensorimotor control was added to provide muscle-like reflexes and tracking of the prescribed joint configuration. Also, a kinematic transformation was implemented in software to convert the prescribed virtual leg configurations output from the network controller to equivalent configurations for the mechanical hexapod's legs.

The robot hexapod walks successfully, exhibiting a continuum of statically stable insect-like gaits. The walking gait and corresponding speed depend on a single external input to the central command neuron. For each value of the command input, a steady-state stepping pattern and walking speed ensue. The controller also behaves well during transients. When the command input is changed abruptly, there is an orderly transition in gait and speed as the robot accelerates.

Due to its functional distribution, the neural net controller is robust to the removal of any central or sensory individual connection (Chiel *et al.*, 1992) and lends itself to physical distribution in much the same way as in animals. This functional and physical distribution of the controller could lead to a system that is robust to physical as well as neural damage.

Using the neural network to control a robot revealed two new results. First, we found that its formation of metachronal wave gaits is robust to symmetric time delays of feedback from the angle sensors up to about 60% of the pacemaker time constant. Second, the 180° across-the-body phasing of the legs swings is sensitive to small asymmetric time delays in this feedback. Note that if we had chosen to proceed with more accurate computer simulations instead of implementing the controller for a mechanical hexapod, it is doubtful that the controller's sensitivity to asymmetric time delays would have been apparent.

The computer simulation of the neural network eased its implementation, and the effects of hardware on the controller were more readily identified and addressed.

Miniaturizing the present neural net and applying it to this hexapod would result in an autonomous walking vehicle but with relatively limited capabilities. The locomotion controller needs to be developed further to incorporate true leg kinematics, force control, more sensory information, and more leg degrees of freedom for maneuverability. In future work, we will continue to look to biology to gain insights into controller development because this approach has proved very fruitful.

This work has also led us to some broader conclusions about strategies for research into the development of biologically based controllers: Simulation is an important tool for the timely development of models that enhance our understanding of a system. A computer model permits hypotheses to be tested by allowing easy access to and alteration of system parameters, and the complexity of the model can be varied to meet the needs of the experimenter. However, implementation in hardware is essential for testing models of nervous systems and for the development of practical controllers for robots. Even in the most exhaustive simulation some potentially important effects may be neglected, overlooked, or improperly modeled. It is often not reasonable to attempt to account for the complexity and unpredictability of the real world. Hence, implementation in hardware is often a more straightforward and accurate approach for rigorously testing models of nervous systems.

Acknowledgments

This work was supported by NASA Goddard Space Flight Center grant NGT-50588, Office of Naval Research grant N00014-90-J-1545, and the Cleveland Advanced Manufacturing Program (CAMP) through the Center for Automation and Intelligent Systems Research (CAISR) at CWRU.

References

BEER, R. D., CHIEL, H. J., and STERLING, L. S. (1989). In *Advances in Neural Information Processing Systems 1;* D. S. Touretzky, ed. (pp. 577–585). Morgan Kaufmann, San Mateo, CA.

BEER, R. D. (1990). *Intelligence as Adaptive Behavior: An Experiment in Computational Neuroethology.* Academic Press, Cambridge, MA.

BROOKS, R. A. (1989). *Neural Comput.* **1,** 253–262.

BIZZI, E., and MUSSA-IVALDI, F. A. (1990). In Osherson, D. N., Kosslyn, S. M., and Hollerbach, J. M. (eds.), *An Invitation to Cognitive Science,* Vol. 2: *Visual Cognition and Action* (pp. 213–242). MIT Press, Cambridge, MA..

CHIEL, H. J., BEER, R. D., QUINN, R. D. and ESPENSCHIED, K. (1992). *IEEE Trans. Robot. Automat.* **8,** 293–303.
DELCOMYN, F. (1985). *Annu. Rev. Entomol.* **30,** 239–256.
GRAHAM, D. (1985). *Adv. Insect Physiol.* **18,** 31–140.
MCKERROW, P. J. (1991). *Introduction to Robotics.* Addison-Wesley, Reading, MA.
NARENDRA, K. S., and PARTHASARATHY, K. (1990). *IEEE Trans. Neural Networks* **1,** 4–27.
QUINN, R. D., and LIN, N. J. (1989). *Proc. AIAA Guidance, Navigation and Control Conference,* Part 2, AIAA 89-3562, pp. 1150–1157.
SEJNOWSKI, T. J., and ROSENBERG, C. R. (1987). *Complex Syst.* **1,** 145–168.
WILSON, D. M. (1966). *Annu. Rev. Entomol.* **11,** 103–122.

Chapter XVII   Modeling Neural Function
at the Schema Level:
Implications and Results
for Robotic Control

RONALD C. ARKIN
Mobile Robot Laboratory
College of Computing
Georgia Institute of Technology
Atlanta, Georgia

I. Introduction

Roboticists have long sought to provide their machines with the simplest capabilities of animals: the ability to perceive and act within the environment in a meaningful and purposive manner. Although it seems intuitively obvious that a study of existing biological systems that already have the ability to conduct these tasks successfully is a reasonable way to achieve that goal, it has been largely resisted by the robotics community. Why is that?

There are two principal reasons. First the underlying hardware is fundamentally different. Biological systems bring a large amount of evolutionary baggage that is unnecesary to support intelligent behavior in their silicon-based counterparts. Second, our knowledge of the functioning of biological hardware is often inadequate to support its migration from one system to the other. For these and other reasons, most roboticists ignore biological realities and seek purely engineering solutions.

We argue that there is much that *can* be gained for robotics through the study of neuroscience, ethology, and psychology. But given sparse knowledge of the functioning of even the simplest biological systems, how can we use this information to produce smart robots? Artificial intelligence (AI) frequently advocates the use of abstraction as a means for representing and controlling problem solving. By abstracting the limited knowledge available on biological systems, the problem and its solution can be viewed independently of the underlying neurological substrate and converted into a methodology that is more conducive for computer-based systems. In doing this, we relinquish any claims of cognitive or neuroscientific validity, but we now have modular models and tools derived from biological systems that can demonstrate intelligent robotic behavior. We need to be concerned not with the details of synaptic connections or membrane potentials but rather with a more abstract model of neural functioning. Schemas, as presented by psychologists and neuroscientists, serve as this neural basis within our model of robotic control.

A fundamental problem with much of the underlying artificial neural network research as applied to robotics lies in its myopia: *one cannot see the forest that lies above the dendritic trees.* Our use of schemas models this behavioral forest without concern for specific neural structure. Function is abstracted into motor behaviors and perceptual strategies (referred to as motor and perceptual schemas respectively) which provide for flexibility in configuring a robotic system and adaptability in dealing with a dynamic and potentially hostile world. Reiterating: *function rather than structure is central to this approach.*

In presenting our work on schema-based robotics we first review the neuroscientific basis for schema theory in Section II. These concepts are then demonstrated in the domain of robot navigation in Section III. Methods for effectively integrating perception with motor control are presented in Section IV. A cornerstone of intelligence, learning, is demonstrated within the context of schema-based robotics in Section V, followed by mechanisms that support system survivability in Section VI. This chapter concludes with a discussion of future work in schema-based systems in our laboratory and a summary of the overall methodology.

II. Neuroscientific Motivation for Schema-Based Systems

A. Background

The use of schemas as a *philosophical* model for the explanation of behavior dates as far back as Immanuel Kant (eighteenth century). *Neurophysiological* schema theory emerged early in this century. The first application was by Head in an effort to explain postural mechanisms in humans (Head and Holmes, 1911). Later work by Bartlett (1932) and Piaget (1971) used schema theory as a mechanism for expressing models of memory and learning.

More recently, Neisser (1976) presented a cognitive model of interaction between motor behaviors in the form of schemas interlocking with perception in the context of the *perceptual cycle*. Norman and Shallice (1986) used schemas as a means of differentiating between two classes of behavior, willed and automatic, and proposed a cognitive model that utilizes contention scheduling mechanisms as a means of cooperation and competition between behaviors.

Arbib (1981) was the first to consider the applications of schema theory to robotic systems. Extensions of these principles have also been applied to computer vision systems (Riseman and Hanson, 1987). An overview of schema theory appears in Arbib (1992). Just what is a schema anyway? Many definitions exist for the term, often strongly flavored by its application area (computational, neuroscientific, psychological, etc.). Some representative examples are cited below:

- A pattern *of* action as well as a pattern *for* action (Neisser, 1976).
- An adaptive controller that uses an identification procedure to update its representation of the object being controlled (Arbib, 1981).
- A functional unit receiving special information, anticipating a possible perceptual content, matching itself to the perceived information (Koy-Oberthur, 1989).
- A perceptual whole corresponding to a mental entity (Piaget, 1971).

Our working definition is as follows: *A motor schema is the basic unit of motor behavior from which complex actions can be constructed. It consists of both the knowledge of how to act and the computational process by which it is enacted.*

Why are schemas attractive as a means for expressing computational models of perception and action?

- They afford fairly large grain modularity, in contrast to neural network models, for expressing the relationships between motor control and perception.
- They act as individual agents in a cooperative yet competing manner and as such are readily mappable onto distributed processing architectures.
- They provide behavioral and perceptual primitives out of which more complex emergent behaviors can be readily constructed.
- They have a cognitive and neuroscientific basis that can provide future insights as new data and models become available from those fields.

B. Vector-Based Behaviors

Simply having a tool such as schemas for expressing motor behaviors and perceptual activity is inadequate to construct a working robotic system. Somehow a translation from the behavioral requirements encoded in the schema must be made into a form that a robot can execute. This encoding could be somewhat arbitrary, but we have chosen deliberately to see what data are available from biological experiments which have potential for use in robotic systems. Two neuroscientific models have influenced our work in this regard, suggesting a vector-based representation readily implementable on a robot.

Arbib and House's (1987) model for explaining detour behavior in toads describes path planning behavior in terms of generating divergence fields (directional vectors) based on the animal's needs. In particular, repulsive fields surrounding obstacles, attractive forces leading to food sources, and directional vectors based on the frog's spatial orientation give rise to a computational model of path planning in toads that is consistent with observed experimental data.

We have taken great liberties with this model in translating it to the robotics domain. Although we have retained the notion of similar behavioral primitives (motor schemas), we have reformulated them considerably (e.g., eliminating entirely the dependence on spatial orientation of the agent). Using an analog of a well-established robotic technique, potential fields (Khatib, 1985; Krogh, 1984), an instantaneous output vector is generated by each active motor behavior which is combined through simple vector addition. No path planning at all is conducted; rather *reactive* control mechanisms (see later) provide rapid real-time response within the context of a stimulus–response relationship. A complete exposition of the relationship between our work and the Arbib-House model is presented in Arkin (1989c).

A more recent and ongoing project involves not only navigation but also mobile manipulation. Here we are studying control of a mobile vehicle with an arm attached as a single entity. In the ballistic phase of this motion we are concerned with how the arm can preshape as the robot is simultaneously translating through space. Studies by Bizzi *et al.* (1991) have shown that limb control in amphibians has a correlation with a vectorial mapping of the spinal cord. We anticipate using a variation on this model as the basis for providing coordinated movement during navigation of a manipulator through a cluttered and dynamic environment.

C. Robotic Reactive Control

The general format of our schema-based control system fits within the paradigm of reactive control. This recently emerged artificial intelligence approach is fundamentally different from the more deliberative methods that preceded it.

The hallmarks of this method are:

- Tasks are decomposed into a collection of low-level primitive behaviors.
- Global representations are avoided—a direct coupling between perception and motor action is utilized.
- Due to the rapid sensory sampling of the world, the methodology is well suited for dynamic changes in the environment.

Chapter XVII Modeling Neural Function at the Schema Level

There are many variations on this approach, each differing in the ways of expressing the behaviors and combining or selecting the resulting action. Brooks' (1986) subsumption architecture is an early example of this method. Other representative examples include Payton (1986) and Kaelbling and Rosenschein, (1990). Most of these robotics efforts have largely ignored a cognitive and neuroscientific basis and rather note similarities between their results and the biological domain. Our methodology, at the schema level, has been different however. We first look to existing cognitive, ethological, and neuroscientific data and models from which we can extrapolate into the robotics domain. Beer *et al.* (1990) and a few others follow a similar tack, although they are more often concerned with neural network models rather than behavioral ones.

III. Schema-Based Robot Navigation

Mobile robot navigation has served as the primary domain for testing our schema-based methodology. To build a robotic schema-based system we use the basic procedure described below:

1. Characterize the problem domain by developing an understanding of the task the robotic system is going to undertake and the constraints imposed on its achievement by the environment.
2. Enumerate the necessary motor behaviors to accomplish the task. Decompose these behaviors into primitive stimulus–response reactions. Here, ethological studies often prove fruitful in providing insights into the types of behaviors and the nature of the motor response evoked by a particular stimulus.
3. Formulate the motor behaviors into motor schemas which encode a specific algorithmic response to an environmental condition. We use a variant on the potential fields methodology (Khatib, 1985; Krogh, 1984) for this purpose.
4. Test the motor schemas in simulation.
5. Develop perceptual algorithms that are capable of delivering the required environmental information directly to the motor schemas.
6. Implement the entire schema-based system on the robot.

Experimentation and simulation are applied iteratively until successful task-achieving behavior is attained.

A major advantage of the methodology from a robotics perspective is that the perceptual and motor schemas themselves are often reusable while also allowing for incremental system development and testing. In complex environments the

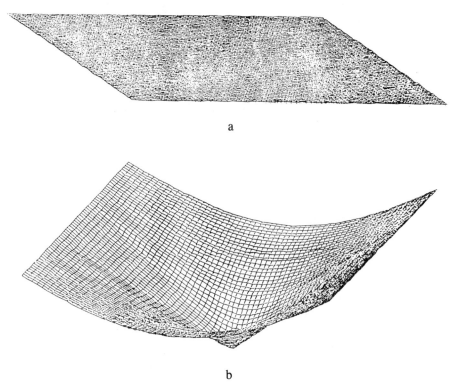

FIGURE 1. Representative motor schemas. (a) Move-ahead (tilted plane). (b) Move-to-goal (well).

schemas can also be reconfigured using a more traditional artificial intelligence planner (Arkin, 1990).

Some of the schemas that we have developed for robots include:

- **move-ahead:** the agent moves in a specified direction at a given velocity.
- **move-to-goal:** There are two forms of this schema, ballistic and controlled, based on biological evidence for the coexistence of these two forms of motion (Brooks *et al*, 1973). In the ballistic form, the robot moves at a constant velocity toward the goal. Typically, feedforward control is used to provide the estimate of the goal's location to the schema. The controlled form reduces the velocity of the agent as it approaches the goal proportionately with the distance from the goal.
- **avoid-static-obstacle:** Any nonthreatening obstacle (even if it is moving) produces a repulsive field around it directing the agent away from it.
- **stay-on-path:** Move rapidly at a constant velocity toward the path if the agent is located off it. If on the path, move more slowly toward its center.

Chapter XVII Modeling Neural Function at the Schema Level

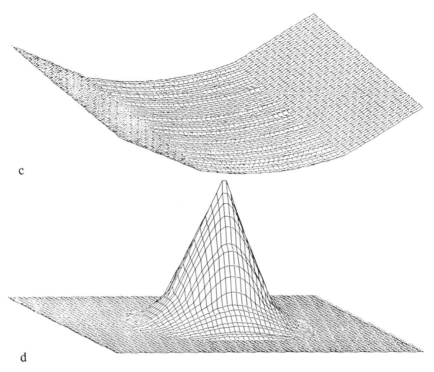

FIGURE 1. (cont.) (c) Stay-on-path (valley). (d) Avoid-static-obstacle (peak). Reprinted by permission of the MIT Press from the *International Journal of Robotics Research,* Volume 8, Issue 4.

- **escape and dodging:** For threatening obstacles, take evasive action as rapidly as necessary based on an estimated time to collision. Dodging involves sidestepping a ballistic object, and escape incorporates fleeing from a potential predator.
- **docking:** Approach a workstation within a specified approach zone. This behavior applies constraints to the trajectory the robot can take above and beyond the requirements of **move-to-goal.**
- **follow-the-leader:** Maintain a safe distance behind a moving object.
- **move-up, move-down, maintain-altitude:** Used for navigating over undulating terrain, it provides the robot with the ability to climb hills, seek valleys, or travel at a constant height using inclinometer data.
- **noise:** A random noise vector important for exploration in unknown terrain and coping with problems such as local maxima and minima.

It should be noted that many of these schemas have been reformulated for three-dimensional applications in aerospace and undersea navigation. The mathematical formulations for these schemas appear in Arkin (1989a,b), Arkin and Murphy (1990), Arkin and Gardner (1990), and Arkin and Carter (1992).

Each of the schemas generates an instantaneous response to its perception of the world. These responses are encoded as velocity vectors and are computed, as stated earlier, using an analog of the potential fields method. The computations themselves are extremely rapid and can be computed in parallel, in a manner consistent with the parallelism found in the brain.

A representation of the robot's reaction to its perceived environment for some of these schemas appears in Fig. 1. The gravitational analog of the force exerted on the robot at any point within the field is represented. In Fig. 1a the **move-ahead** schema results in the robot having an instantaneous force exerted on it in the direction of the tilted plain at a magnitude represented by the steepness of the plain. Similarly, for the **move-to-goal,** the goal is viewed as a potential well to which the robot is attracted. The robot's reaction can be readily interpreted at any location within the field. It should be noted that the entire field is never computed during the navigational process, only the instantaneous velocity at the point at which the robot is currently located. This yields a rapid real-time response. The entire field is illustrated only to aid the reader's comprehension.

By utilizing simple combination mechanisms (i.e., simple vector addition), complex emergent behaviors arise from these simple responses. Fig. 2 depicts such an example. An example of an actual robot run is shown in Fig. 3. This methodology has been tested in a wide range of domains, both in simulation and on robotic hardware, including navigation in campus settings (indoor and outdoor) (Arkin, 1989a), over rough terrain (Arkin and Gardner, 1990), manufacturing environments (Arkin and Murphy, 1990), multiagent settings (Arkin, 1992a), and undersea and aerospace environments (Arkin, 1989b).

From a robotics perspective, there are many advantages to this methodology.

- It allows software reuse.
- It affords parallel computation.
- It is highly modular, allowing incremental development.
- It can cope with uncertainty in perception through frequent resampling of the environment.
- The motor computation itself is economical, permitting real-time response.
- Perceptual processing is inherently more efficient as it is directed on a need-to-know basis rather than satisfying the goal of task-independent global world model construction.

In addition, the schema-based methodology is biologically plausible and thus ties in with existing neuroscientific research. This wealth of literature can provide fodder for designing new and more complex systems.

Chapter XVII Modeling Neural Function at the Schema Level

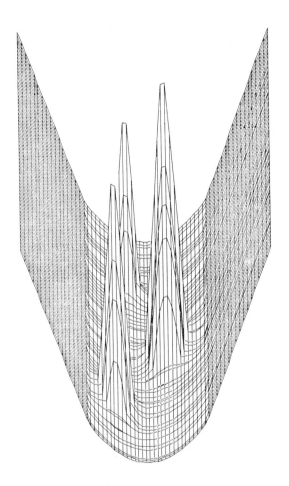

FIGURE 2. Combined motor schemas. In this example three **avoid-static-obstacle** schemas are combined with **stay-on-path** and **move-ahead** schemas.

We would like to think that we are encouraging ethologists, psychologists, and neuroscientists to seek new models or to refine existing models, in both their expression of behaviors through function and the control of these behaviors for potential use by roboticists. Often biological scientists do not take the additional step from the analysis of their data to the development of a model that accounts for it. These models are crucial, in our estimation, not only for an understanding of animal systems but

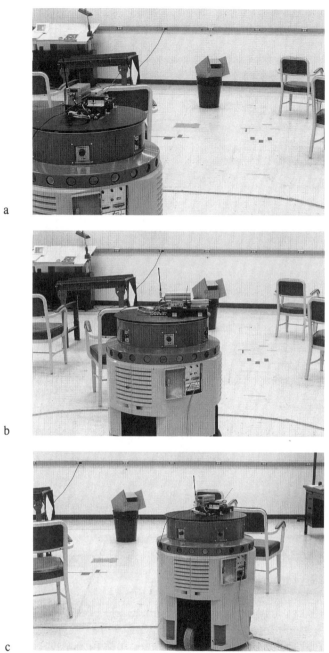

FIGURE 3. Robot run sequence showing robot approaching workstation using *docking* and *avoid-static-obstacle* schemas.

Chapter XVII Modeling Neural Function at the Schema Level

d

e

f

FIGURE 3. (cont.)

also for developing intelligent machines. Although the models themselves may be deficient in accounting for all of the biological data, it is important that they be generated nonetheless so that we, as roboticists, may develop a better understanding of neural function. Brain scientists often use iterative models (Arbib *et al.,* 1989), and this is greatly appreciated by many of us who trying to build intelligent machines, although it is perhaps of less importance within their home discipline.

IV. Perceptual Support and Integration

A means for effectively integrating perceptual information with motor behaviors is crucial for the success of schema-based methods. The general framework that we have used is *action-oriented perception,* a model drawn from cognitive psychology (Neisser, 1976). Action-oriented perception essentially dictates that perception is conducted on a top-down, as-needed basis, where perceptual control and resources are determined by behavioral needs. This is in contrast to more traditional computer vision work, which to a large extent views perception as an end in itself or whose sole purpose is to construct a model of the world without any understanding of the need for such a model. These non–action-oriented strategies burden a robotic system with unnecessary processing requirements, resulting in sluggish performance.

There are three basic components to our view of action-oriented perception: sensory channeling, context-dependent sensor fusion, and temporal sequencing. Space prevents an in-depth discussion of each of these; an overview with additional references follows.

A. Sensory Channeling

Sensory channeling goes under many names: task-dependent perception, action-oriented perception, selective perception, and sensor fission are a few. It has been shown (Anderson and Donath, 1990; Arkin, 1989a; Brooks and Connell, 1986) that by directly connecting consuming behavioral processes with perceptual strategies that are customized for the particulars of the motor behavior, real-time response can be provided for robotic systems. By avoiding the use of mediating global representations in the pathway between perception and action, a time-constraining bottleneck is removed from robotic systems.

Our schema-based methodology utilizes perceptual schemas which are customized for particular motor behaviors and are instantiated at run time by a higher-level planner. This mapping of perception onto action occurs only when motor requirements are known. There is psychological and neurophysiological evidence for the co-existence of two distinct perceptual systems (Neisser, 1989) and planning systems (Norman and Shallice, 1986). By segregating the perceptual functions required for real-time execution from those that are involved in recognition and high-level planning, we can partition planning and perception cleanly. Schema-based models derived by Neisser (Neisser, 1976) and Norman and Shallice 1986) provide inspiration for our overall robotic control strategy in the context of the Autonomous Robot Architecture (Arkin, 1990).

B. Action-Oriented Sensor Fusion

There are times when two or more sensors can contribute to a correct motor action. In the past, more often than not, sensor fusion occurred in the context of a global task-independent representation. In our laboratory, Murphy has recently constructed and demonstrated a methodology for conducting action-oriented sensor fusion (Murphy and Arkin, 1992). This model draws heavily on psychological theories of sensor fusion (Bower, 1974; Lee, 1978). It uses a state-based mechanism to control important sensor interactions such as cooperation, competition, recalibration, and suppression. The use of contributing perceptual subschemas dedicated to individual sensors and funneled into a controlling parent perceptual schema provides the format for expressing this model.

In the robotic analog of the underlying psychological model, three states are present:

1. **State$_1$**: *Complete sensor fusion.* In this state all the sensors are cooperating with each other in determining a valid percept.
2. **State$_2$**: *Fusion with the possibility of discordance and resultant recalibration of dependent perceptual sources.* Recalibration of suspect sensors occurs rather than the forced integration of their potential spurious readings into the derived percept.
3. **State$_3$**: *Fusion with the possibility of discordance and a resultant suppression of discordant perceptual sources.* In this state, spurious readings from suspect sensors are entirely ignored by suppressing the output stream of the sensor(s) in question.

Results demonstrating this action-oriented fusion process are presented in Murphy and Arkin (1992).

C. Temporal Sequencing

The final aspect of action-oriented sensor fusion requires the recognition that the perceptual requirements for even a single motor behavior change over time and space. For example, different perception is required to recognize an object that is far away from when it is close. Entirely different perceptual cues may be used over the course of a single behavior. When looking for a specific object initially, model-based strategies may be used for the identification, after which adaptive tracking methods based on models constructed from previous sensings may be utilized.

Drawing from a recent paper (Arkin and MacKenzie, 1992), we illustrate by example. In this instance, to recognize a workstation in a manufacturing environment, four distinct perceptual algorithms are coordinated. The initial long-range algorithm to support the ballistic phase of the docking motor schema uses a phototropic (light-seeking) algorithm. A temporal activity (motion) detection algorithm is also available if the requirements imposed by the specific workstation warrant its use. [See MacKenzie and Arkin (1991) for the details of these algorithms.] After long-range discovery, a model-based algorithm (Vaughn and Arkin, 1990) is used for explicit identification. As soon as a positive identification is obtained, an adaptive tracking methodology is used, based on region segmentation (Murphy and Arkin, 1990). At the final stages of docking, ultrasound positions the robot. The robot traverses a distance from 30 feet to one-half of a foot utilizing four different perceptual schemas during the process.

The example depicted in Fig. 4 shows an example run using this methodology. A finite state acceptor model (Fig. 5) is used to express the relationships between the individual perceptual strategies in the context of the single motor behavior. Robust perceptual processing occurs by allowing failure transitions to be present within the model.

V. Learning

According to some, an AI system must have the capability to learn in order to be deemed intelligent (Schank, 1987). Although this restriction is perhaps overzealous, we certainly agree with the importance of adaptability and learning for a system to be useful in a complex and dynamic environment.

We note that schemas have long been used as a model for expressing learning in biological systems. Piaget's work (1971) on the assimilation and accommodation of new knowledge has been expressed in a schema-based methodology that provides an alternative paradigm to neural network–based systems. Instead of utilizing fine-grained pattern classification techniques, larger chunks of information are encoded and modified within a schema-based system.

Chapter XVII   Modeling Neural Function at the Schema Level

Although there are many ways in which learning can be applied within our Autonomous Robot Architecture (Arkin, 1990), we focus on the motor schema execution phase for this chapter. Schemas, since they are active concurrently, require gains that express the relative dominance of the set of behaviors. Each schema has a gain, which is a floating-point value used to express the strength of that particular behavior's contribution to the global behavior of the system. The set of gains expresses the cooperation and competition between these behaviors as they strive to influence the robot's activities. These values correlate with *activation levels* typically used to express biological schema models, (e.g., Arbib, 1981; Norman and Shallice, 1986; Neisser, 1976).

In addition, certain schemas have specific variables that are used to express the reaction of the agent with the environment. In obstacle avoidance, the *sphere of influence* encodes the distance from which an obstacle will start exerting a repulsive force on the robot. For the random noise schema, how long the robot maintains one particular random heading before changing its direction is another example of a variable.

We have developed two schema-based learning systems which adapt the robot's performance by selecting or altering gains to best fit the current environment. The first is an on-line method that continuously adjusts the robot's set of schema gains and variables based on a short history of performance. The second is an off-line method involving the use of genetic algorithms for evolving initial sets of gains and parameters when an estimate of the characteristics of the robot's environment is known *a priori*.

A.   On-Line Adjustment of Schema Gains and Variables

The on-line learning method is concerned with dynamically modifying behavioral gains and variables during execution. The underlying philosophy behind this approach to learning is that if something is working well, then keep doing it and do it a bit more, and if it isn't effective then change what you're doing a little. By permitting run-time behavioral changes, some of the problems associated with pure potential fields navigation, such as local maxima, local minima, and cyclic behavior (Arkin, 1989a), can be handled.

Our implementation of this approach (Clark *et al.*, 1992) adopts a rule-based methodology that monitors a short history of the robot's performance (typically 10 time steps). Factors such as progress toward a goal, total distance traveled, and number of sensible obstacles affect the behaviors. Four rules have been articulated that yield behavioral change when certain conditions occur during execution: **no-movement,** where the robot is basically stationary but not yet at the goal; **movement-toward-goal,** the robot is making satisfactory progress; **no-progress-with-obstacles,** the robot is moving but

a

b

c

FIGURE 4. Temporal sequencing example. (a-d) Photos of robot path taken during run. (e) Phototropic and model-based recognition. Cross marks show result of phototropic algorithm;

Chapter XVII Modeling Neural Function at the Schema Level

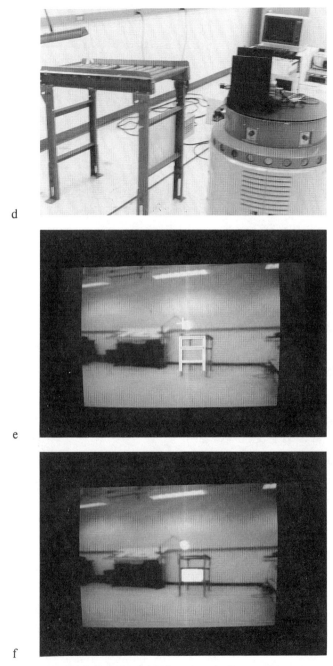

d

e

f

FIGURE 4. (cont). superimposed model shows result of recognition process. (f) Adaptive tracking. Region extracted by robot during navigation.

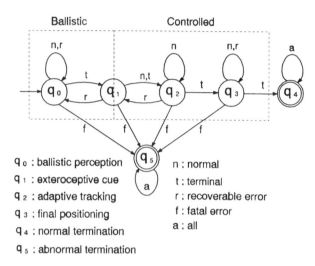

FIGURE 5. Finite State Acceptor model for Fig. 4.

not making progress toward the goal; and **no-progress-no-obstacles,** the robot is moving too randomly, not making progress toward the goal even when no obstacles are present. For each of these conditions, the gains and variables of the individual schemas are adjusted in a manner that enables the robot to cope with the current situation more effectively. For example, when the condition **no-progress-with obstacles** is detected, the noise gain and persistence are increased, the goal gain is decreased, and the obstacle gain and sphere of influence are increased slightly for each time step the robot remains in this condition. The net effect is that the robot is forced out of the local minimum (box canyon) which typically is the cause of this condition. The robot moves around the obstacle formation, solving the problem without any explicit recognition of the box canyon itself. As soon as the robot starts to make progress, the **movement-to-goal** condition is detected, which reverses the trend by gradually reversing the changes to the schema gains and variables that were established by the previous condition.

Fig. 6 depicts the results obtained with this method for another strategy, called the squeezing strategy, which turns down the obstacle avoidance gain and sphere of influence and increases the goal gain when robot progress is impeded (in an effort to let the robot squeeze through tight places). Fig. 7 shows a typical run, with and without on-line learning turned on.

Chapter XVII Modeling Neural Function at the Schema Level

Course	% Complete	Steps	Total Distance	Avg Step Size	Contacts
No Obstacles	100	49.30	42.09	0.86	0.00
25% Obstacles	100	192.70	109.74	0.66	1.40
50% Obstacles	90	868.70	431.87	0.54	24.60
75% Obstacles	80	1134.75	545.94	0.61	32.00

a

Course	% Complete	Steps	Total Distance	Avg Step Size	Contacts
No Obstacles	100	44.20	40.07	0.91	0.00
25% Obstacles	100	54.20	45.21	0.81	0.00
50% Obstacles	90	95.67	61.41	0.69	1.33
75% Obstacles	90	210.43	133.06	0.68	21.57

b

FIGURE 6. On-line learning results. (a) Results without adjustment over 10 randomly generated worlds. (b) Results with learning over same worlds as in (a) but using squeezing strategy.

B. Genetic Algorithms for Schema-Based Systems

Another strategy for learning is to enable the robot to develop characteristics that are suitable for the particular environment it finds itself in. A particularly useful method for accomplishing this is through genetic algorithms (Grefenstette, 1988; Goldberg, 1989). The goal of this form of off-line learning is to allow a robot to develop a suitable set of schema gains and variables that match it well to its current world. We refer to this as having the robot find its *ecological niche,* i.e., developing a set of behaviors that best fit a particular environment.

Space prevents a thorough discussion of the genetic algorithm (GA) methodology we use. Suffice it to say that the agent is presented in simulation with many example worlds that typify the characteristics of a particular environment (e.g., cluttered). By encoding the gains and variables in such a manner that the genetic operators of reproduction, crossover, and mutation can be applied, families of robots are produced. Based on an arbitrarily defined fitness function (it can be defined over values such as safety, speed, timeliness), the agents that perform best in the test worlds are allowed to reproduce while those that do not meet the fitness criteria are removed (i.e., survival of the fittest). The reader is referred to Pearce *et al.* (1992) for the details of our method.

Fig. 8 shows a steady improvement in the performance of the system. In this particular case the robot is evolving effective control parameters to operate in a relatively sparsely cluttered environment. It can be seen that the

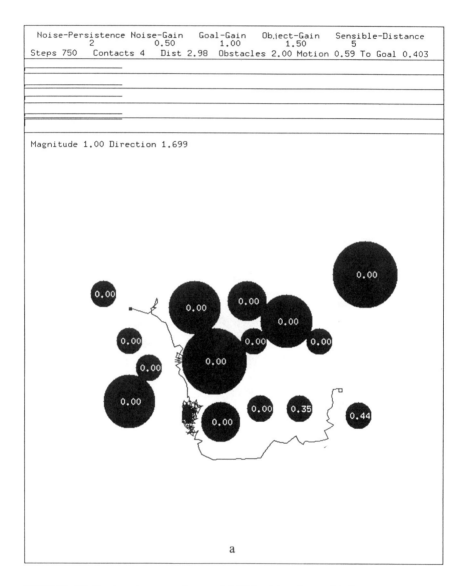

FIGURE 7. On-line learning example runs. (a) Without learning enabled.

number of steps to reach a goal is significantly reduced over the evolutionary time scale, while the number of virtual collisions (intrusions into a safety margin surrounding obstacles) drops down to zero. Similar results have been produced for other environments as well (Pearce *et al.* 1992).

Chapter XVII Modeling Neural Function at the Schema Level 403

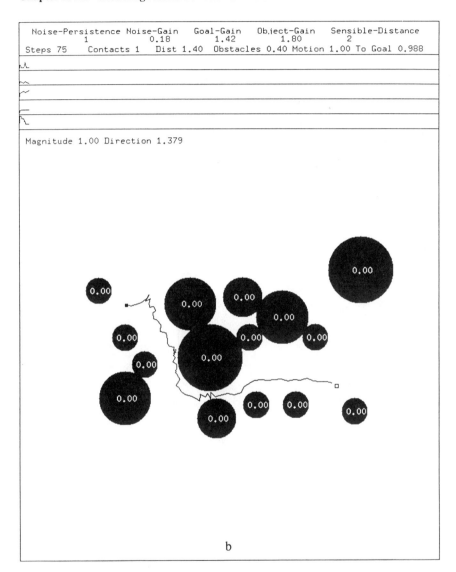

FIGURE 7. (cont.) (b) With learning enabled (squeezing strategy). (a) and (b) reprinted from "Learning Momentum: On-Line Performance Enhancement for Reactive Systems," by R. J. Clark, R. C. Arkin, and A. Ram, in the *Proceedings of the IEEE International Conference on Robotics and Automation,* Nice, France. © 1992 IEEE.

VI. System Survivability—Homeostatic Control

For a robot to be considered truly autonomous it is necessary to provide it with the ability to cope not only with environmental events but also with changes in its own

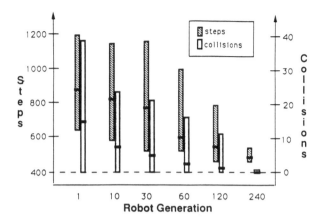

FIGURE 8. Evolution of time and virtual collision measures for a robot operating in a sparsely cluttered environment. Each box represents the range of values for the robot family, with the middle bar being the median. Reprinted from "Learning Momentum: On-Line Performance Enhancement for Reactive Systems," by R. J. Clark, R. C. Arkin, and A. Ram, in the *Proceedings of the International Conference on Robotics and Automation,* Nice, France. © 1992 IEEE.

internal state. By supplementing the schema-based control system with a new class of schemas, called signal schemas, we provide our behavior-based methodology with the flexibility to deal with changes in available fuel, internal temperature, and other potential stresses induced by the robot's own limitations. Fig. 9 depicts the relationships between all the schema classes.

For our purposes we define homeostatic control as the maintenance of a safe internal environment for the robotic agent. We have exploited an analog of the mammalian endocrine system as a means for providing communication and control between the various behaviors. This approach is characterized as follows:

- Reliance is placed on broadcast communication methods.
- Targetability is embedded within the behaviors themselves and is not a product of the communication mechanisms.
- Negative feedback control is used.
- The information transmitted is concerned with regulation of the robot's internal state.
- Behavioral (schema) gains and variables are affected by the transmitted information.

This approach is in contrast to the neural control paradigm, which requires specific interconnections between communicating units. Based on our knowledge that hormonal (endocrine) control is a natural supplement to biological neural control, we feel it has a significant role in robotic control as well. *Fight-or-flight* behavior

Chapter XVII   Modeling Neural Function at the Schema Level

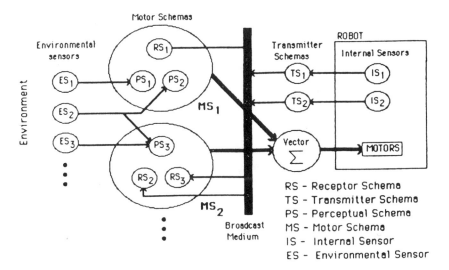

FIGURE 9. Interschema relationships. Reprinted from "Learning Momentum: On-Line Performance Enhancement for Reactive Systems," by R. J. Clark, R. C. Arkin, and A. Ram, in the *Proceedings of the IEEE International Conference on Robotics and Automation*, Nice, France. © 1992 IEEE.

and *feast-or-famine* metabolism can be emulated with this methodology. The details of the homeostatic control methodology are presented in Arkin (1992b).

The system is currently demonstrated only in simulation. Fig. 10 shows the change in navigational behavior produced as a robot's fuel reserves change. The longer yet faster path reflects satiety, with the robot having more than ample fuel to conduct its mission and thus performing it in the fastest manner available to it. As the fuel reserves dwindle, however, as shown by the remaining paths, the robot's behavior becomes slower and somewhat less cautious. These new paths now come closer to clipping obstacles while the robot's speed approaches the most efficient level available for the given motor configuration. This results as a product of both a reduction in the overall speed threshold and the obstacle repulsion as the fuel supplies lessen. It can be seen that the path taken is qualitatively different, solely as a result of the internal condition of the robot and not as a property of any changes within the environment.

VII.   Summary and Conclusions

This chapter has presented an overview of a behavior-based approach to robotic control, inspired largely by cognitive and neuroscientific models of behavior. Different classes of schemas have been described, each with its own specific role in managing a robotic system:

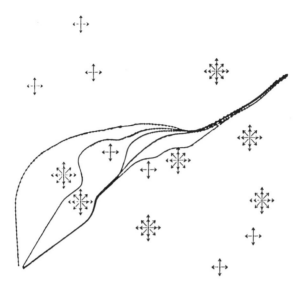

FIGURE 10. Collection of paths representing robot behavior given different available fuel reserves. Note how the paths become slower and shorter as the fuel supply dwindles.

- Motor schemas—the basic unit of motor behavior for a robot encoding the way in which a robot responds to environmental stimuli.
- Perceptual schemas—perceptual strategies which embody the action-oriented perception paradigm and deliver perceptual information to the motor schemas on a *need-to-know* basis.
- Signal schemas—a class of schemas concerned with monitoring the internal environment of the robot and modifying the robot's behavior in a manner consistent with maintaining a survivable system.

Our research has demonstrated the many advantages of schema-based systems: they are highly modular, allowing incremental development; they provide for software reuse; they are cognitively plausible; they are naturally parallel; and they have demonstrated robust robotic performance in both simulation and real-world robots.

Learning can also be readily incorporated into these systems. This is no surprise, as schemas have been used by Piaget (1971) and others to express models of human learning. We have developed both on-line adaptation techniques using rule-based modification of schema values and off-line discovery of ecological niches for robotic systems using genetic algorithms.

Our future work involves the extension of these principles to multirobot systems (Arkin, 1992a), studying the role of communication between agents and its impact on team accomplishment of goals. We are also extending our work to include mobile manipulation, focusing on the coordinated control of a robotic arm mounted on a mobile base.

In conclusion, it is important to restate the role of abstraction in building intelligent systems. We need not have neural roadmaps of the interconnections of animal systems to produce robotic intelligence. Even if we did have such maps, it is unclear how they could be used effectively when dealing with such disparate hardware. It is more important in our estimation to encourage psychologists, neuroscientists, and ethologists to abstract the results of their experiments into descriptive models that are of potential use for roboticists. This is done with the knowledge that as more experimental data become available the proposed model may need further refinement to be consistent with the organism it was derived from. The models used by roboticists, however, do not and should not have to *duplicate* the biological systems whose *behavior* they are trying to emulate. Nonetheless, the inspiration and guidance these biological models can provide in developing robotic systems that can function in the same world as biological life does should not be underestimated.

Acknowledgments

The author would like to thank the graduate students who helped with the realization of these ideas: Russell Clark, Doug MacKenzie, Robin Murphy, Michael Pearce, and David Vaughn. Prof. Ashwin Ram also contributed in the area of learning systems. Finally, the support of the National Science Foundation (grant IRI-9113747) and Georgia Tech's Material Handling Research Center and Computer Integrated Manufacturing Systems Program is gratefully acknowledged.

References

ANDERSON, T., and DONATH, M. (1990). "Animal Behavior as a Paradigm for Developing Robot Autonomy," in Maes, P. (ed.), *Designing Autonomous Agents* (p. 145). MIT Press, Cambridge, MA.

ARBIB, M. A. (1981). "Perceptual Structures and Distributed Motor Control," in Brooks, V. (ed.), *Handbook of Physiology — The Nervous System, II* (p. 1449).

ARBIB, M. A. *et al.,* (1989). "Experimentation and Modeling: An Introductory Discussion," in Ewert, J. P., and Arbib, M. A. (eds.), *Visuomotor Coordination: Amphibians, Comparisons, Models, and Robots* (p. 3). Plenum, New York.

ARBIB, M. A. (1992). "Schema Theory," *Encyclopedia of Artificial Intelligence,* 2nd ed., in press.

ARBIB, M. A., and HOUSE, D. (1987). "Depth and Detours: An Essay on Visually Guided Behavior," in Arbib, M. A., and Hanson, (eds.), *Vision, Brain, and Cooperative Computation* (p. 139). MIT Press, Cambridge, MA.

ARKIN, R. C. (1989a). "Motor Schema Based Mobile Robot Navigation," *Int. J. Robotics Res.* **8,** 4, 99.

ARKIN, R. C. (1989b). "Three Dimensional Motor Schema Based Navigation," *Proc. NASA Conference on Space Telerobotics,* (p. 291). Pasadena, CA.

ARKIN, R. C. (1989c). "Neuroscience in Motion: The Application of Schema Theory to Mobile Robotics," in Evert, P., and Arbib, M. (eds.), *Visuomotor Coordination: Amphibians, Comparisons, Models, and Robots* (p. 649). Plenum, New York.

ARKIN, R. C. (1990). "Integrating Behavioral, Perceptual, and World Knowledge in Reactive Navigation," *Robotics Auton. Syst.* **6,** 105.

ARKIN, R. C. (1992a). "Cooperation without Communication: Multi-Agent Schema-Based Robot Navigation," *J. Robotic Syst.* **9,** 3.

ARKIN, R. C. (1992b). "Homeostatic Control for a Mobile Robot: Dynamic Replanning in Hazardous Environments," *J. Robotic Syst.,* **9,** 2.

ARKIN, R. C., and CARTER, W. C. (1992). "Active Avoidance: Escape and Dodging Behaviors for Reactive Control," in *Proc. Applications of Artificial Intelligence X,* Orlando, FL.

ARKIN, R. C., and GARDNER, W. (1990). "Reactive Inclinometer-Based Mobile Robot Navigation," *Proc. IEEE International Conference on Robotics and Automation,* (p. 936). Cincinnati, OH.

ARKIN, R. C., and MACKENZIE, D. (1992). "Temporal Coordination of Perceptual Algorithms for Mobile Robot Navigation," working paper, College of Computing, Georgia Institute of Tecnology, Atlanta, GA.

ARKIN, R. C., and MURPHY, R. R. (1990). "Autonomous Navigation in a Manufacturing Environment," *IEEE Trans. Robotics Automat.* **6,** (4), 445.

BARTLETT, F. C. (1932). *Remembering: A Study in Experimental and Social Psychology.* Cambridge at the Univ. Press, London.

BEER, R., CHIEL, H., and STERLING, L. (1990). "A Biological Perspective on Autonomous Agent Design," in Maes, P. (ed.), *Designing Autonomous Agents* (p. 169). MIT Press, Cambridge, MA.

BIZZI, E., MUSSA-IVALDI, F. and GISZTER, S. (1991). "Computations Underlying the Execution of Movement: A Biological Perspective," *Science* **253,** 287.

BOWER, T. (1974). "The Evolution of Sensory Systems," in MacLeod, R., and Pick, H. (eds.), *Perception: Essays in Honor of James J. Gibson* (p. 141). Cornell Univ. Press, Ithaca, NY.

BROOKS, R. (1986). "A Robust Layered Control System for a Mobile Robot," *IEEE J. Robotics Automation* **RA-2,** (1), 14.

BROOKS, R., and CONNELL, J. (1986). "Asynchronous Distributed Control System for a Mobile Robot," *Proc. Mobile Robots,* SPIE 727, (p. 77).

BROOKS, V.B., COOKE, J., and THOMAS, J. S. (1973). "The Continuity of Movements," in Stein, R. B., et al. (ed.), *Control of Posture and Locomotion* (p. 257). Plenum, New York.
CLARK, R. J., ARKIN, R. C. and RAM, A. (1992). "Learning Momentum: On-line Performance Enhancement for Reactive Systems," *Proc. IEEE International Conference on Robotics and Automation,* Nice, France.
GOLDBERG, D. (1989). *Genetic Algorithms in Search, Optimization, and Machine Learning,* Addison-Wesley, Reading, MA.
GREFENSTETTE, J. (1988). "Credit Assignment in Rule Discovery Systems Based on Genetic Algorithms," *Machine Learning,* **3,** (p. 25).
HEAD, H., and HOLMES, G. (1911). "Sensory Disturbances from Cerebral Lesions," *Brain* **34,** 102.
KAELBLING, L., and ROSENSCHEIN, S. (1990). "Action and Planning in Embedded Agents," in Maes, P. (ed.), *Designing Autonomous Agents* (p. 35). MIT Press, Cambridge, MA.
KOY-OBERTHUR, R. (1989). "Perception by Sensorimotor Coordination in Sensory Substitution for the Blind," in Ewert, J. P., and Arbib, M. A. (eds.), *Visuomotor Coordination: Amphibians, Comparisons, Models, and Robots* (p. 397). Plenum, New York.
KHATIB, O. (1985). "Real-Time Obstacle Avoidance for Manipulators and Mobile Robots," *Proc. IEEE Int. Conf. Robotics and Automation,* (p. 500), St. Louis.
KROGH, B. (1984). "A Generalized Potential Field Approach to Obstacle Avoidance Control," SME-RI *Technical Paper* MS84-484.
LEE, D. (1978). "The Functions of Vision," in Pick, H., and Saltzman, E. (eds.), *Modes of Perceiving and Processing Information* (p. 159). Wiley, New York.
MACKENZIE, D., and ARKIN, R. C. (1991). "Perceptual Support for Ballistic Motion in Docking for a Mobile Robot," *Proc. SPIE Conference on Mobile Robots VI,* Boston, MA.
MURPHY, R. R., and ARKIN, R. C. (1990). "Adaptive Tracking for a Mobile Robot," *Proc. Fifth IEEE International Symposium on Intelligent Control,* (p. 1044). Philadelphia.
MURPHY, R. R., and ARKIN, R. C. (1992). "SFX: An Architecture for Action-Oriented Sensor Fusion," submitted to *1992 International Conference on Intelligent Robotics and Systems (IROS).*
NEISSER, U. (1976). *Cognition and Reality: Principles and Implications of Cognitive Psychology.* Freeman, San Francisco.
NEISSER, U. (1989). "Direct Perception and Recognition as Distinct Perceptual Systems," text of address delivered to the Cognitive Science Society, August.
NORMAN, D., and SHALLICE, T. (1986). "Attention to Action: Willed and Automatic Control of Behavior," in Davidson, R., Schwartz, G., and Shapiro, D. (eds.), *Consciousness and Self-Regulation: Advances in Research and Theory,* (**4,** p. 1)> Plenum Press, New York.
PAYTON, D. (1986). "An Architecture for Reflexive Autonomous Vehicle Control," *IEEE Conference on Robotics and Automation,* (p. 1838).
PEARCE, M., ARKIN, R. C., and RAM, A. (1992). "The Learning of Reactive Control Parameters through Genetic Algorithms," submitted to *International Conference on Intelligent Robotics and Systems (IROS).*
PIAGET, J. (1971). *Biology and Knowledge,* Univ. of Chicago Press, Chicago.
RISEMAN, E., and HANSON, A. (1987). "General Knowledge-Based Vision Systems," in Arbib, M., and Hanson, A. (eds.), *Vision, Brain and Cooperative Computation* (p. 287). MIT Press, Cambridge, MA.
SCHANK, R. (1987). "What Is AI Anyway?" *AI Magazine,* winter (p. 59).

VAUGHN, D. L., and ARKIN, R. C. (1990). "Workstation Recognition Using a Constrained Edge-Based Hough Transform for Mobile Robot Navigation," *Proc. SPIE Conference on Sensor Fusion III,* (p. 503). Boston.

Index

A

Acceleration, wind, 115
"Action Units," 349–350
Actuator, 365
Adaptive Suspension Vehicle, 324, 325, 347
Amputation, 30–38
Anemotaxis, 161–168
Antenna, 102, 103, 104, 107, 109, 110, 131, 132, 134, 175–177, 180–181
Antennal lobe, 177–178
Aplysia, 273
Arachnids, 107
Auditory receptors, 145

B

Backpropagation, 251, 255, 258, 263, 276
Balance, 44–46
Body motion-then-footholds paradigm, 346

C

CCCPG model, 288, 297
   amplitude modulating reflexes, 302
   central pattern generator, 298
   coordinating system, 308
   load compensation, 305
   metastable coordination, 300
   omnidirectional walking, 299
   oscillator, 297
   phase modulating reflexes, 302
   recruiter, 298
   speed adaptation, 299
   yaw correction, 310
Central pattern generator, *see* CPG
Cerci, 90, 92, 94, 99, 102, 107, 110, 115, 116, 119, 277–278
Cholecystokinin, 237
Cockroach, 6–18, 21–41, 43–68, 89–111, 113–136, 199–220, 267–284
Command system, 295
   command neuron, 269–271, 366, 372, 373, 374, 379
   coordination modulation, 305

411

excitatory connections, 298
presynaptic inhibition, 296
neuronal oscillator parameters, 295
Comparator, 95, 96, 97
absolute, 96
relative, 96
Computational maps, 89
Computational neuroethology, 267, 283
Computer simulation, 91, 92, 130, 227–316, 366, 379, 380
Conditioning
associative, 199–220
detectability of, 261*ff*
nonassociative, 261*ff*
Constraints on models, physiological, 258, 260, 275–279
Context dependent behavior, 131–135, 148–149
Control, controller
centralized control, 365
distributed control, 273–274, 355, 362, 365, 366
feedback control, 370, 371, 375, 377
homeostatic, 403–405
for locomotion, 29, 31, 38–40, 322, 324, 326, 328, 349–350
neural controller, 365–380
open-loop control, 375, 377
reactive, 386–387
reflex position controller, 369, 370, 371, 375, 378
for robotics, 38,3940
sensorimotor control, 371, 379
tracking controllers, 371
vibration controllers, 371
Coordination, 28, 29, 38–39, 305, 366
Corpora Pedunculata;
see Mushroom Bodies
Counterturning, 164–169, 172–175, 188
CPG, 28, 32, 38, 39, 230, 293
entrainment of, 270–271

for insect walking, 268–271
neuronal oscillator, 293
endogenous pacemaker, 293
role of sensory and central mechanisms, 274
for six-legged locomotion, 268–271
Crabs, 7, 9, 11, 12, 14, 16
Crayfish, 135
Crickets, 90, 145

D

Deadband, 370
Deafferentation, 28
Descending neurons, 179, 181–184
flip-flopping, 183–184
long-lasting excitation, 182–184
Distributed processing, 251, 258, 259, 261, 264, 128, 130
Duty factor, 372, 373
Dynamics
forward, 3–6
inverse, 3–6
leg, 16

E

Energy, mechanical, 9–12, 17
Engram, 261*ff*
Escape response, 90, 101, 102, 106, 113–136, 275–276
algorithms, 92, 94, 95
Evasive behavior, 90, 101, 106, 107, 139–156
Extensor muscle, 25
Exteroceptive reflexes, 106, 309
yaw correction, 310

F

Feedback, 28–29, 35–38, 39, 82, 369, 370, 371, 375, 377, 378, 379
negative, 44
peripheral, 231
positive, 50

Index

Filiform hairs, *see* Cerci
Finite state machines
  augmented, 356, 396
  kangaroo jumping, 344
  running, 336
Flight, 139–156, 159–189
  casting, 169, 171–174
  motor outflow, 173, 184–188
  zigzagging, 160–162, 164–169, 171–174, 184–187
Force
  balancing, 361
  ground reaction, 6, 7, 9, 11, 14, 16–18
  muscle, 5

G

Gain control, 76–77
Gait, 9–15, 18, 22–25, 38, 366, 371, 372, 373, 374, 375, 377, 378, 379
  alternating tetrapod, 308
  change of, 25, 30–36
  metachronal wave, 268, 270–271, 273
  quasi-dynamic gait, 327
  tetrapod, 30, 33, 116
  tripod, 23–25, 30, 33, 36, 268, 270–271, 273, 361, 372, 373
Gastric mill model, 243
  reduced model, 245
Genetic algorithm, 255, 401–403
Giant interneurons, 90, 91, 106, 117–130, 135
  model, 278–279
Gradient descent, 251, 253, 255
Gymnastic maneuvers, 320, 338

H

Habituation, 151, 262
Hodgkin–Huxley model, 252

I

Input resistance, 71–75, 258
Input-output function, 256, 259, 260, 261
Insect
  amputee, 31, 37
  flight, 145
  freely walking, 29, 30, 36, 37
  nymphal cockroach, 97
  simulated, 96, 273
  suspended, 35–38, 116–117
Integration, 73–75, 92–97, 117–134
Interneurons, 90, 93, 99, 104, 110
  auditory, 148
  descending giants, 106
  dGI, 91, 92
  DMI, 104, 105, 106, 110
  GI, 90, 91, 93, 95, 98, 100, 102, 105, 106, 110, 117–130, 135
  local, 127, 128, 134
  model, 278–278
  moth descending, 179, 181–184
  premotor, 89
  sensory, 89, 104
  thoracic, 90, 97, 119, 121–136, 276
  vGI, 91, 92, 96, 99, 100
Interneurons-1, 148
Inverted pendulum model, 14, 15, 18, 326–328

K

Katydid, 144
Kangaroo
  jump, 341, 343–345
  simulation, 339, 340

L

Lacewings, 146
Learning
  algorithms, 366
  robot, 396–403
  see also Conditioning
Leg (insect), 21–39
  amputation, 29–38

coordination, 28
disturbance of, 29
joints, 116, 117, 129, 133
pattern of movement, 22, 116–118
segment, 31
Lesions, 92, 271, 273–274, 374
cercal, 281–282
adaptation, 282
electrocautery, 100–102
ventral giant interneurons, 282
Load compensation, 44–49, 54–64
stepping strategies, 45–47, 56–64
swaying strategies, 45, 47, 55–56, 59–64
Lobster
robotic implementation, 288
sense organs
cuticular stress detector, 302, 307
funnel canal organ, 302, 307
thoracocoxal joint receptors, 306
stomatogastric ganglion, 229–249
walking behavior, 289
walking joints
coxobasal joint, 290
merocarpopodite, 290
thoracocoxal joint, 290
Local bending reflex, 255*ff*
Local circuits, 70–71, 81–83, 127, 128, 134
Locomotor system, 22, 29, 31, 35, 39
Locust, 126, 134

M

Mechanical horse, 323
Mechanosensory, 100, 102
Metachronal wave, 373, 374, 375, 377, 379. *See also* Gait.
Modulation, 135, 234*ff*
of reflexes, 64–65
Monte Carlo method, 79, 253
Motor burst, 26, 27, 31–38
double, 31

multiple, 31, 33
pattern, 22, 24–29, 31, 35
period, 33, 38
Movements
arm, 89
eyes, 89
head, 89
preparation by mushroom bodies, 214
turning, 91, 94, 96, 102, 107, 108, 110
Multimodal, 106, 131–134
Muscle, 3–18, 25, 28, 39, 43–66
coxal, 31
extensor, 24
flexor, 24
Mushroom bodies
electrical activation in, 211
history of, 200
hymenopteran
functional properties of, 201, 207, 217
microablations of
techniques for, 203, 216
results of, 209–211
neuron organization of, 206–207, 217–218, 220
position in sensorimotor pathway, 206, 214, 218
recordings from, 205

N

Neural code, 89, 91, 128–130, 258–260, 275–276
ensemble coding, 90
Neural network models, 28, 227–316, 365–380
minimal, 260
Neuromodulators, *see* Modulation
Non-GI, 102
Nonspiking interneurons, 71–75, 126–128
Numerical differentiation, 255

Index

O

Odor
  plumes, see Plumes
  quality coding, 180
  quantity coding, 180
Olfactory search strategies, 165
Omnidirectional cell, 97, 119–121
Omnidirectional walking, 293
Optimization, 251, 253, 260
Optomotor anemotaxis,
    see Anemotaxis

P

Pacemaker neuron 269
Parameter space, 252, 253, 258
Pattern generator, see CPG
Perception
  action oriented, 394–396
  robot, 394–396
Perceptual cycle, 384
Phase, definition of, 25
  plots, 26, 27, 32–34, 36, 37
Pheromone, 135
Physiological state, 134–135
Pitch stabilization, 362
Place memory
  behavioral tests for, 203–204
  deletion of, 203
Plumes
  temporal structure, 180–181
Postural control, 45–46, 54–59, 335, 337–338, 361–362
Praying mantis, 107, 146
Predators, 106–108
Proctolin, 237
Projection neurons, 178
Pronase, 92, 97, 98, 99, 100
Proprioceptive reflexes, 43–66, 301
  amplitude modulating, 302
  load compensation, 43–66, 305
  passive traction, 303
  phase modulation, 302

Proprioceptors, 49–54, 106, 133–134, 269–271
Prowling, 362

Q

Quantal release, 79–81
Quiescent state, 134–135

R

Receptors, stress, 35
Receptive field, 81–82, 91–93, 120–121, 126, 275–279
Reciprocal inhibition, 241, 269–270
Reflexes, 50–54, 64–65, 90, 111 69–83, 269, 359–360
  resistance reflex, 51–52, 59
  flexion reflex, 51–52, 59
Relaxation oscillator, 244
Rheotaxis, 160
Robot, 3, 7, 8, 12, 13, 16–18, 111
  angle ranges, 368, 375, 378
  autonomy, 355ff, 380
  behavior, 386
  Blatta electromagnetica, 111
  control architecture, 287, 288
  endocrine system as a model, 403–405
  hexapod, 271, 358, 365–380
  learning, 396–403
  legged, 21–22, 38–40, 312, 319–354
    control, 322, 324, 326, 328, 349—350
      algorithms, 319, 331–338, 345
      hopping, 333–334
      posture, 335, 337–338
      speed, 334–335
    dynamic runners, 330–345
    dynamic walkers, 326–330, 347
    milestones, 321
    passive dynamic, 329–330
    Statically stable, 320–326
  mobile, 388–394
  navigation, 387–394

schema-based, 384
survivability, 403–405
Robustness, 271, 274, 362, 366, 374, 379
Running, 11, 12, 14–16, 18

S

Schema
　activation level, 397
　learning, 396–403
　motor, 385, 388, 406
　perceptual, 394, 406
　signal, 404, 406
　theory, 384–385
Sense organ, 29, 35, 39, 50–54
　campaniform sensilla, 35, 50
　chordotonal organs, 50–54, 59–60
　disturbance of, 39
　force receptors, 50
　hair plate, 50
　joint angle receptors, 50–54
　see also Cerci
Sensitization, 262
Sensor fusion, 395–396
Sensor reliability, 365
Sensorimotor processing, 110, 69–83
Sensory channelling, 394–395
Sensory modality, 90, 103, 131–134
Signal sorting, 92
Simulated annealing, 255
Spatial memory, 216
Specification problem, 89, 90, 110
Spiders, 107, 108
Spike trains, 95
Spiking local interneuron, 77–81
Spring-mass model, 7–10, 16–18
Stability, 12–15, 18, 268–271, 320–326, 365, 366, 375, 378
Stance, 23, 25, 26, 291, 369, 372
Static crawlers, 326
Stilt biped, 328
Stepping, 29
　cycle, 25, 29
　diagram, 22–24
　pattern, 368, 371, 372, 373, 374, 379
　rate, 23, 31
Stiffness, 369, 377, 378
Stomatogastric network, 234
Stomatogastric system, 230, 232
Subsumption architecture, 355
Summation
　spatial, 77
　temporal, 73–74
Swing, 23, 26, 291, 369, 375
Synapse
　development, 77–78
　quantal analysis, 79–81
Synaptic plasticity, 261*ff*
Synaptic shunt, 71–72
Synthetic neuroethology, 362

T

Tactile stimulation, 103, 131–134
Temporal patterning, 95, 100, 396
TI_As, 119–136, 276
Time constant, 258, 374, 378, 379
Time delays, 367, 374–379
Toad, 106, 107
Tonotopy, acoustic, 145
Trochanter, 31, 35
Turn
　amplitude, 94, 95
　bias, 92, 97, 99, 110
　direction, 94, 96

U

Ultrasound, 147

V

Ventral median branches,
　　see VM branch
Video analysis, 116, 117, 129, 130
Virtual leg, 369, 374, 377, 379

Visual cues, 89, 106
VM branch, 122–130
Voltage-gated currents, 72–73

W

Walking 6, 7, 9, 11, 13, 16
   cockroach, 21–40, 60–64
   fast, 22, 35, 39
   freely, 30–35
   high speed, 23, 28, 30, 31
   neural basis of, 28–29
   normal, 22–29
   pattern, 28, 29
   perturbation of, 29–30
   resetting effects on walking, 47–48, 64
   rough terrain, 319, 345–348
   slow, 22, 31, 35, 39
   suspended, 35–38
   truck (General Electric), 322, 324
Weight space, 252, 253, 259, 261, 264
Wind field, 120, 121, 126
Wind-induced drift, 170–173
Wind-sensory 90, 96, 100, 103, 106, 110

Y

Yaw aerodynamics, 145